ADVANCED STUDIES IN EXPERIMENTAL AND CLINICAL MEDICINE

Modern Trends and Latest Approaches

ADVANCED STUDIES IN EXPERIMENTAL AND CLINICAL MEDICINE

Modern Trends and Latest Approaches

Edited by

P. Mereena Luke, MSc, MTech

K. R. Dhanya, PhD

Didier Rouxel, PhD

Nandakumar Kalarikkal, PhD

Sabu Thomas, PhD, DSc, FRSC

AAP | APPLE ACADEMIC PRESS

First edition published 2021

Apple Academic Press Inc.
1265 Goldenrod Circle, NE,
Palm Bay, FL 32905 USA
4164 Lakeshore Road, Burlington,
ON, L7L 1A4 Canada

CRC Press
6000 Broken Sound Parkway NW,
Suite 300, Boca Raton, FL 33487-2742 USA
2 Park Square, Milton Park,
Abingdon, Oxon, OX14 4RN UK

Apple Academic Press exclusively co-publishes with CRC Press, an imprint of Taylor & Francis Group, LLC

Library and Archives Canada Cataloguing in Publication

Title: Advanced studies in experimental and clinical medicine : modern trends and latest approaches / edited by P. Mereena Luke, MSc, MTech, K.R. Dhanya, PhD, Didier Rouxel, PhD, Nandakumar Kalarikkal, PhD, Sabu Thomas, PhD, DSc, FRSC.

Names: Luke, P. Mereena, 1983- editor. | Dhanya, K. R., 1983- editor. | Rouxel, Didier, editor. | Kalarikkal, Nandakumar, editor. | Thomas, Sabu, editor.

Description: Includes bibliographical references and index.

Identifiers: Canadiana (print) 20200327208 | Canadiana (ebook) 20200327348 | ISBN 9781771889063 (hardcover) | ISBN 9781003057451 (ebook)

Subjects: LCSH: Clinical trials. | LCSH: Clinical medicine. | LCSH: Medicine, Experimental. | LCSH: Medicine—Research. | LCSH: Biology—Research. | LCSH: Nanostructures. | LCSH: Biomedical materials.

Classification: LCC R853.C55 .A38 2021 | DDC 610.72/4—dc23

Library of Congress Cataloging-in-Publication Data

Names: Luke, P. Mereena, 1983- editor. | Dhanya, K. R., 1983- editor. | Rouxel, Didier, editor. | Kalarikkal, Nandakumar, editor. | Thomas, Sabu, editor.

Title: Advanced studies in experimental and clinical medicine : modern trends and latest approaches / edited by P. Mereena Luke, K.R. Dhanya, Didier Rouxel, Nandakumar Kalarikkal, Sabu Thomas.

Description: First edition. | Palm Bay, FL : Apple Academic Press, 2021. | Includes bibliographical references and index. | Summary: "This volume, Advanced Studies in Experimental and Clinical Medicine: Modern Trends and Approaches, provides a selection of chapters on new developments in various areas of clinical medicine, including dental, surgery, and general practice. These scientific papers and chapters analyze the diagnostic processes and inform of new and novel diagnostic techniques. This book is divided into two sections; the first section contains review papers and includes an overview of experimental and clinical medicine, explaining its history from ancient medicine to modern medicine. The second section presents a selection of original research papers from respected authors on a variety of topics. This book is recommended to immunologists, physiologists, and medical practitioners"-- Provided by publisher.

Identifiers: LCCN 2020040092 (print) | LCCN 2020040093 (ebook) | ISBN 9781771889063 (hardcover) | ISBN 9781003057451 (ebook)

Subjects: MESH: Biomedical Research | Clinical Laboratory Techniques | Nanostructures | Biocompatible Materials

Classification: LCC R853.C55 (print) | LCC R853.C55 (ebook) | NLM W 20.5 | DDC 610.72/4--dc23

LC record available at https://lccn.loc.gov/2020040092

LC ebook record available at https://lccn.loc.gov/2020040093

ISBN: 978-1-77188-906-3 (hbk)
ISBN: 978-1-00305-745-1 (ebk)

About the Editors

P. Mereena Luke
Research Scholar, Department of Polymer Technology, Chemical Faculty, Gdansk University of Technology, Gdansk, Poland

P. Mereena Luke is a Research Scholar in the Department of Polymer Technology, Chemical Faculty, at Gdansk University of Technology, Poland. She recently completed a two-year project sponsored by the University Grants Commission on "Preparation and Evaluation of Dendrimeric Nanosystems for Drug Delivery Applications" at the International and Inter University Center for Nanoscience and Nanotechnology, Mahatma Gandhi University, Kottayam, India. She has published articles in international and national professional journals and has also presented her work at different international and national conferences. She has written many chapters for several books. Her research interests include dendrimer-based drug delivery, nanomedicine, natural polymers, phytochemistry synthesis, and applications of dendrimer drug complexes for biomedical applications.

K. R. Dhanya, PhD
Postdoctoral Fellow, International and Inter University Center for Nanoscience and Nanotechnology, Mahatma Gandhi University, Kottayam – 686560, Kerala, India

K. R. Dhanya, PhD, is a Postdoctoral Fellow at the International and Inter University Center for Nanoscience and Nanotechnology, Mahatma Gandhi University, Kottayam, India. She has published articles in professional international and national journals and has presented her work at various international and national conferences as well. She has authored and co-authored several book chapters. Her research interests include polymer synthesis, hydrogels, water purification, polymer-nanocomposites, and photocatalytic degradation.

Didier Rouxel, PhD

Professor, Institute Jean Lamour, Université de Lorraine, Nancy Cedex – 50840–54011, France

Didier Rouxel, PhD, is a full professor at the Institut Jean Lamour, Université de Lorraine, France. His current research areas include elastic properties of polymeric materials studied by Brillouin spectroscopy, development of polymer-nanoparticle nanocomposite materials, development of micro-devices based on electroactive polymers, piezoelectric nanocrystals, and microsensor development for surgery. He has over 100 publications to his name, including over 60 in international journals, plus book chapters, invited conference presentations, seminars, and other communications. He is a member of the French Society of Nanomedicine; the PhD thesis director of the C' Nano Thesis National Prize 2013 (category Interdisciplinary Research) (Van Son Nguyen, "Development of P (VDF-TrFE); and Nanocrystal Hybrid Nanocomposite Films and Integration into Micro-Structured Devices" from the University of Lorraine, 2012); and thesis director of the winner of the French-Indian Scholarship Raman-Charpak 2015 (Camille Thevenot, "Modulation of Elastic Properties of Nanocomposites," Mahatma Gandhi University, Kottayam University). Dr. Rouxel was awarded the Scientific Excellence Award (PES) in 2013 and is a holder of the doctoral supervision and research bonus (PEDR) 2007–2011.

Nandakumar Kalarikkal, PhD

Director, International and Inter University Center for Nanoscience and Nanotechnology, Mahatma Gandhi University, Kottayam – 686560, Kerala, India

Nandakumar Kalarikkal, PhD, is the Director of the International and Inter University Center for Nanoscience and Nanotechnology as well as the Director and Assistant Professor in the School of Pure and Applied Physics at Mahatma Gandhi University, Kerala, India. His current research interests include synthesis, characterization, and applications of various nanostructured materials, laser plasma, and phase transitions. He has published more than 160 research articles in peer-reviewed journals and has edited 15 books. He has supervised many PhD, MPhil, and master's theses. He has four patents to his credit. Very recently, Dr. Nandakumar was honored with "Professor at Lorraine" by the University of Lorraine, France, for his academic excellence.

Dr. Nandakumar obtained his master's degree in physics with a specialization in industrial physics and his PhD in semiconductor physics from Cochin University of Science and Technology, Kerala, India. He was a postdoctoral fellow at NIIST, Trivandrum, and later joined Mahatma Gandhi University, Kerala, India.

Sabu Thomas, DSc
Vice Chancellor, Mahatma Gandhi University; Professor of Polymer Science and Technology and Founding Director of the International and Inter University Center for Nanoscience and Nanotechnology, Mahatma Gandhi University, Kottayam – 686560, Kerala, India

Sabu Thomas, PhD, DSc, FRSC, is the Vice Chancellor of Mahatma Gandhi University, Kottayam, Kerala, India, and Professor of Polymer Science and Engineering at the School of Chemical Sciences as well as Founder Director of the International and Inter University Center for Nanoscience and Nanotechnology at Mahatma Gandhi University, Kottayam, India. He has supervised many PhD, MPhil, and master's theses. He has five patents to his credit. He received the coveted Sukumar Maithy Award proclaiming him the best polymer researcher in the country for the year 2008. Professor Thomas received a number of awards, including MRSI and CRSI medals; the Dr. APJ Abdul Kalam Award for Scientific Excellence–2016; and the Lifetime Achievement Award from the Indian Nano-Biologists Association for his excellent work. He was the recipient of a Fulbright-Nehru International Education Administrators Award in 2017. He received TRiLA Academician of the year 2018 award. He has over 800 publications to his credit, including over 70 books. He was cited over 35,765 citations and has an H Index – 94 (Google Scholar) and 81 (Scopus), and listed in one of the most cited researchers in materials science and engineering by Elsevier Scopus Data 2016. Dr. Thomas has been ranked fifth in India as one of the most productive scientists.

Contents

Contributors

Efrén D. J. Andrades
Faculty of Pharmacy and Bioanalysis, Universidad de Los Andes, Mérida – 5101, Venezuela

Mohammad Zaheer Ansari
International School of Photonics, Cochin University of Science and Technology, Kochi – 682 022, Kerala, India, E-mails: mohamedzaheer1@gmail.com; mdzaheer@cusat.ac.in

Sanjay Kumar Bharti
Institute of Pharmaceutical Sciences, Guru Ghasidas Vishwavidyalaya (A Central University), Bilaspur, Chhattisgarh – 495009, India, E-mail: skbharti.ggu@gmail.com

Humberto Cabrera
Optics Laboratory, The Abdus Salam International Center for Theoretical Physics (ICTP), Strada Costiera 11, Trieste – 34151, Italy, E-mail: hcabrera@ictp.it

Sutapa Som Chaudhury
Research Scholar, Center for Healthcare Science and Technology, Indian Institute of Engineering Science and Technology, Shibpur, Howrah – 711103, West Bengal, India, E-mail: somchaudhurysutapa@gmail.com

K. R. Dhanya
Postdoctoral Fellow, International and Inter University Center for Nanoscience and Nanotechnology, Mahatma Gandhi University, Kottayam – 686560, Kerala, India, E-mail: k_r_dhanya@yahoo.co.in

Suja George
Professor, Department of Chemical Engineering, Malaviya National Institute of Technology Jaipur, Jaipur, Rajasthan – 302017, India, E-mail: sgeorge.chem@mnit.ac.in

Sandhya Gopalakrishnan
Associate Professor, Department of Prosthodontics, Government Dental College, Kottayam, Kerala, India, E-mail: sandhya_gopal@rediffmail.com

R. Gopinath
Research Scholar, Virology and Immunology Laboratory, Department of Microbiology, Dr. ALM PG IBMS, University of Madras, Taramani, Chennai – 600 113, Tamil Nadu, India

Akshara Goyal
Department of Chemical Engineering, Malaviya National Institute of Technology Jaipur, Jaipur, Rajasthan – 302017, India

Hilda C. Grassi
Faculty of Pharmacy and Bioanalysis, Universidad de Los Andes, Mérida – 5101, Venezuela

A. B. Gupta
Department of Chemical Engineering, Malaviya National Institute of Technology Jaipur, Jaipur, Rajasthan – 302017, India

Sayan Dutta Gupta
Department of Pharmaceutical Chemistry, GokarajuRangaraju College of Pharmacy, Hyderabad, Telangana – 500090, India, E-mail: sayandg@rediffmail.com

Nurit Hadad
Department of Clinical Biochemistry and Pharmacology, Faculty of Health Sciences Ben-Gurion University of the Negev and Soroka Medical University Center, Beer Sheva, Israel

Józef T. Haponiuk
Chemical Faculty, Polymers Technology Department, Gdansk University of Technology, Gdansk – 80233, Poland, E-mail: jozeph.haponiuk@pg.edu.pl

J. M. Jeffrey
Department of Genetics, Dr. ALM PG IBMS, University of Madras, Taramani, Chennai – 600 113, Tamil Nadu, India

M. Jeyadevasena
Department of Microbiology, Dr. ALM PG IBMS, University of Madras, Taramani, Chennai – 600 113, Tamil Nadu, India

Neetha John
Associate Professor, Central Institute of Plastics Engineering and Technology, JNM Campus, Udyogamandal, Kochi – 683501, Kerala, India, E-mail: neethajob@gmail.com

Tomy Muringayil Joseph
Research Scholar, Chemical Faculty, Polymers Technology Department, Gdansk University of Technology, Gdansk – 80233, Poland, E-mail: say2tomy@gmail.com

Nandakumar Kalarikkal
International and Inter University Center for Nanoscience and Nanotechnology, Mahatma Gandhi University, Kottayam – 686560, Kerala, India

Kesenia Kasianov
Department of Clinical Biochemistry and Pharmacology, Faculty of Health Sciences Ben-Gurion University of the Negev and Soroka Medical University Center, Beer Sheva, Israel

Rachel Levy
Professor, Department of Clinical Biochemistry and Pharmacology, Faculty of Health Sciences Ben-Gurion University of the Negev and Soroka Medical University Center, Beer Sheva, Israel, E-mail: ral@bgu.ac.il

P. P. Lizymol
Professor, Scientist F and in Charge, Division of Dental Products, Biomedical Technology Wing, Sree Chitra Tirunal Institute for Medical Sciences and Technology, Poojappura, Thiruvananthapuram – 695012, Kerala, India, Tel.: +91-471-2520221, Fax: +91-471-2341814, E-mails: lizymol@rediffmail.com; lizymol@sctimst.ac.in

P. Mereena Luke
Research Scholar, International and Inter University Center for Nanoscience and Nanotechnology, Mahatma Gandhi University, Kottayam – 686560, Kerala, India; Chemical Faculty, Polymers Technology Department, Gdansk University of Technology, Gdansk – 80233, Poland, E-mail: merinaluke@gmail.com

Debarshi Kar Mahapatra
Department of Pharmaceutical Chemistry, DadasahebBalpande College of Pharmacy, Nagpur, Maharashtra – 440037, India, E-mails: mahapatradebarshi@gmail.com; dkmbsp@gmail.com

Yafa Malada-Edelstein
Department of Clinical Biochemistry and Pharmacology, Faculty of Health Sciences Ben-Gurion University of the Negev and Soroka Medical University Center, Beer Sheva, Israel

Elanchezhiyan Manickan
Department of Microbiology, Dr. ALM PG IBMS, University of Madras, Taramani, Chennai – 600 113, Tamil Nadu, India, E-mail: emanickan@yahoo.com

Mayank Mehta
Department of Chemical Engineering, Malaviya National Institute of Technology Jaipur, Jaipur, Rajasthan – 302017, India

A. Mujeeb
International School of Photonics, Cochin University of Science and Technology, Kochi – 682 022, Kerala, India

Chitrangada Das Mukhopadhyay
Center for Healthcare Science and Technology, Indian Institute of Engineering Science and Technology, P.O., Botanical Garden, Shibpur, Howrah, West Bengal – 711103, India, E-mails: chitrangadadas@yahoo.com; chitrangadam@chest.iiests.ac.in

Suthandira Munisamy
IRT Perundurai Medical College, Perundurai, Erode District, Tamil Nadu – 638053, India

G. Sathya Narayanan
Department of Microbiology, Dr. ALM PG IBMS, University of Madras, Taramani, Chennai, Tamil Nadu, India

Pugalendhi
Voluntary Health Science, Taramani, Chennai, Tamil Nadu, India

G. Reena
Department of Microbiology, Dr. ALM PG IBMS, University of Madras, Taramani, Chennai, Tamil Nadu, India

Didier Rouxel
Institute Jean Lamour, Université de Lorraine/UMR-CNRS 7198, Nancy Cedex – 50840–54011, France

Bhuban Ruidas
Center for Healthcare Science and Technology, Indian Institute of Engineering Science and Technology, Shibpur, Howrah – 711103, West Bengal, India

Prasanta Kumar Sarkar
Department of Rasashastra, J. B. Roy State Ayurvedic Medical College and Hospital, Kolkata – 700004, West Bengal, India

G. Sathyanarayanan
Department of Microbiology, Dr. ALM PG IBMS, University of Madras, Taramani, Chennai, Tamil Nadu, India

Koel Sinha
Research Scholar, Center for Healthcare Science and Technology, Indian Institute of Engineering Science and Technology, P.O., Botanical Garden, Shibpur, Howrah, West Bengal – 711103, India, E-mail: ksinha2110@gmail.com

Yulia Solomonov
Department of Clinical Biochemistry and Pharmacology, Faculty of Health Sciences Ben-Gurion University of the Negev and Soroka Medical University Center, Beer Sheva, Israel

Sabu Thomas
School of Chemical Sciences and International and Inter University Center for Nanoscience and Nanotechnology, Mahatma Gandhi University, Kottayam – 686560, Kerala, India, E-mail: sabuthomas@mgu.ac.in

Kannan Vaidyanathan
Professor, Department of Biochemistry and Head Molecular Biology,
Amrita Institute of Medical Science and Research Center, Kochi, Kerala, India;
Pushpagiri Institute of Medical Science and Research Center, Tiruvalla – 689101,
Kerala, India, E-mail: drkannanvaidyanathan@gmail.com

K. Vasanthi
Research Scholar, Virology and Immunology Laboratory, Department of Microbiology,
Dr. ALM PG IBMS, University of Madras, Taramani, Chennai – 600113, Tamil Nadu, India

Ana Velásquez
Faculty of Pharmacy and Bioanalysis, Universidad de Los Andes, Mérida – 5101, Venezuela

Abbreviations

ABS	acrylonitrile-butadiene-styrene
ACV	acyclovir
AD	Alzheimer's disease
AiBNP	*A. indica* bimetallic nanoparticle
AIF	apoptosis-inducing factor
AiGNP	*A. indica* gold nanoparticle
AiSNP	*A. indica* silver nanoparticle
ALDH	aldehyde dehydrogenase
ALS	amyotrophic lateral sclerosis
ANOVA	analysis of variance
AVD	absolute values of difference
BBB	blood-brain barrier
BCAA	branched-chain amino acids
Bis-GMA	bisphenol-A glycidyl methacrylate
BM	Bacopa monnieri
BNCT	boron neutron capture therapy
CAD	computer-aided design
CaF_2	calcium fluoride
CAH	congenital adrenal hyperplasia
CapG	capping protein
CAT	computerized axial tomography
CBS	corticobasal syndrome
CDG1a	congenital disorder of glycosylation type 1a
CFTR	cystic fibrosis transmembrane conductance regulator
CL	*Curcuma longa*
CLIC4	chloride intracellular channel
CMS	carbon molecular sieving
CNS	cashew nutshell
CNS	central nervous system
CNSL	cashew nut shell liquid
COM	co-occurrence matrix
COX-2	cyclooxygenase-2
CPE	cytopathic effects

CsBNP	*C. sinensis* bimetallic nanoparticle
CSF	cerebrospinal fluid
CsGNP	*C. sinensis* gold nanoparticle
CsSNP	*C. sinensis* silver nanoparticle
CUL3	cullin 3
CVD	chemical vapor deposition
DA	dopamine
DCFHDA/DCFDA	2′,7′-dichlorofluorescein diacetate
DDAH1	dimethylarginine dimethylaminohydrolase 1
DHCR7	7-dehydrocholesterol reductase
DMEM	Dulbecco's modified eagle media
DMSO	dimethyl sulfoxide
DNA	deoxyribonucleic acid
DTS	diametral tensile strength
EGCG	epigallocatechin-3-gallate
EGF	epidermal growth factor
ETC	electron transport chain
FDM	fused deposition modeling
FE-SEM	field emission scanning electron microscopy
FT-IR	Fourier transform infrared
FWHM	full width at half maximum
GAG	glucosamine glycans
GB	*Gingko Biloba*
GD	generalized differences
GM2AP	GM2 activator protein
GM-CSF	granulocyte-macrophage-colony-stimulating factor
GS	glycogen synthase
GSTO1	glutathione-S-transferase omega 1
HATs	histone acetyltransferase
HB	Hirak Bhasma
HD	Huntington's disease
HGP	*human genome project*
HGPS	Hutchinson-Gilford progeria syndrome
HIV	human immunodeficiency virus
HR/AM	high-resolution accurate mass
HR-SEM	high resolution scanning electron microscopy
HR-TEM	high resolution transmission electron microscopy
Hsp90	heat shock protein 90
HSV	herpes simplex virus

ICP-OES	inductively coupled plasma optical emission spectra
IFN-γ	interferon-gamma
IgG	immunoglobulin
IL-1β	interleukin-1beta
IMS	imaging mass spectrometry
iNOS	inducible nitric oxide synthase
KLHL3	Kelch-like 3
LAA	leucocyte ascorbic acid
LAT	latency-associated transcripts
LB broth/agar	Luria Bertani broth/agar
LDHB	lactate dehydrogenase
LIT	liver infusion tryptose
LMWLs	low-molecular-weight-ligands
LOX-1	lipoxygenase
MADD	multiple acyl-CoA dehydrogenase deficiency
MBC	minimal bactericidal concentration
MDR	multi-drug resistant
MEEVD	Met-Glu-Glu-Val-Asp
MIC	minimal inhibitory concentration
MLP	multilayer perceptron
MPCR	multiplex PCR
MS	mass spectrometry
MSUD	maple syrup urine disease
MTT	3-(4,5-dimethylthiazol-2-yl)-2,5-diphenyltetrazolium bromide
NBS	newborn screening
NCIs	National Cancer Institutes
NCL	neuronal ceroid lipofuscinosis
NF-κb	nuclear factor kappa B
NK cells	natural killer cells
NKC	Nilavembu Kudineer Chooranaum
NKCBNP	Nilavembu Kudineer Chooranaum bimetallic nanoparticle
NKCGNP	Nilavembu Kudineer Chooranaum gold nanoparticle
NKCSNP	Nilavembu Kudineer Chooranaum silver nanoparticle
NMR	nuclear magnetic resonance
NPC	Niemann-Pick C
Nrf-2	nuclear factor erythroid 2

NVP	N-vinylpyrrolidone
ORMOCERS	organically modified ceramics
PAA	polyacrylic acid
PAH	phenylalanine hydroxylase
PBLG	poly(y-benzyl-Z,-glutamate)
PBS	poly(butylene succinate)
PBSA	poly(butylene succinate-co-adipate)
PCB	printed circuit boards
PCL	poly-ε-caprolactone
PCOS	polycystic ovarian syndrome
PCR	polymerase chain reaction
PD	Parkinson's disease
PDT	photodynamic therapy
PE	polyethylene
PEAs	poly(ester amides)
PEG	polyethylene glycol
PF2D	two-dimensional protein fragmentation
PGA	poly(glycolic acid)
PgJBNP	*P. granatum* juice bimetallic nanoparticle
PgJGNP	*P. granatum* juice gold nanoparticle
PgJSNP	*P. granatum* juice silver nanoparticle
P-gp	p-glycoprotein
PgPBNP	*P. granatum* peel bimetallic nanoparticle
PgPGNP	*P. granatum* peel gold nanoparticle
PgPSNP	*P. granatum* peel silver nanoparticle
PHAII	pseudohypoaldosteronism Type II
PHB	poly(hydroxyl butyrate)
PHBV	*poly(3-hydroxybutyrate-co-3-hydroxyvalerate)*
PKU	phenylketonuria
PLA	polylactic acid
PLGA	poly(L-lactic acid-co-glycolic acid)
PMBV	p-vinylphenylboronic acid
PP	polypropylene
PPF-DA	poly(propylene fumarate)-diacrylate
ppm	parts per million
PPMC	Pearson product-moment coefficient
PPP	poly(p-phenylene)
PTH	parathyroid hormone
PTT	poly(trimethylene terephthalate)

PU	polyurethane
PVA	polyvinyl alcohol
RBCs	resin-based composites
ROI	region of interest
ROS	reactive oxygen species
RRM	reduced retention model
SAA	ascorbic acid in serum
SAP	serum alkaline phosphatase
SBCADD	short/branched-chain acyl-CoA dehydrogenase deficiency
SBDD	structure-based drug design
SC5D	5-desaturase
SE	standard error
SERCA1	sarcoplasmic reticulum calcium ATPase 1
SFE	supercritical fluid extraction
SI	selectivity index
SILS	single-incision laparoscopic surgery
SIP	serum inorganic phosphorus
SLOS	Smith-Lemli-Opitz syndrome
SOD	superoxide dismutase
SQRDL	sulfide: quinone oxidoreductase
SSA	serum albumin
SUAC	succinyl acetone
TB	tuberculosis
TEGDMA	triethylene glycol dimethacrylate
THSP	time history speckle pattern
Tm	temperature
TMS	tandem mass spectrometry
TPR	tetratricopeptide repeat
UDMA	urethane dimethacrylate
UPJ	ureteropelvic junction
UV-Vis	ultraviolet-visible
VHN	Vickers hardness number
WBC	white blood cells
WHO	World Health Organization
WS	*Withania somnifera*
XDR	extremely drug-resistant
XRD	x-ray diffraction
YPD	yeast extracts peptone dextrose

Preface

I hope this book will establish inspiration, powerful insight, and constant motivations in the field of medical research both in clinical as well as experimental medicine. Developments in the field of healthcare based on innovative technologies and novel findings enable a change in health research that could reinforce studies that were not feasible in the past, leading to new ideas on treatments and medicines. Research in healthcare has significant social value. It can provide accurate data on the tendencies to disease and risk factors, public health interventions, organizational capacity, treatment methods, and expenses of various health technologies and their uses. Clinical trials can provide significant information on the effectiveness and undesirable side effects of medical events by regulating factors that might affect the research outcomes, but the response from actual clinical experience is also essential for comparing and improving the use of drugs, vaccines, medical devices, and also different diagnostics methods. Thus it is essential to spot and track each and every experience or action within the biomedical field to identify and compare the adverse effects and also to determine efficacy under distinct circumstances. To establish proper guidelines for effective and safe clinical practices and to guarantee high-quality patient care, it is also essential to record and evaluate the experience in clinical practice from the preliminary stages onwards. Various types of health research has generally contributed to substantial findings, the development of innovative therapies, and significant improvements in healthcare and public health.

This book, *Advanced Studies in Experimental and Clinical Medicine: Modern Trends and Latest Approaches*, provides a framework for learning from information on medical history and developments. The volume includes present trends, inventive conclusions, and useful medical research opportunities in the area of medical research, including internal medicine, infectious disease, the dental field, and surgery.

This book is divided into two different sections. The first section includes reviews of recent trends in innovative medicine and biological aspects. It can be read by people with a basic scientific background.

The second section presents original research works from the conference ICECM 2017. The chapters in this clinical and biomedical section may seem overwhelming with their complexity, but we hope that readers won't be easily discouraged. Even if it is difficult to understand the specific content of a chapter, scientific proof, methodical data discussion, citations from the supporting documents, suggestions, and the precise conclusions will facilitate an effective comprehension of the chapters.

The current state of biomedical affairs has reached our expectations by the technological and scientific advancements, and the field of experimental and clinical medicine continues to require improvement and communication. We are resolved to avoid and learn from the mistakes of the past, to overcome the limits, and to start benefiting from the knowledge bases created by the great scientists and experienced physicians, as well as from the discoveries of modern equipment and innovative technologies in this era that has seen the revolution of the medical and pharmaceutical sectors.

We hope this book will be widely read and will enable readers to understand how much we have improved in medicine, treatment, and diagnosis. This book also offers information on new perspectives and opportunities in the field of advanced medical research and healthcare.

Part I
Review Papers

CHAPTER 1

Developments of Health Care: A Brief History of Medicine

P. MEREENA LUKE,[1,2] K. R. DHANYA,[1] TOMY MURINGAYIL JOSEPH,[3] JÓZEF T. HAPONIUK,[2] DIDIER ROUXEL,[4] and S. THOMAS[1]

[1]*International and Inter-University Center for Nanoscience and Nanotechnology, Mahatma Gandhi University, Kottayam – 686560, Kerala, India, E-mail: merinaluke@gmail.com (P. M. Luke)*

[2]*Department of Polymer Technology, Chemical Faculty, Gdansk University of Technology, Gdansk, Poland*

[3]*Research Scholar, Chemical Faculty, Polymers Technology Department, Gdansk University of Technology, Gdansk – 80233, Poland, E-mail: say2tomy@gmail.com*

[4]*Institute Jean Lamour, Université de Lorraine/UMR-CNRS 7198, Nancy Cedex – 50840–54011, France*

ABSTRACT

This chapter focusing the important areas in the history of medicine discusses additional investigation of facts, and concepts that influenced clinical practice in past decades and in contemporary clinical practices. This chapter also highlighting the medical traditions and cultures of various societies around the world, as well as the excellent work of renowned scientists and physicians who have contributed to our existing knowledge of medicine, diagnosis, treatments and even health.

1.1 INTRODUCTION

Disease and injury are as old as mankind; the people in the ancient period strongly believed that the diseases may be due to the natural or the supernatural causes such as the wrath of the Gods, natural causes, and the imbalances within the human body [1]. At this time the people's convictions about the causes of disease was purely based on superstitious. The rituals and magic spells were used to deflect sickness. The historical evidences from the stone ages support the proofs of various diseases such as smallpox, arthritis, inflammations, dental defects, leprosy, bone tumors, scurvy, and tuberculosis (TB), etc., and also it brings out the evidences that there were attempt for treating diseases and injuries [2].

The invention of the microscope led to the discovery of microorganisms and finally identified as the source of infectious diseases [3]. Detection of the particular microorganism causing a disease leads to vaccination or immunization against the infectious disease [4]. Modern surgery developed only after the permission of human dissection and it made tremendous awareness to human anatomy physiology and circulatory system [5]. Eventually, the need for modern anesthesia and antiseptic practices were raised. The discovery of x-rays and the development of medical scanners led to Modern diagnostic practices [6]. Present medicine is much more advanced than ancient medicine in terms of scientific knowledge. Contribution of each invention or discovery and improved technologies led to significant progress in the field of advanced healthcare thus it is possible to identify the causes of disease, provide the proper diagnosis, suitable treatments, and also to evaluate the efficacy of treatments. The technological advances and the modern studies being carried out around the world make medicine a more advanced industry [7]. Traditional medicine systems have always played a major part in global health care needs. Presently, it proceeds to do so and will also play a significant role in the future. Diseases, medicine, and treatments are quite familiar today. It can be seen that medicine as science, research, and practice and it covers diagnosis, treatment, and preserving one's health and life through suitable medication. This Chapter outlines the importance of medical advances throughout history and how modern medicine developed and improved methods of diagnosis.

1.2 HISTORY OF MEDICINE

1.2.1 EGYPTIAN MEDICINE

The rise of Egyptian civilization was in about 3000 BC. Sekhet-Eanach was the first doctor known to history. The second doctor was Imhotep (2,600 BC) He seems to have been a successful physician. He started using simple surgery instead of magic [8]. The oldest known medical book is the Ebers

Papyrus, written about 1500 BC, covering 200 illnesses, extracting medicine from crops, and pointing out the Egyptian physicians used a wide range of herbal and mineral drugs medicated steam inhalation had used for the treatment of patients with chest problem, and doctors used ointments for healing wounds [9]. According to the Egyptians concept, the human body was full of passages. They considered that it could cause disease if these passages in a human body were blocked, to open the passages; they used laxatives and caused vomiting. However, they believed that spells and magic would help to cure the sickness and they used amulets to prevent the disease. They were, curious about the fundamental and primary sources of illness and they started to search for a physical cause of disease [10]. The Egyptians had some awareness of anatomy from the experience of making mummies. Egyptian surgery was restricted only to the treatment of injuries, fractures, and the treatment of blisters and cysts or abscesses. They had surgical instruments such as probes, saws, forceps, scalpels, and scissors clamps, sutures, and cauterization [11].

1.2.2 CHINESE MEDICINE

Chinese medical history starts around the second century B.C, older medical methods and techniques are not clearly documented. The first consistent medical treatise is the "*Yellow Emperor's Inner Canon*." The HAN Dynasty considered the most glorious period in Chinese medical history [12]. Chinese medication was based on the Yin and Yang concept [13]. More specifically, Chinese Physicians use the concepts of Yin and Yang to describe the naturally occurring opposing and interdependent physical conditions that exist in a balanced state in the body. Yin is related with tissue of the organ and is feminine, soft, cold, moist, receptive, dark, and associated with water, while yang is associated with function of the organ is masculine, dry, hot, and bright also allied with fire. With a yin deficiency, organs are lacking in nutrition, while a yang deficiency results an insufficient performance of an organ or organ system. Yin and yang are equally distributed onto a healthy body, and disease was considered an imbalanced condition of the Yin-Yang dialectical components. This imbalance of the body is due to the absence of vital energy stream (Qi) that circulates through channels [meridians] in the body [14]. *Huangdi Neijing,* describes the shape and size of the major internal organs it clearly indicates that the ancient Chinese performed primitive dissection and anatomy had never developed beyond this, due to the rigorous ban against on the dissection of the human body [15]. During the period the knowledge of anatomy is very poor and limited hence the idea regarding the internal organs were considerably mistaken. Based on the ancient theory, all internal organs are classified into two major categories, the five firm [zang] organs and are considered to be yin in nature, the heart, spleen, lung, liver, and kidneys belong to this category. The second category considered as yang nature and these six hollow organs [fu] includes gall bladder, stomach, small intestine, large intestine, bladder, and triple burner. There is also another group of tissues and organs, functioning as the zang organs but in the form of fu organs-called extraordinary fu organs, including the marrow brain, the bones, the vessels, and the uterus [16].

The ancient Chinese diagnosis based on the pulse and the procedure of taking the pulse known as the pulse diagnosis [17]; it helps to recognize very subtle pulse variation. The categories of pulse classification used in traditional Chinese medicine have expanded to 51 different varieties of pulse which were to be taken from 11 different areas of the body. A strong, steady pulse would indicate to the practitioner that the person is healthy while the scattered pulses are the indication of the illness and the critical condition may be close death. Acupuncture is one of the important treatments in ancient china and is the practice of inserting needles into the superficial skin, subcutaneous tissue, and muscles at particular acupuncture points [18]. In traditional Chinese medicine, the human body has as many as 2,000 acupuncture points linked by 12 major meridians. These meridians carry energy, or "Qi," between exterior parts of the body and its internal organs. Acupuncture is believed to maintain the balance between Yin and Yang, thus enabling the normal flow of "Qi" throughout the body and restoring physical and mental health [19]. Moxibustion is another traditional Chinese therapy which consists of burning dried moxa [mugwort root] made from dried *Artimesia vulgaris* on particular points on the body to facilitate healing. Moxibustion facilitated to warming and refreshing the blood, it enhances the stimulation of the flow of Qi. Moxibustion utilized to treat the diseases such as Arthritis, Back pain, Headaches, Migraines, Muscle stiffness, Menstrual cramps, Digestive problems, Ulcers, Cancer, Infertility, Tendonitis [20].

1.2.3 INDIAN MEDICINE

In India, the history of traditional medicines and its health-care record goes back to 5000 years BCE, when health-care needs and diseases were described in ancient manuscripts such as "Charaka Samhita" (990 BCE), "Sushruta Samhita" (660 BCE), and "Dhanwantari Nighantu" (1800 CE), where the use of plants and polyherbal formulations was emphasized and widely practiced. As per Indian belief a good health, requiring equilibrium between air, bile, and mucous component [21]. In Ayurveda *Charaka Samhita* is the oldest and the most authentic manuscript it was written by Charaka he was a well-known Ayurvedic physician in ancient India. This comprehensive text describes various primordial theories on the human body, etiology, symptomology, and of wide range medicines for a varieties of diseases and contains 120 chapters arranged in eight books. The Sarira-Sthaka is one of the significant books, and are mainly discusses the anatomy, embryology, and technique of dissection [22]. The Susruta Samhita was written by another prominent physician and surgeon Susruta. He was a proponent of human dissection his texts include a systematic method for the dissection of the human cadaver. The prohibition on human dissection was not existed in India, this facilitated the Indian physicians to acquire a good knowledge of human anatomy especially bones, muscles, blood vessels, and joints. Thus, India has established to a higher standard in surgery than any other ancient civilization [23].

1.2.4 ARAB MEDICINE

Islamic medicine was exceedingly developed during the post-classical era by incorporating the concept of other ancient medicine such as Greek, Roman, and Persian medicine and traditional Indian medicine-Ayurveda, while at the same time it helped to making numerous advances and innovations. Islamic medicine was later embraced in Western Europe's medieval medicine together with knowledge of classical medicine after European physicians got to know Islamic medical authors during the 12th century Renaissance.

Al-Razi was the first major Persian physician (865–925 C.E). He was the first person to distinguish measles from smallpox, and he found the chemical kerosene and number of other compounds including alcohol and Sulphuric acid [24]. He became the chief physician in Baghdad and Rayy hospitals. Al-Razi is known as the father of pediatrics and his book "The *Diseases of Children*" was considered as the first document to define pediatrics as a separate sector of medicine. He was an expert in ophthalmology also, he found allergic asthma. Al-Razi was the first doctor to study and write about immunology and allergy. According to available records, Al-Razi recognized fever as a mechanism of protection against disease and infection [25]. Abu ' Ali al-Husayn ibn Sina Born around 980 in Afshana near Bukhara in Central Asia was another great doctor in the Islamic world. He was better known by the Latin name "Avicenna" in Europe.

He is renowned as a polymath, and as a philosopher as a doctor whose significant work was the "Canon of Medicine" [*al-Qanunfi'l-Tibb*] continued to teach in Europe and the Islamic world as a medical textbook until the early modern age. The Canon of Medicine" put standards in the Middle East and Europe; it formed the basis for the traditional type of Unani medicine in India [26].

1.2.5 GRECO-ROMAN MEDICINE

Hippocrates codified, systematized, and located Greek Medicine into its classical form. The fundamental principles of natural healing in Greek Medicine based on the medical philosophy of Hippocrates (460–360 BCE). He was one of the authors of the Hippocratic Corpus. According to the Hippocratic physiology, the body consists of four fluids or humors-Blood, Black bile, Yellow bile, Phlegm, and health was accomplished when all these four fluids were in equilibrium. Anatomical knowledge was not a much stronger point in Hippocratic medicine [27]. There was a religious restriction on dissecting cadavers in ancient Greece. The impact of classical Greek medicine was to understand physiology or how the living, respiratory human organism as a whole relates and reacts to its surroundings and how it works to guarantee its health, survival, and well-being [28]. This gave a holistic approach to Greek medicine. Hippocrates laid the theoretical foundation for Greek medicine, further developed, expanded, and introduced by other physicists and philosophers-Plato, Aristotle, and Galen. Galen mainly focused on anatomy and physiology. Both Galen and Dioscorides were pioneering innovators who made significant contributions to Greek Medicine's theory and practice. Galen was the Roman Empire's main physician, and Dioscorides was an herbalist and known as the Father of Pharmacy [29]. Galen's anatomical observations were based on animal dissections and vivisections then he tried to interpret human anatomy. His first human anatomical observations were started when he worked as a gladiator surgeon; the wounds from fighting provided the basic awareness about the internal structure of the human body. His concepts were streamlined after Galen's death and became the foundation of medicine until the Renaissance. His focus on the four humors [blood, yellow bile, black bile, and phlegm] was particularly combined with the other fours, including the elements, qualities, seasons, and age groups [30].

1.2.6 THE MEDICAL RENAISSANCE

The renaissance is the period between the 14th and 17th centuries in European history at this period new involvement has arisen and prospered in the cultural and scientific area. In the early 1400s, the Medical Renaissance began and ended in the late 1600s. During this time, great physicians and humanists have made unique progress in medicine and surgery during this time [31]. In the early classic Renaissance period, Linacre, Erasmus, Leonicello, and Sylvius were considered as the first, contributors in the field of medicine. Andreas Vesalius and Ambroise Paré were made exceptional anatomical contributions by the publication of the "Human Factory" in 1543 (Vesalius), and Ambroise Paré Published "The Apologie and Treatise" which describing inimitable surgical developments. The period of Medical Renaissance, included a great number of gifted physicians and surgeons who made exceptional contributions to human anatomy; Vesalius collected detailed anatomical information; Paré focused on advanced surgical techniques; and Harvey, revealed anatomy and physiology of the circulatory system [32].

Ambroise Paré (1510–1590) was most praised doctor during Renaissance. He was a French military barber surgeon who delivered many discoveries. He is considered the father of surgery and modern forensic pathology and an innovator in surgical techniques especially in battlefield medicine, and wound treatments [33]. Paré devised a clamp, it helped to control of bleeding vessels at the amputated sites. Paré becomes aware of improving his techniques and advancing the surgical treatments, with the improved result and eventually better survival. The expanded collection of Par'e also included the use of artificial limbs and artificial eyes to replace losses on the battlefield [34]. Techniques have also been developed for bladder stone operations. Par'e also invented some ingenious approaches for the suturing of wounds on the face. Paré was also an influential personality in the progress of obstetrics during the middle of the 16th century [35]. He revived the procedure of podalic version for the safe birth of the child by turning the fetus into a viable position at the womb [36].

Hieronymus Fabricius (1537–1619) was an anatomist and surgeon he was known as "The Father of Embryology" in medical science [37]. He prepared an atlas of human and animal anatomy known as Tabulae Pictae. This work involves illustrations from many distinct artists and these anatomical illustrations are credited to Fabricius. At the end of the Renaissance, William Harvey (1578–1657), a British medical doctor and cardiovascular researcher discovered the general circulation and published his findings in "The Motu Cordis" in 1628. This discovery replaced old theories regarding the blood circulation with evidence-based on experiments. The studies of Harvey on blood circulation are fundamental facts to understanding the role of the heart within the human body [38]. However, his work was not accepted. Despite the support of the Royal School of Physicians, several found it hard to accept his ideas they supported the theories behind bloodletting that was fundamental to the practice of the time. Harvey was the most intellectual physician of the very late Renaissance and the beginning of the New Science.

1.3 VACCINATION

Edward Jenner (1749–1823) his revolutionary contribution to vaccination and ultimate eradication of smallpox is well renowned around the world. The work of Jenner is widely regarded as the basis of immunology. Between the fifth and sixth centuries, smallpox was introduced in Europe, and later it was spread in South America by Spanish and Portuguese conquistadors and it destroyed the native populations [39]. The sweeping of smallpox to the continent was eventually led in the fall of the Aztec and Inca empires. In the 18th century, smallpox was widespread in Europe around 60% of all infected persons, and 80% of infected children died. Survivors frequently experienced some degree of enduring scar and loss of organs (such as lips, nose, or ear tissue) smallpox is responsible for blindness also. Jenner's method of vaccination against smallpox became popular and it eventually replaced variolation, which had been the standard before his demonstration [40]. At the end of the 20th century (around 150 years, after Jenner's death in 1823), after a huge surveillance and vaccination program, smallpox would eventually be eradicated.

1.4 THE BIRTH OF ANESTHESIA

In 1846, an American dentist, William Morton, demonstrated that ether causes complete absolute insensitivity to pain while an operation performed in front the doctors and students at the Massachusetts General Hospital [41] Morton used ether vapor to sedate the patient, without pain he was able to remove a tumor from the patient's neck. Ether was one of the main anesthetics, but it caused the patients to be choking and also had administration difficulties. Ether anesthesia, however, supplied the patient with an enhanced surgical experience. It also enabled physicians to develop more advanced surgical skills. In November 1847, Chloroform was established by James Simpson who was a Professor of Obstetrics in Edinburgh [42]. This was a more effective agent but it had serious side effects such as sudden death [in the case of n very anxious patients] and it also caused severe liver damage. In 1860, Albert Niemann isolated cocaine, which thus became the first local anesthetic. Newer, less toxic, local anesthetic agents were established in the early 1900s [43].

1.5 19TH CENTURY: THE RISE OF SCIENTIFIC MEDICINE

1.5.1 INTRODUCTION OF BACTERIOLOGY AND GERM THEORY OF DISEASE

Research on disease prevention has grown enormously in the 19th century. Instruments such as the stethoscope and machines such as the electrocardiogram were also invented. Antoni van Leeuwenhoek (1632–1723) developed the most powerful microscopes that had discovered micro-organisms [protozoans and bacteria] [44]. Consequently, his work resulted in the understanding of disease causes such as Black Death. In the late 1800's and early 1900's many scientific developments especially in the field of microbiology, this translational period led by the discovery of Louis Pasture (1850) later in 1880 this research was successively expanded by Robert Koch as the germ theory of diseases [45].

Pasture discovered aerobic and anaerobic organisms and initiate to think about the possibility of a causal relationship between germs and diseases. The germ theory facilitated the detection of actual microorganisms that causes several diseases. The great achievement of Pasteur was the discovery of a vaccine for rabies [46]. Pasteur developed a vaccine for rabies which was capable to be injected in the period after the dog bite and before the onset of symptoms. He showed how the diseases in both animals and people could be cured by vaccination. Pasteur's work has been implemented as the germ theory of disease and put a conclusion to different speculations of sickness, such as the humoral theory [47].

Robert Koch was a scientist from Germany. In molecular biology, he developed essential techniques used to find treatments and to look for the cause of TB. In 1905, this work got the Nobel Prize. Koch discovers the bacillus bacteria responsible for anthrax; also, he was able to demonstrate that the disease was transmissible in mice. Koch also established an innovative method to producing pure cultures of different types of bacteria by placing the bacteria on a solid culture medium. He used new aniline dyes to distinguish between different kinds of bacteria. The improved microscope and better techniques for producing pure cultures of bacteria, Robert was able to find a tiny bacterium which he called the tubercle bacillus. The tubercle bacillus was much tinier than anthrax the bacteria. Later he discovered other bacteria that caused Cholera [48].

Pasteur instigated the germ theory of disease and Koch, turn it into science as bacteriology. Koch developed the techniques for the study of microorganisms and identified their relationship with specific diseases. Koch formalized four postulates to confirm an organism was the cause of a disease. These postulates of Koch are essential because they were one of the first techniques used by physicians to determine the cause of a disease [49].

1. The same organism must be present in every case of the disease;
2. The organism must be isolated from the diseased host and grown in pure culture;
3. The isolate must cause the disease, when inoculated into a healthy, susceptible animal;
4. The organism must be re-isolated from the inoculated, diseased animal. The postulates provided a group of procedures for the analysis of diseases. It helps to identify the causes of diseases and which allow to finding the proper treatments for the particular diseases. During 1879 and 1906, the micro-organisms causing several diseases were found and are listed in Table 1.1.

The modified germ theory opens the new prospects for the infectious diseases control with improved diagnostic methods and provides an insight into the significance of vectors with respect to the transmission of diseases and understanding of the carrier states.

1.5.2 ANTI-SEPTICS

Doctor Joseph Lister (1827–1912), a surgeon at the Royal Infirmary in Glasgow, Scotland, used antiseptics for the first time. In 1867, Lister initiated to clean the surgical tools with carbolic acid [phenol], and soaked bandages were also used directly on wounds. At that period, even a tiny wound that got infected could cause death. Patients who had surgery also suffered with the risk of infection. The advancement of Dr. Lister started this new practice

TABLE 1.1 Disease-Causing Microorganism and the Year of Identification

Sl. No.	Diseases	Disease-Causing Microorganism	Year of Identification
1.	Gonorrhea	*Neisseria gonorrhoeae*	1879
2.	Typhoid fever	*Salmonella enterica serotype Typhi bacteria, Salmonella paratyphi*	1880
3.	Suppuration	*Staphylococcus aureus*	1881
4.	Glanders	*Burkholderia mallei*	1882
5.	Tuberculosis	*Mycobacterium tuberculosis*	1882
6.	Pneumonia	*Streptococcus pneumoniae*	1882–1883
7.	Cholera	*Vibrio cholerae*	1884
8.	Erysipelas	*Streptococcus pyogenes*	1884
9.	Diphtheria	*Corynebacterium diphtheriae*	1883–1884
10.	Tetanus	*Clostridium tetani*	1884
11.	Cerebrospinal meningitis	*Neisseria meningitidis*	1887
12.	Influenza	*Influenza virus*	1892
13.	Food poisoning	*Clostridium perfringens bacteria*	1898
14.	Plague	*Yersinia pestis*	1896
15.	Pseudo-tuberculosis of cattle	*Yersinia pseudotuberculosis*	1889
16.	Botulism	*Clostridium botulinum*	1896
17.	Bacillary dysentery	*Shigella bacterium*	1898
18.	Paratyphoid fever	*Salmonella Typhi, Salmonella Paratyphi*	1900
19.	Syphilis	*Treponema pallidum subspecies pallidum*	1905
20.	Whooping cough	*Bordetella pertussis*	1906

that led to clean and sterile surgery and wound care that saved millions of lives. Lister inspired by Louis Pasteur's work proving "germ theory," and the causes of infections, in 1865. Lister published his work in 1867 in a paper entitled *On the Antiseptic Principle in the Practice of Surgery*. The germ theory was not widely accepted until the late 1890s. By 1900, it was finally routine practice for doctors to wash their hands, sterilize their tools, clean wounds, and keep operating rooms clean [50].

1.6 20TH CENTURY: ADVANCES IN CLINICAL PRACTICE

More progress has been produced in medicine and health improvement in the 20th century than in the past 5,000 years. The causes of so many diseases were found in the 20th century and the large number of diseases could be able to prevented or cured by the 21 century. The discoveries and developments that changed the face of medicine in all respect. The improvement in the standard of living, education, diet, and nutrition, and the development of health care services the establishment of a free National Health Service, etc., has helped improve public health. Progress in science and technology led to the innovations of new techniques and utensils for diagnosing and treating disease. Disease prevention was the prime reason of the average life expectancy increased enormously from 47 to 75 years [51].

Initial focus persisted on disease control during the first half of the 20th century, as well as significant achievements in endocrinology, nutrition, and other related areas were also accomplished. Fundamental concepts of the disease process altered in the years after World War II by knowledge of cell biology. New biochemical and physiological achievements opened the way for accurate diagnostic tests, more effective therapies, and treatments. Stupendous developments in biomedical engineering enabled the doctor and specialist to test the structures and functions of human body using non-invasive imaging methods, for example, ultrasound (sonar), computerized axial tomography (CAT), and nuclear magnetic resonance (NMR). With each new logical improvement, therapeutic practices of only a couple of years sooner ended up outdated.

1.6.1 INFECTIOUS DISEASES AND CHEMOTHERAPY

Continuing research in the first half of the 20th century concentrated mainly on the nature of infectious diseases and their mode of transmission [52]. Different types of pathogenic organisms were discovered and classified. Diseases like typhus fever caused by rickettsias, are highly pleomorphic bacteria that may occur in the forms of cocci 0.1 μm in diameter, rods 1–4 μm long, or threads up to about 10 μm long and are smaller bacteria. Other pathogenic organisms like protozoans are unicellular eukaryotes relatively complex internal structure and carry out complex metabolic activities. Protozoans causing tropical diseases including malaria [there are four species that infect humans: *P. vivax, P. ovale, P. malariae,* and *P. falciparum*] The viruses are smallest pathogenic organisms causing many diseases, German measles mumps, measles, polio, and malignant tumor [*Rous sarcoma* Virus]. Chemotherapy was launched in the early 20th century. Chemotherapy was pioneered by German scientist Paul Ehrlich (1854–1915) [53]. He introduced the term chemotherapy and studied substantial quantities of chemicals to evaluate their efficacy in treating infectious diseases and he defined the successors as "magic bullets."

1.6.2 DEVELOPMENT OF ANTIBIOTICS

With his invention of enzyme lysozyme (1921) and antibiotic penicillin (1928), Sir Alexander Fleming, a Scottish biologist, defined new horizons for modern antibiotics [54]. In 1929, Fleming published his results and then in a briefer report in 1932. Fleming's work was largely ignored at the time because as per the main scientific view, the use of antibiotic drugs would not work against infectious disease and would be toxic to use on humans. In 1944, penicillin was commonly used to treat infections of soldiers. It was widely used in the battlefield as well as in hospitals across Europe. Penicillin had been nicknamed 'the wonder drug' by the end of World War II and had saved many lives. Howard Florey and Ernst Chain (Scientists in Oxford) have succeeded in effectively purifying and the large scale manufacture of penicillin. The emergence of penicillin was considered as the beginning of "golden era" of antibiotics [55]. In 1945, Ernst Boris Chain and Sir Howard Walter Florey they shared the Nobel Prize in Medicine with Alexander Fleming for their role in making the scale-up production of antibiotic. The invention of penicillin from *Penicillium notatum* was convenient for the treatment of bacterial infections including syphilis, gangrene, and TB. Eventually, semi-synthetic penicillins and edible penicillins were produced. The systematic search began for other antibiotics and the new antibiotics were produced and commercialized.

1.6.3 INVENTION OF SULFONAMIDE DRUGS

German bacteriologist Gerhard Domagk found the second magic bullet against streptococcal infection after years of systematic research in 1932. This was a red dye called Prontosil. With the aid of the new, powerful electron microscopes which had been in use since the early 1930s, scientists found that the active component was a sulfonamide from coal tar. Soon after French workers found that the Prontosil could be metabolized in the patient to sulfanilamide, which was the active antibacterial molecule. After 1935, this idea was to change when it was found that Prontosil could use against streptococcal infection. In 1936, British physician Leonard Colebrook and co-workers confirmed the effectiveness of both Prontosil and sulfanilamide in streptococcal septicemia [56]. The major befits of sulfonamide drugs was the rapid action, superior potency, broad antibacterial range, with lesser toxicity. The discovery of sulfonamides led to the development of drugs which cured gonorrhea, pneumonia, meningitis, and scarlet fever. In subsequent years many derivatives of sulfonamides, or sulfa drugs, were synthesized and tested for antibacterial and other activities. Most of the antibiotic classes [Penicillins, Macrolides, Fluoroquinolones, Tetracyclines, Aminoglycosides, Sulfonamides, Cephalosporins, Glycopeptides, Carbapenems, Lincosamide, etc.], were discovered and introduced to the market between 1940 and 1962. There are generally several antibiotics in each class that have been found over time or are improved versions of earlier types.

1.7 IMMUNIZATION AGAINST VIRAL DISEASES

Cell biology progressed in the second half of the 20th century, providing a much deeper insight of both normal and anomalous conditions in the body. The availability of electron microscopes make possible to examine to the structures and the components of the cell, and chemical exploration revealed evidences to their functions and complex metabolism in the cells. The introduction of tissue culture also helps to growing viruses in the laboratory atmosphere. It led to the fast growth in progress of effective viral vaccines. Microbiologist Max Theiler developed the first viral vaccine for yellow fever in the late 1930s [57]. The first comparatively efficient vaccine for influenza was produced around 1945 [58]. In the 1950s and 1960s, polio, measles, and rubella vaccines were developed by American physician Jonas Salk in 1954; and in 1960, an oral polio vaccine was developed by virologist Albert Sabin which helped wipe out polio from the western world in the 20th century, and which may help to make the eradication of polio worldwide in 21st century [59]. Effective vaccines for measles and rubella (German measles) were used during the 1960s. A better measles vaccine was developed in 1968. More vaccines have been created to control illnesses of the childhood. Children's health was usually better after the war [60].

1.8 ENDOCRINOLOGY

The study of endocrine functions in its modern form was implemented in the latter half of the nineteenth and the first decades of the twentieth century [61].

Endocrinology is the study of disorders on endocrine system. It helps to diagnose a wide range of symptoms and variants and the management of disorders by the deficiency or excess of one or more hormones in fact, the key events that led to identification of endocrine functions took place between 1890 and 1905, in this period, and word “hormone” began to be institutionalized. Secretin, a chemical messenger secreted by the intestinal mucosa, was found in 1902 by Bayliss and Starling [62]. For this class of internal secretions, Earnest Starling and Edward Sharpey-Schafer (1905) proposed the name “hormone.” Endocrinology was introduced as a fresh scientific branch [63]. English physician George Redmayne Murray was effectively treated myxedema by using thyroid gland extract (1891). Sharpey-Schafer and George Oliver later identified the substance found in the adrenal gland extracts [Adrenaline, also called epinephrine] which was the cause of high blood pressure [64]. In 1901, Jokichi Takamine, a Japanese chemist isolated the adrenaline [65].

1.9 *IN VITRO* FERTILIZATION

Dr. Robert Edwards and Dr. Patrick Steptoe perform effective IVF treatment in England for the first time in November 1977. Eggs are collected from Lesley Brown and fertilized with sperm samples from John Brown [66]. Today, IVF is considered a medical procedure for infertility. In some instances, children have been conceived with donor eggs and sperm. An approximately 6.5 million IVF-conceived kids were born across the world. These babies of the so-called test tube are as healthy and normal as kids that are typically conceived.

1.10 ADVANCES IN CANCER TREATMENT

In the 20th century, both chemotherapy and radiotherapy were developed as treatments for cancer. Before the development of these treatments, the large number of individuals who have cancer died of the disease.

1.10.1 SURGERY

The discovery of general anesthesia in the mid of the 19th century commences a golden age of surgical innovation. In **1880**s, an American surgeon William Halsted pioneered radical cancer operations. The radical surgery proved that surgeons could remove cancers. The first radical mastectomy to treat breast cancer was conducted by Halsted. This surgical procedure remained as the normal surgery for breast cancer treatment until the latter half of the 20th century [67]. The surgeries have limitations in some cases as, if the tumor has initiated to spread or the tumors are at inaccessible sites.

1.10.2 RADIATION

S.W. Goldberg and Efim (London) used radiation therapy for cancer treatment for the first time (1903). They used radium to treat two skin basal cell carcinoma patients and in both patients, the disease was eradicated. Radiation therapy can be used to treat nearly all types of solid tumors, including the brain, breast, cervix, larynx, liver, lung, pancreas, prostate, skin, stomach, uterus, or soft tissue sarcomas, leukemia, and lymphoma [68]. The dose of radiation at each site depends on several factors of factors, including the radiosensitivity of each type of cancer and surrounding tissues or organs, etc., radiation therapy eliminates cancer cells by damaging their DNA (deoxyribonucleic acid). The significant benefits of radiotherapy have facilitated the implementation of modern, sophisticated methods for treatment, therapy, delivery, and imaging to be executed into regular radiation oncology practice.

1.10.3 INTRODUCTION OF ANTICANCER DRUGS AND CHEMOTHERAPY

At the beginning of the 20th century, chemotherapy was first developed, although it was not originally intended as a treatment for cancer. It was found during the Second World War that individuals exposed to nitrogen mustard had a considerably decreased number of white blood cells (WBC) [68]. This finding prompted scientists to explore the possibility of using mustard agents to avoid the development of rapidly dividing cells such as cancer cells. This discovery dramatically altered the scientific opinion and marked a turning point for cancer research until the researchers presumed that cancer-induced by acting on proteins rather than genes was caused by cancer. In 1949, the US Food and Drug Administration approved the nitrogen mustard as the first chemotherapeutic drug (mechlorethamine) for cancer treatment. Nitrogen mustard have a capacity to alkylate molecules including protein, DNA, and RNA, thus it belongs to the class of drugs called alkylating agents. It facilitates to destroy the cancer cells by chemically modifying their DNA. It promotes the destruction of cancer cells by altering their DNA chemically. In 1953, Roy Hertz and Min Chiu Li succeeded the first complete curing of a human solid tumor by using the drug Methotrexate [70]. Chemotherapy has the ability to harm surrounding healthy tissues, many of the chemotherapy side effects can be attributed to damage to cells that divide rapidly and these anti-mitotic drugs are sensitive to bone marrow cells, digestive tract, and hair follicles. Radiation may cause damages to these tissues. This results in the most prevalent side effects of chemotherapy such as mucositis with immunosuppression (inflammation of the digestive tract lining), hair loss, and autoimmune illnesses including rheumatoid arthritis, systemic lupus erythematosus, various sclerosis, vasculitis, and many others. The primary limitations of chemotherapy are the high toxicity and often does not destroy complete tumor, thus the chance recurrence is high. Researchers also developed chemotherapy by the mid-20th century, and is still used today.

1.10.4 TARGETED THERAPY

Targeted therapy another mode of cancer treatment it primarily assisted with is a drug. But it is distinct from traditional chemotherapy [71]. Targeted therapy operates by targeting the particular genes, proteins, or tissue environment of the cancer that contribute to the development and survival of cancer. These genes and proteins are discovered in cancer cells or cancer growth-related cells. Molecularly targeted therapy has enabling cancer cells to be destroyed and stopping the proliferation of tumor while saving healthy cells [72].

1.10.5 DEVELOPMENT OF MOLECULAR BIOLOGY AND GENETICS

DNA is the genetic sequence that drives the function of the cell and ultimately distinguishes between one life type and another. In the 1950s, Watson, and Crick discovered the DNA structure, medical science opened up new possibilities for all forms of life [73]. Each piece of DNA includes enormous quantities of data. Advanced computer technology permits the DNA codes of human and other life forms to be stored, organized, retrieved, and analyzed. Since this discovery, which laid the foundation for molecular biology, many research areas have changed significantly with new insights and developments and also found their way into our daily lives such as DNA sequencing, genetic fingerprinting, or personalized medicine. However, new sequencing techniques have opened up significant opportunities, especially in the field of medicine.

Genomic medicine has already demonstrated its benefit in cancer diagnosis and directing therapeutic approaches. Since the late 1990s, the cancer treatments including surgery, radiation, and chemotherapy has increasingly been supplemented by therapies aimed at specific molecular pathways in cancer growth and development [74]. Genomic information can now help clinicians to decide treatment strategies by classifying a tumor according to its mutations and related sensitivities to drugs. In some cases, patients were spared expensive and complex procedures based on a molecular diagnosis, such as bone marrow transplants. The genetic features may have an enormous impact on the effectiveness of anti-cancer drugs. More popularly, physicians are starting to use genomic data to refine cancer diagnosis and prognosis and improve the of life quality of person [75]. This technology allows the detection and monitoring of non-invasive tumor responses to therapy that assure an extensive improvement in patient management. In the late 1990s, The *Human Genome Project (HGP)* The HGP was started to record the pattern of coding for a sequence of human DNA; however, by the turn of the century it was not completed.

1.11 DEVELOPMENTS IN SURGERY

Development in the modern surgery was mainly based on the discovery of anesthesia, anti-septic, and a-septic practices. The introduction of intravenous injection agents such as barbiturates helps the patient to sleep rapidly without any obnoxious inhalation agents. Then in the 1940s and early 1950s the introduced muscle relaxants with curare in the initial period and then over the next decades, up an entire series of other agents. Halothane came in the mid-1950s, which was a revolutionary inhalation agent that was much easier to use. All of these groups of drugs have since been refined so there are now much more potent and less toxic intravenous agents, inhalational agents, local anesthetics, and muscle relaxants. Anesthesia has become very secure, and the death rate is very less [1 in 250,000]. However, with current highly advanced monitoring systems and a better understanding of body functions, will makes an improvement in this field. The major benefits of anesthesia are the accuracy of surgical procedures, including complex operations; it also helps to make surgeries more common.

1.11.1 PLASTIC SURGERY

Major developments in plastic surgery did not take place before the 20th century, when victims of war, especially soldiers, had to perform plastic surgery. Indeed, it was the First World War that brought plastic surgery to a unique level in the medical field. Soldiers injured with a number of extensive facial and head injuries induced by modern weapons had to be treated by military doctors. These serious wounds required new innovations in reconstructive surgery. The history of modern plastic surgery was commenced in 1960s and 1970s [76].

During this time, there have also been many important scientific developments. Silicone was a novel substance created at that period and it become more popular as a key ingredient of some plastic surgery procedure. It was initially used to treat defects of the skin. Now they are used as implants. In the 2000s, popularity of cosmetic surgery enhanced, and medical advances have made greatest achievements in plastic surgery, especially in reconstructive or cosmetic surgical treatment. Currently, the most significant trend in plastic surgery is the less invasive processes to eliminate the visible signs of aging.

1.11.2 ORGAN TRANSPLANTATION SURGERY

Scientists had to wait until the 19th and 20th centuries to perform successful organ transplants by the science and surgical techniques that made modern transplant medicine possible. Between 1900 and 1920, successful bone, skin, and cornea transplantation took place. The progresses in the transplantation of solid organs were started in the 1950s. The kidney was the first organ to be transplanted effectively in humans. Dr. Joseph E Murray [Peter Brent Brigham Hospital, Boston, Massachusetts]. The first effective kidney transplant between the identical twins was carried out in 1954 and Dr. Joseph E Murray received the Nobel Prize in Medicine in 1990 [78].

In the late 1960s, liver, heart, and pancreas transplantations were successfully performed, while in the 1980s lung and intestinal transplanting surgeries were started. The main problem was the body's tendency to activate the immune system against the "foreign" organ [rejection]. Patients were given strong medicines to restrain their whole immune system in order to prevent rejection, which may cause them life-threatening infections. Transplantation surgeries were restricted until around the early 1980s due to difficulties with organ rejection. Presently, kidney, liver, pancreas, lungs, heart, intestines, bone marrow and bones, corneas, skin, and heart valves are being transplanted.

A South African surgeon, Dr. Christian Barnard (1922–2001) transplanted a human heart from one person into another's body in 1967 at Cape Town. This first experience of clinical heart transplantation stimulated worldwide publicity and the procedure was rapidly co-opted by many surgeons. However, many patients died soon after the surgery, the number of cardiac transplants fell from 100 in 1968 to 18 in 1970 [79]. The main issue in the case of transplantation surgery was the normal tendency of the body to reject the new tissues. Significant developments in tissue typing and immunosuppressive drugs over the next 20 years have enabled more transplant procedures and enhanced recipient survival rates. In 1983, the Columbia University Medical Center launched clinical trials with cyclosporine-an immunosuppressive drugs originating from soil fungus (Discovered by Jean Borel in 1970) which was approved for commercial use (November 1983) and it is the most frequently prescribed immunosuppressant in organ transplantation [80]. When Cyclosporin was introduced as an immunosuppressive drug, many of the rejection problems were controlled. Advanced medical technology prevents organ rejection, has led to more efficient transplantation and increased demand.

1.11.3 LAPAROSCOPY (KEYHOLE SURGERY)

Laparoscopy was invented by George Kelling in 1901, in Germany. In 1910, Hans Christian Jacobaeus of Sweden performed the first laparoscopic operation in humans. It is a surgical procedure performed in the abdomen or pelvis using minute cut (usually 0.5–1.5 cm) with the aid of a camera [81]. Laparoscopic surgery includes operations within the abdominal or pelvic cavities, thoracic or chest cavity [thoracoscopic surgery] using endoscopes and ultrasound scanning. Laparoscopic and thoracoscopic surgery belong to the broader field of endoscopy. Laparoscopic surgery has been extensively accepted technical innovation. Laparoscopic surgery has also been generally used in hepatic, pancreatic, gynecological, and urological surgery. Advanced single-incision laparoscopic surgery (SILS) and robotic surgery are promising an excellent diagnosis and treatments [82]. A surgical robot is a computer-controlled system that can help a surgeon to use and control surgical tools. These robots significantly enhanced the quality of the surgeries, even for experienced laparoscopic surgeons. The development of highly sophisticated surgical robots helps to perform single-port surgery for complex procedures.

1.12 MEDICAL ADVANCES IN THE 21ST CENTURY

The rapid medical advances in this era have been encouraged by communication among scientists around the world. They freely exchanged thoughts and concepts and reported on their research efforts through journals, meetings, and later computers and electronic media. The technology thus became much developed in the 21st century, particularly in the case of medical science, offering excellent diagnosis, treatments, and almost completely wiping out and controlling life-threatening diseases, etc. It provides a healthy social life than in the past decades.

1.12.1 COMPLETION OF HUMAN GENOME PROJECT (HGP)

One of the most exceptional medical breakthroughs of the 21st century was the HGP. It was complete on 14 April 2003 [83]. The project achieved its goal in sequencing and mapping a 99.99% human genome within the intended target. This is a door that opens to more medical inventions than a breakthrough. Genome sequencing currently has the greatest effect in the treatment of cancer, the identification of genetic disease, and the provision of information on the probable reaction of an individual to medication.

1.12.2 IMPROVEMENTS IN SURGERY

Another result of new technologies generated by information science is the modern trend in surgery toward minimally invasive and non-invasive therapeutic procedures. Robotic surgeries allow minimal invasive procedure and less pain and blood loss and avoid the need for blood transfusions, also reduces the risk of infections and other complication thus the patient cured with a potentially lesser recovery time. It has demonstrated to be more effective in treating cancerous and non-cancerous tumors than standard surgical methods [84]. Face transplantation is an innovative achievement of modern reconstructive surgery and is about to become a prevalent chance for surgery. Partial face transplantation was first performed in France in 2005, whereas in Spain in 2010 complete face replacement was carried out [85]. Turkey France, the United States, and Spain are the leading countries in research and successful face transplant procedures. Face transplantation is performed to enhance the worth of life for a person seriously disfigured by facial trauma, burns, disease, or birth defects. It is intended to enhance both appearance and functional abilities, such as chewing, swallowing, talking, and breathing through the nose.

1.12.3 ADVANCEMENT IN STEM CELL RESEARCH, ARTIFICIAL ORGANS, AND 3D PRINTING

Stem Cell Research another innovative and active field in biomedical research. Embryonic stem cell research can lead to more efficient treatments for severe illnesses such as juvenile diabetes, Parkinson's disease (PD), heart failure, and spinal cord injuries. Stem cells differ from other body cells because they are capable of differentiating into other types of cells/tissues. This capacity enables them to replace dead or faulty cells and has been used in patients with certain disorders or defects to replace erroneous cells/tissues [86]. In 2013, Japanese researchers at Yokohama City University succeeded in creating a functional human liver from stem cells with the incredible breakthrough [87]. Similarly, Researchers at Duke University have grown bundles of muscle fibers that twitch and respond to electrical stimuli [88].

3D printing could also bring enormous advantages to patients waiting for organ transplantation. Research on using 3D printers to produce artificial heart, kidney, and liver structures, as well as other significant organs, is presently being performed. 3D printing for cellular production was first launched in 2003, while Thomas Boland [Clemson University] patented the use of cell inkjet printing [89]. This method used a modified spotting scheme to place cells in integrated 3D matrices on a substrate. Organs that have been successfully printed and implemented in a medical environment, organs that have been effectively printed and applied either flat [skin], vascular, [blood vessels], or hollow[bladder]. They are often generated with the recipient's own cells when artificial organs are ready for transplantation. More complicated organs are being researched, those consisting of strong cellular structures, including the heart, pancreas, and kidneys [90]. Organovo Company manufactured a human liver using 3D bio-printing in 2013, although not suitable for transplantation, and was mainly used as a medium for drug testing [91].

1.12.4 PROGRESS IN TREATMENTS

The progress of immunotherapy and gene therapy improves cancer treatment. Immunotherapy was developed in the 1990s. The pioneering research in this field by P. Allison and Tasuku Honjo helped to inflict the intrinsic capacity of the immune system to attack tumor cells. (This work awarded with Nobel Prize in Physiology or Medicine in 2018). It helped to develop a completely new therapeutic concept against cancer [92]. It Works for many types and stages of cancer, Durable in many individuals, less toxic and Works well with other treatments. The use of gene therapy technology to treat blood cancers such as leukemia is one of the most exciting medical developments in recent history [93]. Recent studies have also disclosed the ability to use gene therapy to reverse other cancers such as breast cancer. The gene therapy can be used to eliminate the need for traditional therapies such as radiation, chemotherapy, or surgery.

AIDS remains one of the most deadly diseases of the modern age. Apparently, with the recent combination therapy involving various antiretroviral active drugs, the survival rate of HIV positive patients has risen considerably. The Highly Active Antiretroviral Therapy maintains the function of immune system and prevents infections that cause death in a patient [94]. Death from cardiovascular diseases has decreased by an enormous margin over the past decade. Cardiac issues can be handled with genetically engineered tissue plasminogen activator, frequently known as t-PA, by busting blockage and stopping blood circulation. Blocks in the blood vessel can be opened through an artery with a stent [95]. Another artery is substituted for the damaged artery by the bypass surgery thus it rising the heart patients' survival rate.

1.12.5 INTRODUCTION OF NANOTECHNOLOGY

The implementation of nanotechnology is another inventive sector that flourishes in the 21st century. The new word nanomedicine relates to the use of nanomaterials, nano, electronic biosensors, and nanoparticles to initiate the molecular level corrective procedure. Nanotechnology promotes diagnoses at the cellular and subcellular levels and also permits the rapid identification of diseases. It is considered that nanomedicine will have the greatest effect on drug delivery and regenerative medicine. It enables the targeted drug delivery, which enhancing effectiveness and significantly reducing side effects. It also provides efficacy for therapeutic materials being released under-regulated manner. Nanoparticles are also used to enhance the inherent repairing processes of the body; artificial activation and control of adult stem cells is a significant part of these studies [96].

1.13 CONCLUSION

Every century has brought revolutionary changes in healthcare and medicine throughout human history. Today, owing to findings such as penicillin inventions such as x rays and accomplishments such as the eradication of smallpox, we have expanded the average rate of human life expectancy by about 25 years. We have learned to repair hearts during this time span, mapped DNA, and even carried out a partial brain transplant. New developments in science, engineering, and computer technology over the next decade will revolutionize health care and medicine. Adopting digital technology in healthcare and medicine has resulted in fresh practices, treatments, and techniques that have saved our lives. Maybe nothing in modern history has had such an impact on healthcare and medicine fields by the emergence of the digital age.

KEYWORDS

- **bacteriology**
- **computerized axial tomography**
- **digital age**
- ***human genome project***
- **nuclear magnetic resonance**
- **vaccination**

REFERENCES

1. Jackson, M., (2014). *The History of Medicine: A Beginner's Guide.*
2. Dobanovački, D., Milovanović, L., Slavković, A., Tatić, M., Mišković, S. S., Škorić-Jokić, S., & Pećanac, M., (2012). Surgery before common era (B.C.E.*). *Arch. Oncol., 20*, 28–35. doi: 10.2298/AOO1202028D.
3. Kriss, T. C., & Kriss, V. M., (1998). History of the operating microscope: From magnifying glass to micro neurosurgery. *Neurosurgery, 42*, 899–908. doi: 10.1097/00006123-199804000-00116.
4. Bloom, D. E., (2011). The value of vaccination. *Adv. Exp. Med. Biol., 697*, 1–8. doi: 10.1007/978-1-4419-7185-2_1.
5. Elliot, H., (2001). A history in surgery. In: *A Hist. Surg.* (pp. 4, 5).
6. Bosque, O. G., & Hsiang, W., (2018). Medical technology. *Yale J. Biol. Med., 91*, 203–205.
7. Afr, S., & Cam, J. T., (2007). *A. J. Traditional Review, 4*, 319–337.
8. Dawson, W. R., (1929). Egyptian medicine. *Nature, 124*, 776–777. doi: 10.1038/124776a0.

9. Meyerhof, M., (2002). *Ancient Egyptian Medicine* (Vol. 8, pp. 198–200). Hasan Kamal, Isis. doi: 10.1086/358371.
10. Saber, A., (2010). Ancient Egyptian surgical heritage. *J. Investig. Surg., 23*, 327–334. doi: 10.3109/08941939.2010.515289.
11. El-Zawahry, M. D., Ramzy, A. F., El-Sahwi, E., Bahnasy, A. F., Khafaga, M., Rizk-Allah, M. A., & El-Hoda, M. F. A., (1997). Surgery in Egypt. *Arch. Surg., 132*, 698–702. doi: 10.1001/archsurg.1997.01430310012001.
12. Ashworth, U. E., (2014). History of medicine: China. *Encycl. Br.*
13. Zhang, W. B., (2013). Analysis on the concepts of qi, blood, and meridians in huangdi neijing (Yellow Emperor's Canon of Internal Classic). *Zhongguo Zhen Jiu.*, 33.
14. Yu, F., Takahashi, T., Moriya, J., et al., (2006). Traditional Chinese medicine and kampo: A review from the distant past for the future. *J. Int. Med. Res., 34*, 231–239, doi:10.117 7/147323000603400301..
15. Guo, R., (2010). The foundations of Chinese medicine. *Focus Altern. Complement. Ther., 11*, 71–71. doi: 10.1111/j.2042-7166.2006.tb01251.x.
16. Churton, M., Brown, A., Churton, M., & Brown, A., (2016). Traditional theory. *Theory and Method*, 7–83. doi: 10.1007/978-1-137-02163-2_2.
17. Bedford, D. E., (1951). The ancient art of feeling the pulse. *Br. Heart J., 13*, 423–437.
18. Chang, S., (2012). The meridian system and mechanism of acupuncture: A comparative review. Part 1: The meridian system. *Taiwan. J. Obstet. Gynecol. 51*, 506–514. doi: 10.1016/j.tjog.2012.09.004.
19. Quiroz-González, S., Torres-Castillo, S., López-Gómez, R. E., & Jiménez, E. I., (2017). Acupuncture points and their relationship with multireceptive fields of neurons. *JAMS J. Acupunct. Meridian Stud., 10*, 81–89. doi: 10.1016/j.jams.2017.01.006.
20. Li, A., Wei, Z. J., Liu, Y., et al. (2016). Moxibustion Treatment for Knee Osteoarthritis. Med. (United States), *95,* doi:10.1097/MD.0000000000003244.
21. Reddy, D. V. S., (1936). Ancient texts of Indian medicine. *Br. Med. J., 2*, 1056–1057. doi: 10.1136/bmj.2.3959.1056-d.
22. Pandav, C., Desai, S., Singh, J., & Desai, M., (2012). Contributions of ancient Indian physicians-implications for modern times. *J. Postgrad. Med., 58*, 73. doi: 10.4103/0022-3859.93259.
23. Loukas, M., Lanteri, A., Ferrauiola, J., Tubbs, R. S., Maharaja, G., Shoja, M. M., Yadav, A., & Rao, V. C., (2010). Anatomy in ancient India: A focus on the susruta samhita. *J. Anat., 217*, 646–650. doi: 10.1111/j.1469-7580.2010.01294.x.
24. Band, I. C., & Reichel, M., (2017). Al rhazes and the beginning of the end of smallpox. *JAMA Dermatology, 153*, 420. doi: 10.1001/jamadermatol.2017.0771.
25. Edriss, H., Rosales, B. N., Nugent, C., Conrad, C., & Nugent, K., (2017). Islamic medicine in the middle ages. *Am. J. Med. Sci., 354*, 223–229. doi: 10.1016/j.amjms.2017.03.021.
26. Sobhani, Z., Nami, S. R., Emami, S. A., Sahebkar, A., & Javadi, B., (2017). Medicinal plants targeting cardiovascular diseases in view of Avicenna. *Curr. Pharm. Des., 23*. doi: 10.2174/1381612823666170215104101.
27. Gupta, S., (2015). Hippocrates and the Hippocratic oath. *J. Pract. Cardiovasc. Sci., 1*, 81. doi: 10.4103/2395-5414.157583.
28. Ravenel, M. P., (2008). Greek medicine. *Am. J. Public Heal. Nations Heal., 27*, 294–295. doi: 10.2105/ajph.27.3.294.
29. Grube, G. M. A., (2006). Greek medicine and the Greek genius. *Phoenix., 8*, 123. doi: 10.2307/1086122.

30. Glouberman, S., & Glouberman, S., (2018). Galen's four humors: The first medical model. In: *Mech. Patient* (pp. 15–23). doi: 10.4324/9780429490330-3.
31. Weakland, J. E., (2002). Medieval and early renaissance medicine. *Hist. Eur. Ideas., 14*, 302–303. doi: 10.1016/0191-6599(92)90273-f.
32. Toledo-Pereyra, L. H., (2015). Medical renaissance. *J. Investig. Surg., 28*, 127–130. doi: 10.3109/08941939.2015.1054747.
33. Banerjee, A. D., & Nanda, A., (2011). Ambroise Paré and 16th century neurosurgery. *Br. J. Neurosurg., 25*, 193–196. doi: 10.3109/02688697.2010.544786.
34. Tershakowec, M., & Bagwell, C. E. (1981). Ambroïse Paré and the renaissance of surgery. *Surg Gynecol Obstet. 152*(3), 350-354.
35. Gibson, T., (1955). The prostheses of Ambroise Paré. *Br. J. Plast. Surg., 8*, 3–8. doi: 10.1016/S0007-1226(55)80003-3.
36. Baskett, F., (2019). Paré, Ambroise (1510–1590). In: *Eponyms Names Obstet. Gynaecol.* (pp. 312–313). doi: 10.1017/9781108421706.253.
37. Oppenheimer, J., (2004). The embryological treatises of *Hieronymus fabricius* of aquapendente. In: Howard, B. A., (ed.), *Q. Rev. Biol., 43*, 329–330. doi: 10.1086/405851.
38. O'rourke, B. M., (2013). Harvey, by Hercules! The hero of the blood's circulation. *Med. Hist., 57*, 6–27. doi: 10.1017/mdh.2012.78.
39. Plotkin, S., (2014). History of vaccination. *Proc. Natl. Acad. Sci. U.S.A. 111*, 12283–12287, doi:10.1073/pnas.1400472111.
40. Riedel, S., (2005). Edward Jenner and the history of smallpox and vaccination. *Proc. (Bayl. Univ. Med. Cent), 18*, 21–25.
41. Forrester, R., & Forrester, C. R., (2013). The history of medicine revised. *JAMA, 309*, 528. doi: 10.1001/jama.2012.145217.
42. Pai-Dhungat, J. V., & Parikh, F., (2015). James Young Simpson and painless labor. *J. Assoc. Physicians India, 63*, 50.
43. Wawersik, J., (1991). History of anesthesia in Germany. *J. Clin. Anesth., 3*, 235–244. doi: 10.1016/0952-8180(91)90167-L.
44. Lane, N., (2015). The Unseen World: Reflections on Leeuwenhoek, 1677. "Concerning little animals." *Philos. Trans. R. Soc. B Biol. Sci. 370*, doi:10.1098/rstb.2014.0344.
45. Bastian, H. C., (1875). The germ-theory of disease. *Br. Med. J., 1*, 469476. doi: 10.1136/bmj.1.745.469.
46. Fu, Z. F., (1997). Rabies and rabies research: Past, present and future. *Vaccine, 15*. doi: 10.1016/S0264-410X(96)00312-X.
47. Gaynes, R. P., (2014). Louis Pasteur and the germ theory of disease. In: *Germ Theory* (pp. 143–171). doi: 10.1128/9781555817220.ch9.
48. Harries, A. D., (2008). Robert Koch and the discovery of the tubercle bacillus: The challenge of HIV and tuberculosis 125 years later. *Int. J. Tuberc. Lung Dis., 12*, 241–249.
49. Grimes, D. J., (2016). Koch's postulates: Then and now. *Microbe Mag., 1*, 223–228. doi: 10.1128/microbe.1.223.1.
50. Pennington, T. H., (1995). Listerism, its Decline and its Persistence: The Introduction of aseptic surgical Techniques in three British Teaching Hospitals, 1890–1899. *Med. Hist., 39*, 35–60. doi: 10.1017/s0025727300059470.
51. Riley, J. C., (2015). *Rising Life Expectancy: A Global History*. doi: 10.1017/CBO9781316036495.

52. Cruickshank, R., (1947). Control of infectious diseases. *Health Educ. J., 5*, 174–177. doi: 10.1177/001789694700500408.
53. Aminov, R. I., (2010). A brief history of the antibiotic era: Lessons learned and challenges for the future. *Front. Microbiol., 1*, 1–7. doi: 10.3389/fmicb.2010.00134.
54. Bennett, J. W., & Chung, K. T., (2001). Alexander Fleming and the story of penicillin. *Adv. Appl. Microbiol., 49*, 56. sci-hub.tw/10.1016/s0065-2164(01)49013- (accessed on 24 June 2020).
55. Kathrin I. Mohr., (2016). History of antibiotics research. *Curr. Top. Microbiol. Immunol., 398*, 237–272. doi: 10.1007/82_2016_499.
56. Turk, J. L., (1994). Leonard Colebrook: The chemotherapy and control of streptococcal infections. *J. R. Soc. Med., 87*, 727–728.
57. Frierson, J. G., (2010). The yellow fever vaccine: A history. *Yale J. Biol. Med., 83*, 77–85.
58. Cobey, S., & Hensley, S. E., (2018). Immune history and influenza vaccine effectiveness. *Vaccines, 6*, 28. doi: 10.3390/vaccines6020028.
59. Blume, S., & Geesink, I., (2000). A brief history of polio vaccines. *Science 288*(80), 1593, 1594. doi: 10.1126/science.288.5471.1593.
60. Davidson, T., (2017). *Vaccines: History, Science, and Issues, 1*. ABC-CLIO.
61. Levy, M., & Coll, T., (2015). A history of endocrinology: The fantastical world of hormones. *Endocrinologist*, 1128.
62. William, M. B., & Ernest, H. S., (1998). The discovery of secretin. *Endocrinologist, 8*, 1–5. doi: 10.1097/00019616-199801000-00001.
63. Henriksen, J. H., & Schaffalitzk, O. B., (2002). Secretin, its discovery, and the introduction of the hormone concept. *Scand. J. Clin. Lab. Invest., 60*, 463–472. doi: 10.1080/003655100448446.
64. The Treatment of Myxedema with Thyroid Extract, (2011). *Jama J. Am. Med. Assoc., 75*, 36. doi: 10.1001/jama.1920.02620270040013.
65. Rao, Y., (2019). The first hormone: Adrenaline. *Trends Endocrinol. Metab.*, doi: 10.1016/j.tem.2019.03.005.
66. Bavister, B. D., (2002). Early history of *in vitro* fertilization. *Reproduction, 124*, 181–196. doi: 10.1530/rep.0.1240181.
67. Ozmen, V., (2017). Paradigm shift from Halstedian radical mastectomy to personalized medicine. *J. Breast Heal., 13*, 50–53. doi: 10.5152/tjbh.2017.312017.
68. Mould, R. F., (2007). Priority for radium therapy of benign conditions and cancer. *Curr. Oncol., 14*, 118–122. doi: 10.3747/co.2007.120.
69. American Cancer Society, (2014). *Evolution of Cancer Treatments: Chemotherapy* (pp. 1–6). Cancer.org.
70. ElBagoury, M., & Kotb, M., (2018). Chemotherapy over the years. *J. Pharm. Sci. Res., 10*, 316–318.
71. Charlton, P., & Spicer, J., (2016). Targeted therapy in cancer. *Med. (United Kingdom), 44*, 34–38. doi: 10.1016/j.mpmed.2015.10.012.
72. Padma, V. V., (2015). An overview of targeted cancer therapy. *Biomed., 5*, 1–6. doi: 10.7603/s40681-015-0019-4.
73. Fry, M., (2016). Discovery of the structure of DNA. In: *Landmark Exp. Mol. Biol.*, (pp. 143–247). doi: 10.1016/b978-0-12-802074-6.00005-9.
74. Goodman, D. M., Lynm, C., & Livingston, E. H., (2013). Genomic medicine. *JAMA-J. Am. Med. Assoc., 309*, 1544. doi: 10.1001/jama.2013.1927.

75. Norton, M. L., (2009). Genome medicine: The future of medicine. *Genome Med.,* 1. doi: 10.1186/gm1.
76. Saad, M. N., Lechtveld, P., & Lewis, E., (1976). Reviews in plastic surgery and general plastic and reconstructive surgery. *Plast. Reconstr. Surg., 57,* 91. doi: 10.1097/00006534-197601000-00020.
77. Jabbari, B., Grunzweig, K., & Totonchi, A., (2018). Botox in plastic surgery. In: *Botulinum Toxin Treat* (pp. 147–155). doi: 10.1007/978-3-319-99945-6_12.
78. Haghighat, R., (2011). *375 Years of Science at Harvard,* 8–13.
79. Hakim, N. S., Papalois, V. E., Kroshus, T. J., & Kshettry, V. R., (2010). The history of heart transplantation and heart valve transplantation. In: *Hist. Organ Cell Transplant* (pp. 194–208). doi: 10.1142/9781848160057_0009.
80. Calne, R. Y., (1985). Cyclosporin a as an immunosuppressive agent in transplantation. In: *Chronic Ren. Dis.,* (pp. 497–505). doi: 10.1007/978-1-4684-4826-9_50.
81. Hatzinger, M., Kwon, S. T., Langbein, S., Kamp, S., Häcker, A., & Alken, P., (2006). Hans Christian Jacobaeus: Inventor of human laparoscopy and thoracoscopy. *J. Endourol., 20,* 848–850. doi: 10.1089/end.2006.20.848.
82. Udwadia, T., (2010). Single-incision laparoscopic surgery: An overview. *J. Minim. Access Surg., 6,* 91. doi: 10.4103/0972-9941.72354.
83. Kanavakis, E., & Xaidara, A., (2001). The human genome project. *Arch. Hell. Med., 18,* 475–484. doi: 10.1258/jrsm.98.12.545.
84. Sodergren, M. H., & Darzi, A., (2013). Robotic cancer surgery. *Br. J. Surg., 100,* 3, 4. doi: 10.1002/bjs.8972.
85. Barret, J. P., Gavaldà, J., Bueno, J., Nuvials, X., Pont, T., Masnou, N., Colomina, M. J., et al., (2011). Full face transplant: The first case report. *Ann. Surg., 254,* 252–256. doi: 10.1097/SLA.0b013e318226a607.
86. Baharvand, H., & Aghdami, N., (2012). *Advances in Stem Cell Research.* doi: 10.1007/978-1-61779-940-2.
87. Taniguchi, H., (2015). *Organ Generation Using iPS Cells* (pp. 23–24).
88. Madden, L., Juhas, M., Kraus, W. E., Truskey, G. A., & Bursac, N., (2015). Bioengineered human myobundles mimic clinical responses of skeletal muscle to drugs. *Elife.,* 4, doi: 10.7554/elife.04885.
89. Boland, (2006). *Ink-Jet Printing of Viable Cells.*
90. Boland, T., Mironov, V., Gutowska, A., Roth, E. A., & Markwald, R. R., (2003). Cell and organ printing 2: Fusion of cell aggregates in three-dimensional gels. *Anat. Rec.-Part A Discov. Mol. Cell. Evol. Biol., 272,* 497–502. doi: 10.1002/ar.a.10059.
91. Choudhury, D., Anand, S., & Naing, M. W., (2018). The arrival of commercial bioprinters-towards 3D bioprinting revolution! *Int. J. Bioprinting,* 4. doi: 10.18063/ijb.v4i2.139.
92. Allison, J. P., (2018). *Discovery of Cancer Therapy by Inhibition of Negative Immune Regulation: The 2018 Nobel Prize in Physiology or Medicine.* Nobel Assem. Karolinska Institutet.
93. Mulherkar, R., (2001). Gene therapy for cancer. *Curr. Sci., 81,* 555–560.
94. Arts, E. J., & Hazuda, D. J., (2012). HIV-1 antiretroviral drug therapy. *Cold Spring Harb. Perspect. Med., 2.* doi: 10.1101/cshperspect.a007161.
95. Manuscript, A., (2013). *Introduction : Plasminogen Activators in Ischemic Stroke,* 41. doi: 10.1161/STROKEAHA.110.595769.
96. Saini, R., Saini, S., & Sharma, S., (2010). Nanotechnology: The future medicine. *J. Cutan. Aesthet. Surg., 3,* 32. doi: 10.4103/0974-2077.63301.

CHAPTER 2

Immunomodulatory Effect of Plant-Based Extracts on Neurodegeneration

KOEL SINHA and CHITRANGADA DAS MUKHOPADHYAY

Center for Healthcare Science and Technology, Indian Institute of Engineering Science and Technology, P.O., Botanical Garden, Shibpur, Howrah, West Bengal – 711103, India, E-mails: ksinha2110@gmail.com (K. Sinha), chitrangadadas@yahoo.com (C. D. Mukhopadhyay)

ABSTRACT

Mechanisms including blood-brain barrier (BBB) integrity, oxidative stress, chronic stress, hippocampal neurogenesis, microglial activation and chronic low-grade neuroinflammation, have shown to be related to cognitive changes across age. Supplementation with one or more plant based extracts or nutraceuticals that act on these mechanisms may be an important next step toward preventing age associated cognitive decline. Dopamine (DA) and serotonin (5-HT) are the main neurotransmitters found in the central nervous system (CNS) which possess immunomodulatory functions. Various experimental evidences support that these molecules are involved in modified cytokine secretion, apoptosis and cytotoxicity. Extracts from *Withania somnifera, Bacopa monnieri, Gingko biloba,* and *Curcuma longa* are being studied for their anti-inflammatory and immunomodulatory properties as well as their cognitive enhancing effects. Extraction is considered to be the important step in the analysis of constituents and the following mentioned extraction techniques such as soxhlet, room temperature, super critical fluid extraction are used. A neurotransmitter profiling findings are useful in pharmacological regulation of the serotonergic and dopaminergic system modulated immune function and may help in establishing therapeutic alterations. The translational gap between *in vitro*, *in vivo* and clinical studies is still a major issue especially with respect to immunomodulatory effects. So this study aims to join that gap in a holistic approach.

2.1 INTRODUCTION

Neurodegenerative diseases such as Alzheimer's disease (AD), Parkinson's disease (PD), Huntington's disease (HD), and amyotrophic lateral sclerosis (ALS) are generally characterized by progressive loss as well as the death of neurons [1]. One of the main causes of neurodegenerative disease is aging, which carries mitochondrial dysfunction, chronic immune-inflammatory response, loss of blood-brain barrier (BBB) integrity, enhancement of oxidative stress, and reduction of hippocampal neurogenesis along with anchorage of chronic low-grade neuroinflammation [2]. Besides, neuronal loss interplay between adaptive and innate immune system also performs an important role in the onset and progression of neurodegenerative diseases. Interestingly, immune in an organism acts as a major defense mechanism for protection against foreign invaders as well as to eliminate diseases. They include the involvement of many types of cells of which some function either as immunostimulants or as immunosuppressors [3]. The process of enhancement of immune reactions is termed as "immunomodulation" which mainly involves the stimulation of non-specific systems. Thus, the suppression of the immune system's elements may allow the pathogenic organisms to surge over the host and may lead to secondary infections [4].

The knowledge about the neurodegenerative disease etiology and immunomodulatory outcomes targeted therapies like neurotransmitter modulators, stem cell-based therapies, hormone replacement therapy, neurotrophic factors, as well as regulators of the mRNA synthesis and its translation into disease-causing mutant proteins have been developed [5–10]. Therefore, the above strategies are great tools to combat neurodegeneration and are often associated with adverse effects and long-term unknown consequences [7, 11]. Neural transplantation and stem cell-based therapies are emerging trends to fight against the neurodegenerative diseases. Despite the advancement in these fields of research, the therapeutic applications of such novel tools are still far from reaching the clinics. Therefore, there is an immense urge towards a long duration alternative therapeutics to combat against this debilitation neurodegenerative disease. Currently, a new approach towards immunomodulation with medicinal plants has been developed for enhancing the host defense mechanism against neurodegenerative diseases. In the Indian medicinal system, a huge number of plants along with their bioactive constituents exhibit immunomodulatory characteristics and hence can be considered as a safer alternative to reduce the adverse effects of modern drugs in the immunological system. Besides, whole plants as well as secondary metabolites can also be used as drugs to treat several neurodegenerative diseases. Not only phytodrugs, but rasayana drugs also plays a crucial role towards the improvement of defense mechanism as well as immunomodulatory activity. Also, many nutraceuticals like proteins, alkaloids, phenolic compounds show immunomodulatory characteristics along with anti-oxidant and anti-inflammatory properties. In this chapter, we will provide a review on some of the plant-based extracts with established immunomodulatory properties and their efficacy in treating neurodegenerative diseases has been discussed along with a broader vision and a better understanding of how these diseases could be related to each other along with the immunomodulatory property of neurotransmitter on various neurodegenerative diseases.

2.2 NEURODEGENERATIVE DISEASES

Neurodegenerative diseases are incurable and debilitating conditions leading to chronic brain damage and neurodegeneration. The etiology of the disease is still elusive, although improved experimental models showed miserable conditions associated with mutated genes, accumulation of abnormal proteins, increased reactive oxygen species (ROS), or destruction of the neurons in a definite area of brain.

2.2.1 ANATOMY AND FUNCTION OF NEURODEGENERATIVE BRAIN

Besides, the classical symptoms, neurodegenerative disease also show other signs and symptoms such as weight loss, abnormal mood, and endocrine perturbations Moreover, major damage is caused in the small brain structure in the ventral side of brain known as hypothalamus symmetrically located on both the sides of the third ventricle. The hypothalamus carries signals from the CNS to the periphery and regulates several functions including reproduction, intake of food, and control of circadian rhythm as well as sleep-wake cycle. Indeed, alteration in normal condition of hypothalamus has been observed in pre-symptomatic carriers or in patients with initial phase of neurodegenerative disease [12]. Furthermore, pathological aggregates and multiple neuropeptidergic populations in the hypothalamus of most neurodegenerative disease patients leading to the observed metabolic phenotype as well as to other non-metabolic symptoms. Despite pathological evidence, to date, a few studies have determined the functional role of hypothalamic alterations in the case of both neurodegenerative disease progression and dysfunction.

2.2.2 MECHANISMS ASSOCIATED WITH NEURODEGENERATION

1. **Mitochondrial-Mediated Neurodegeneration:** Manifestations of neurodegenerative diseases such as AD, PD, HD, MS, and ALS are largely associated with mitochondrial damage [13–16]. Several mitochondrial-dependent mechanisms such as inhibition of mitochondrial electron transport chain (ETC)'s complexes, ROS generation, and impairment of mitochondrial dynamics could contribute to the pathogenesis of neuronal injury as well as neurodegenerative diseases [17, 18]. Mitochondrial abnormalities and defective ETC (Complex I) present in substantia nigra is the foremost cause of neuronal damage in AD, PD, and ALS whereas, neuronal damage in MS is also attributed to Complex I and IV of the ETC along with the loss of mitochondrial membrane potential ($\Delta\Psi_m$) [19, 20]. Furthermore, evidences from clinical studies suggested the active role of mitotoxicity in the neuronal degeneration in HD [13]. Huntingtin, the gene responsible for HD, is reported to directly impair the mitochondrial functions [21]. Based on these findings it can be concluded that mitochondrial damage is the primary event observed in various neurodegenerative diseases. Owing to complex interplay between mitochondrial-toxicity and oxidative stress, it has been unclear whether mitochondrial damage is the main consequence of neuronal damage. Evidences also proved that mitotoxicity-related oxidative stress is also the leading cause for the development of neurodegeneration.
2. **Oxidative Stress-Mediated Neurodegeneration:** Our nervous system is vulnerable to oxidative stress due to several reasons including: (i) high oxygen consumption, (ii) relatively lower level of endogenous antioxidants, (iii) polyunsaturated lipids susceptible to free radical in the plasma membrane of large neuronal cells, and (iv) presence of excitatory neurotransmitters whose metabolism can produce ROS [22–24]. These characteristics make neuronal cells highly prone to ROS-mediated damage. Neuronal degeneration in

AD patients is associated with the oxidative damage to DNA, RNA, proteins, and lipids whereas; oxidative damage to DNA and protein has also been reported in the nigro-striatal region in PD [25–27]. Moreover, oxidative stress-mediated mutations in the gene coding for the ubiquitous Cu/Zn-superoxide dismutase (SOD-1) enzyme and damage to the biomolecules have been associated with the familial forms of the ALS [28, 29]. On the other hand, ROS produced by the activated microglia and mononuclear cells and performs a crucial role in MS pathogenesis. Indeed, oxidative damage to nuclear DNA, mitochondrial DNA has been linked to demyelination and axonal injury in MS [30]. Therefore, all these evidences clearly indicate a causal relationship between the oxidative stress and neuronal degradation.

2.2.3 SYMPTOMS OF NEURODEGENERATION

Although, our understanding about the symptoms involved in neuronal damage has increased in recent years, due to advanced medical technology. The symptoms generally vary from agitation, irritability, and impulsivity followed by apathy and indifference. Other symptoms vary from depression to euphoria, from delusions and hallucinations to anxiety and sleep disturbance, from loss of empathy and socially inappropriate behavior along with changes in eating behavior and stereotyped behaviors such as pacing, wandering, and rummaging. These changes cut across diseases and occur frequently in AD, PD, and a host of other conditions including HD and corticobasal syndrome (CBS) [31].

2.2.4 TREATMENTS AND DRAWBACKS OF NEURODEGENERATIVE DISEASES

Till date, it has been impossible to find a proper cure against neurodegenerative diseases, but medication and therapeutic strategies available nowadays has been addressed to improve the quality of life in patients. Although, there are some treatment options which are effective in enhancing the treatment of AD and PD to some extent. But still now no feasible treatment options have been discovered against ALS and HD. Regarding PD, the association between carbidopa and levodopa represents a potential standard for symptomatic treatment of PD [32]. Levodopa is a natural chemical in brain, when combined with carbidopa can be easily converted to dopamine

(DA). This carbidopa generally prevents conversion of levodopa to DA before entering the brain. Indeed, this is most effective treatment against PD, but due to long-term use, the effect starts to fluctuate. Other various side effects experienced includes nausea, feeling of light-headedness, and sudden involuntary movements. Moreover, the association between levodopa and catechol-O-methyltransferase inhibitors helps to prolong the effects of levodopa by blocking brain enzymes that deplete DA concentration. These side-effects are similar with that of levodopa mainly diarrhea and involuntary movements. With regard to dopaminergic, agonist which generally mimics the DA effect in the brain and are generally not effective as levodopa but the effects are long lasting and can be used in conjunction with levodopa to counter any fluctuation in efficiency. These types of medications can be administered by oral medications or as an injection. The side effects are drowsiness, hallucinations, and compulsive behaviors such as gambling and overeating. This intervention needs careful evaluation due to its adverse side effects and requires a gradual interruption of the therapy. Though already proved effective in amelioration of symptoms but most of the above treatment strategy are unable to undergo modification in terms of disease evolution and progression. Therefore, it is necessary to develop fruitful therapies that provide neuroprotective effects and slows down the progression of neurodegeneration mechanisms involved in PD. With regards to the treatment options for AD, it has no significant effect either on symptoms or disease progression. The current approved treatments against AD utilizes two strategies such as symptomatic and disease-modifying treatment. In this scenario, for symptomatic treatment anticholinesterase inhibitors act as a potent component. On the other hand, for disease-modifying treatment antioxidant and anti-inflammatory agents could be used. Moreover, recent treatments for AD patients temporarily became successful in slowing cognitive deteoriation in AD. The effects of these strategies are at its best marginal and are prescribed as there is not anything better option to be used to fight against AD. Various ongoing clinical trials and the search for effective drug against AD being pursued worldwide based on its pathogenetic mechanisms.

2.2.5 TREATMENT OF NEURODEGENERATIVE DISEASE WITH PLANT EXTRACTS

As the current therapies are symptomatic not curative led to the discovery of new drugs for effective treatment of neurodegenerative diseases.

Furthermore, the treatment for chronic neurodegenerative diseases require long-term drug administration with effective targeted drug delivery system along with reduced side-effects. Based on the above considerations, the plant-based extract could be an interesting candidate as therapeutic agent because of its anti-oxidant and anti-inflammatory properties. In the last two decades, there has been a massive interest for the usage of plant extracts and its metabolites as drug and this is because of the rapid advancement of various techniques such as NMR, HPLC-MS/MS, high-resolution Fourier transform mass spectrometry (MS) and so on [33]. Although, few plant extracts proved to be beneficial due to their *in-vivo* disease protecting features and also turned out to be non-essential nutrients for human beings. Some of these extracts are efficient in regulation of several cellular as well as molecular pathways due to these innumerable properties they are evidenced to play a beneficial role for human health [34]. Hence, the *in-vitro* and *in-vivo* effects of plant extracts have shown intense role in disease prevention along with their ability towards treatment and prevention of neurodegeneration. The association between L-dopa and carbidopa, a peripheral inhibitor of enzyme DOPA decarboxylase represents the gold standard for the symptomatic treatment of PD [13]. Carbidopa is involved in the inhibition of the conversion of L-dopa to DA peripherally, thus, it guarantees that a higher level of the drug reaches the central nervous system (CNS). Currently, L-dopa represents the most effective impossible to assume single patient's response to this drug, some motor symptoms, namely bradykinesia, rigidity, and postural instability, ameliorate 70% from the first weeks of the treatment [18].

2.3 TRADITIONAL HERBS AND THEIR CHEMICAL CONSTITUENTS

2.3.1 WITHANIA SOMNIFERA (WS)

Withania somnifera (WS), locally called "Indian Ginseng" or "Ashwagandha" is the widely used traditional herb in the field of Ayurvedic medicine (Figure 2.1A). It belongs to family Solanaceae and is one of the essential herbs since time immemorial due to its immense medicinal benefits. This classical herb generally improves the function of brain and CNS and is used for age-related cognitive deficits. It also plays a vital role in improving the body's defense mechanism by cell-mediated immunity. This ancient herb also has shown to induce significant immunomodulatory effects [35, 36].

The most active constituents of WS are alkaloids, steroidal lactones such as withanolides, withaferins (Figure 2.2A), and saponins [37]. Many of its active constituents also show immunomodulatory actions [35]. WS can also be used as a general tonic for increasing energy and also for prevention of disease which are related to the immune system [37]. The root extract of WS showed various health-promoting effects like anti-stress, anti-arthiritic, anti-inflammatory, analgesic, anti-pyretic, anti-oxidant, and immunomodulatory properties [38, 39].

FIGURE 2.1 (A) *Withania somnifera*; (B) *Bacopa monnieri*; (C) *Gingko biloba*; (D) *Curcuma Longa.*

2.3.2 *BACOPA MONNIERI (BM)*

Bacopa monnieri (BM), locally known as Brahmi (Figure 2.1B) of family Scrophulariaceae is another important indigenous herb in the Ayurvedic world from centuries. It has been classified under medharasayana. The plant has no distinct odor and is slightly bitter in taste. It is capable of rejuvenating the nerve cells, improve memory, cognitive impairment, as well as concentration, and also act as a therapeutic agent for treating mental illness and epilepsy. BM also plays a beneficial role as a nerve tonic as well as anti-inflammatory and anti-pyretic agents. The different pharmacological activities of BM include active constituents like alkaloids (brahmine, herpestine, and nicotine), saponin (monierin, hersaponin 8, bacoside A and B, bacogenin A1–A4) [40]. Between the two main saponins bacoside-A (Figure 2.2B) plays a role to improve memory power. The BM extract along with its pure constituents also acts as an anti-stress, anti-depressant, anti-convulsant, and anti-dementia agent. The anti-oxidant, anti-cancer, and anti-inflammatory role of BM extract related to immunological activity has been investigated.

2.3.3 *GINGKO BILOBA (GB)*

Gingko biloba (GB) is the ancient Chinese herb which is also known as living fossil (Figure 2.1C) [1]. The leaves of this fossil plant possess several chemical compounds like trilactonic diterpenes (Figure 2.2C) (ginkgolide A-C, ginkgolide J-M), trilactonic sesquiterpene (bilobalide) as well as various flavanoids. Besides, the leaf extract also have compounds well-known for its antioxidant properties and also has the potentiality to inhibit platelet aggregation.

2.3.4 *CURCUMA LONGA (CL)*

Curcuma longa (CL) is a perennial indigenous herb of family Zingiberaceae and is widely distributed throughout India (Figure 2.1D). The rhizome of this plant has also been used as a safe remedy against many ailments. In recent years CL has gained great importance because of its wide range of pharmacological properties including analgesic, anti-inflammatory, wound healing, and immunomodulatory activities. The chief active constituents of CL are curcumin (Figure 2.2D) and its natural derivatives such as demethoxycurcumin and bis-demethoxycurcumin [41].

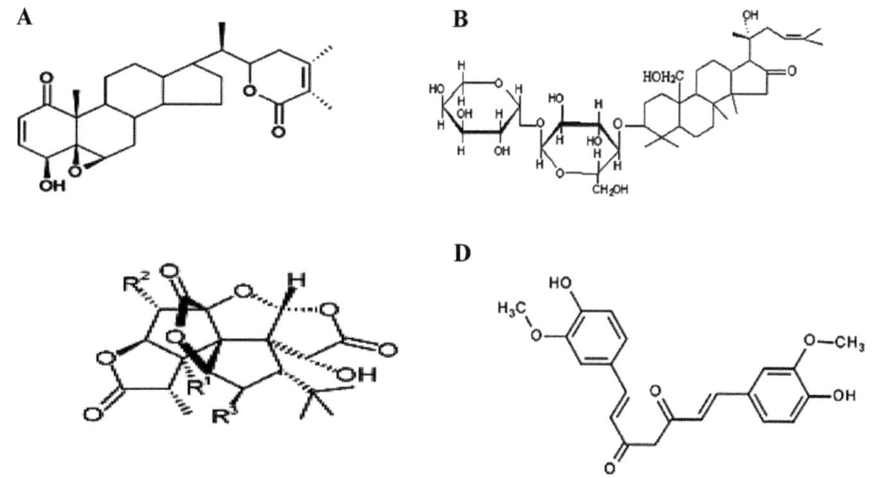

FIGURE 2.2 (A) Withaferin-A; (B) Bacoside-A; (C) Gingkolide-A; (D) Curcumin.

2.4 EXTRACTION TECHNIQUES OF MEDICINAL PLANTS

Extraction involves the separation of active components from their parent compound using selective solvents by using the standard extraction protocols. The obtained products are impure liquids, semisolids, or powder forms applicable only to external or oral use.

2.4.1 EXTRACTION PROCEDURES OF MEDICINAL PLANTS

2.4.1.1 SOXHLET EXTRACTION

This extraction procedure includes finely grounded crude drug which is placed in a porous bag made up of strong filter paper. This bag is placed inside the Soxhlet apparatus for further processing. The solvent is heated in the flask and finally, the vapors condense inside the condenser. Then the condensed extractant drips in the bag with the crude drug and finally, extraction is done by contact. Finally, the liquid level in the chamber rises at the top of the siphon tube and the liquid content remains in the flask. This extraction method is a continuous process and is carried out till a single drop of solvent in the siphon tube doesn't leave residue while evaporated. The method is highly advantageous as large amount of drug can be obtained from small quantity of the solvent and is highly economical and beneficial in terms of time, energy, and economical inputs.

2.4.1.2 ULTRASOUND EXTRACTION (SONICATION)

Ultrasound extraction is the method where ultrasounds with frequencies ranging from 20 kHz to 2000 kHz are involved. This procedure increases cell wall permeability as well efficient in producing cavitation. Though the process is beneficial in few respects but its large scale, intervention is limited due to its high cost. One of the major drawbacks of the procedure is its deleterious effect of ultrasound energy (more than 20 kHz) on the bioactive components of medicinal plants by means of free radical formation along with undesired changes of drug molecules.

2.4.1.3 SUPERCRITICAL FLUID EXTRACTION (SFE)

Supercritical fluid extraction (SFE) an alternative method for sample preparation along with an aim for low use of organic solvents and increased sample throughput. The factors associated with this technique include temperature, pressure, volume of sample, and so on. Basically, cylindrical extraction vessels are utilized for extraction procedure and exhibit good performance whereas, collection of analyte is a significant step. There are many advantages regarding the usage of CO_2 as a potent extracting fluid. Besides, the physical property CO_2 is abundant, cost-effective, and safe but has various drawbacks in terms of polarity.

2.5 IMMUNOMODULATORY NEURONAL DEGRADATION

The alteration of immune response which may increase or decrease immune responsiveness is termed as "immunomodulation." From ancient time plant extracts has been used as a potent medicine to alter the immune system. Many medicinal plants along with their extracts has been applied in various solvents to investigate the immunomodulatory effect on diverse animal model but the result was found to vary from one plant extract to another. Some plant extracts may act as an immunosuppressive agent whereas, others as an immunomodulatory agent and they showed remarkable changes in animal model such as increase in paw value, bone marrow cells, alpha esterase cells, and immunoglobulin (IgG) levels proving the immunomodulatory action of plant extracts. Therefore, these well known medicinal plants of Indian origin are reputed for their effectiveness in immunomodulatory properties on neurodegeneration. A huge number of experiments has been carried out on anti-stress, anti-oxidant, immunomodulatory, anti-aging, and anti-inflammatory actions of the extracts of BM, WS, GB, and CL. It has been evidenced that BM showed a marked decrease in Aβ deposition in the brains of AD mice model whereas, the chemical constituents isolated from root extracts of WS reversed the cognitive defects in ibotenic acid-induced AD mice model [42, 43].

2.5.1 IMMUNOMODULATORY ROLE OF WS

WS, the medicinal herb has been regarded as a traditional herbal medicine for more than 3,000 years. Several body extracts such as leaf, fruit, or root possess beneficial properties along with immunomodulatory property like increase rate in WBC count, phagocytic activities of macrophages, and many others [35, 37]. The extract also acts as a neuroprotective agent in various neurodegenerative diseases like AD, PD, ALS, HD, and many others [44].

AD, the common form of dementia and can be characterized by accumulation of misfolded protein like Aβ and tau (Figure 2.3). Recently, studies have proved that roots of WS showed beneficial effects in symptoms of AD by inhibiting NF-κB activation and by blocking the production of Aβ along with reduction in apoptotic cell death and increasing the antioxidant effect by Nuclear factor erythroid-2 (Nrf-2) molecule migration in the nucleus [45]. In the case of AD brain analysis involvement of immunomodulatory markers such as cytokines Interleukin-1beta (IL-1β), IL-6, and TGF-β) are found to accumulate around the amyloid plaques. This led to the study of various pro-inflammatory and anti-inflammatory cytokines. It has also been observed that there has been a trend of elevated level of pro-inflammatory and anti-inflammatory cytokines in cerebrospinal fluid (CSF) and serum of AD patients [46]. Therefore, the alteration in cytokine level as noticed in body fluid of AD patient resulted in disturbance of immune system.

PD, the second most common neurodegenerative disease after AD. The pathology of PD lies in the accumulation of α-synuclein (Figure 2.3) and the postmortem analysis has evidenced that there is an increase in cytokine level of PD brain such as-TNF-α, IL-1β, IL-2, IL-4, IL-6, epidermal growth factor (EGF), TGF-α and TGF-β1 [47]. Thus, imbalance in immunomodulatory markers in body fluid may lead to neurodegeneration.

ALS, another neurodegenerative disease has got a beneficial effect of WS. The misfolded aggregation of TAR DNA binding protein-43 (TDP-43) is the main feature of ALS pathology. TDP-43 generally binds and coactivates the P-65 subunit of nuclear factor kappa beta (NF-κB). This NF-κB acts as a vital factor in disease pathology. On administration of the potent bioactive constituent withaferin-A in transgenic mouse model showed a marked improvement in ALS pathology whereas, this withaferin-A acts as an antagonist to NF-κB showed a remarkable reduction in the inflammatory response in the transgenic mice model [48]. Other observations suggested a marked improvement in ALS mice model by administration of withaferin-A which delayed the disease onset and progression. Later on, the root extract of WS with human TDP-43 mutation showed clearance of abnormal TDP-43 in neurons and showed improvement in motor as well as cognitive functions. Therefore, this adverse situation can be overcome by the administration of WS extract on neurodegenerative patients.

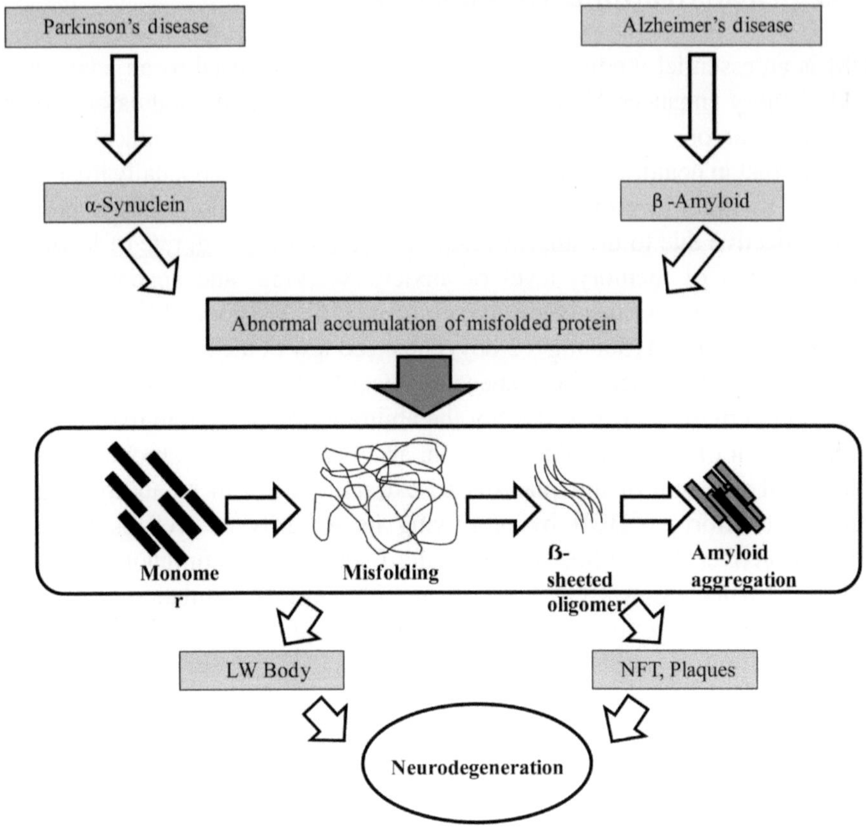

FIGURE 2.3 Molecular basis of neurodegenerative disease pathogenesis.

Presently, the immunomodulatory effect of WS extract showed significant increase in cell-mediated immunity whereas, the root extract of WS elevated the levels of various cytokines like interferon gamma (IFN-γ) and granulocyte-macrophage-colony-stimulating factor (GM-CSF) in mice model showing its immunoprotective and myeloprotective effect. Root extract of WS has been found to induce Th1 polarized cell mediated immune response in BALB/c mice model [49]. The powdered root extract of WS helps in the production of WBC and act as both chemoprotective and immunoregulatory agent in mice. Not only root extract, the leaf extract of WS decreased the production of various pro-inflammatory markers TNF-α, IL-1β, IL-6 along with ROS and also downregulated NF-κB pathway.

2.5.2 IMMUNOMODULATORY ROLE OF BM

BM is an essential medicinal and dietary plant and is used as an adaptogen and memory enhancer. Very few reports about the immunomodulatory effect of this plant on neurodegeneration have been evidenced. Clinical studies showed that in healthy subjects as compared to that of placebo, daily treatment with BM extract (3 months; 300/320 mg/day, 150/160 mg × 2/day) proved to be effective due to the improvement in visual processing, rate of learning, consolidation of memory, level of anxiety, working, and spatial working memory and various other cognitive behaviors [50]. Moreover, for AD patients a higher dose of BM (300 mg × 2/day) enhanced few of the cognitive functions like attention, language, comprehension, etc., through 6 months intervention [51]. Furthermore, BM has also got the ability to improve cognitive function by reduction of inflammation. Later on, an *in-vitro* study has proved several biological potential of this plant-based extract along with its ability to give protective effort on BBB by preserving its integrity as well as function [52]. Moreover, BM extract inhibited the fluctuation in immunomodulatory markers by decreasing COX level along with the downregulation of TNF-α. This extract also inhibits ROS production and reduces DNA damage in rat astrocytes, proving its anti-inflammatory actions [53]. Therefore, the improvement of perturbed immunomodulatory function in neurodegenerative diseases might be an important contributory factor of BM extract.

2.5.3 IMMUNOMODULATORY ROLE OF GB

GB is beneficial for normal cognitive functions due to its proper antioxidant and vascular properties. Presently, *in-vitro* reports in rats evidenced that a vital mechanism plays a crucial role in reducing BBB permeability by enhancing the cognitive effects of Ginkgolide-B [54]. Moreover, other studies discovered that EGb-761 leaf extract, prevented brain endothelial damage by beta-amyloid oligomer. This oligomer plays a key role in AD pathogenesis [55]. The study was conducted along with 216 patients and they received either a regular dose of 240 mg EGb 761 or placebo. Besides, the clinical efficacy was done by responder analysis along with therapy response defined as response in two of the three primary variables. On the other hand, human studies with GB leaf extract decreased IL-6 levels in serum of patients with neurologic disorders [50]. Furthermore, one week treatment randomized controlled pilot study revealed that a combining effect of *Panax ginseng, Ginkgo biloba*, and *Crocus sativus* elevated the working memory in controls [50].

2.5.4 IMMUNOMODULATORY ROLE OF CL

CL is a common perennial herb in India and is well-known for many years for its cosmetic and medical properties. The chief constituent of CL is curcumin, the yellow pigment which possesses potent antioxidant and immunomodulatory properties. The consumption of curcumin in the diet has been associated with reduce AD rates and is neuroprotective in action by decreasing Aβ plaques, metal-chelation, decrease microglial activation, and protection from the breakdown of BBB [56]. The study also revealed the effects of curcumin injection to cerebral ischemia-reperfusion in rats and observed that curcumin has the ability to reduce infarction volume, neurological deficiency, and BBB permeability [56]. Moreover, curcumin also inhibits free radical generation and inflammatory markers such as TNF-α, IL-1α, and IL-6 produced by microglia and reduced the levels of astrocyte marker. Moreover, other studies demonstrated the mode of action of curcumin in clearing the Aβ plaques by treating the macrophages of AD patients [1]. The result demonstrated that curcumin-treated macrophages had a capability of greater uptake and ingestion of plaque in comparison with untreated macrophages. Furthermore, several other experiments have been conducted for examining the effects of curcumin on cognitive function in humans and it decreased deposition of Aβ plaques but did not provide beneficial effects against cognitive functions in AD patients.

2.6 IMMUNOMODULATORY EFFECT OF NUTRACEUTICALS ON NEURODEGENERATION

Some nutraceuticals such as resveratrol, epigallectocatechin-3-gallate (EGCG), Vitamin-D, and Vitamin-E are potent antioxidants and have shown

A single bioactive compound is not sufficient to control the complex nature of various neurodegenerative diseases. Several mechanisms such as BBB integrity, oxidative stress, immunomodulation, inflammation, and aggregation of misfolded protein have shown to be related with neurodegeneration. Since, the translational gap between *in-vitro*, *in-vivo*, and clinical studies is still a major issue especially with respect to immunomodulatory effects of the said plant extracts. Thus, careful consideration should also be made in respect to the immunomodulatory effects of plant-based extracts on neurodegeneration.

Another aspect that should be taken under consideration is that a single compound is not sufficient in controlling neurodegeneration and therefore a synergistic effect of multiple compounds as present in the extract could mutually enhance the condition of this complex disorder. In addition to this, detailed research is needed to explore the immunomodulatory effect of plant extracts along with its bioactive compounds; to elucidate the therapeutic efficacy of this extract in combating the imbalance in immunomodulatory markers in various neurodegenerative diseases. Therefore, further research should be focused on the mechanism and mode of action of herbal compound supplementation and its safe therapeutic role for preventing the imbalance of immunomodulatory markers on neurodegeneration.

KEYWORDS

- **Alzheimer's disease**
- **amyotrophic lateral sclerosis**
- **blood-brain barrier**
- **central nervous system**
- **electron transport chain**
- **epidermal growth factor**

REFERENCES

1. Ratheesh, G., Tian, L., et al., (2017). Role of medicinal plants in neurodegenerative diseases. *Biomanufac. Rev., 2*, 2.
2. Simen, A. A., Bordner, K. A., et al., (2011). Cognitive dysfunction with aging and the role of inflammation. *Ther. Adv. Chronic Dis., 2*, 175–195.

3. Haque, M. R., Ansari, S. H., et al., (2013). Coffea Arabica seed extract stimulate the cellular immune function and cyclophosphamide induced immunosuppression in mice. *Iranian. J. Pharm Res., 12*, 101–108.
4. Mogensen, T. H., (2009). Pathogen recognition and inflammatory signaling in innate immune defenses. *Clin. Microbiol. Rev*., 240–273.
5. Dantuma, E., Merchant, S., et al., (2010). Stem cells for the treatment of neurodegenerative diseases. *Stem Cell Res. Ther., 1*, 37.
6. Dunkel, P., Chai, C. L., et al., (2012). Clinical utility of neuro protective agents in neurodegenerative diseases: Current status of drug development for Alzheimer's, Parkinson's, and Huntington's diseases, and amyotrophic lateral sclerosis. *Expert Opin. Investig. Drugs., 21*, 1267–1308.
7. Dye, R. V., Miller, K. J., et al., (2012). *Hormone Replacement Therapy and Risk for Neurodegenerative Diseases.*
8. Moraes, W. A., (2015). *Current Pharmacological and Non-Pharmacological Therapies for Neurodegenerative Diseases.*
9. Moreno, J. A., Halliday, M., et al., (2013). Oral treatment targeting the unfolded protein response prevents neurodegeneration and clinical disease in prion-infected mice. *Sci. Transl. Med.*
10. Weissmiller, A. M., &, Wu, C., (2012). Current advances in using neurotrophic factors to treat neurodegenerative disorders. *Transl. Neurodegener., 1*, 14.
11. Connolly, B. S., &, Lang, A. E., (2014). Pharmacological treatment of Parkinson's disease: A review. *JAMA, 311*, 1670–1683.
12. Vercruysse, P., &, Vieau, D., (2018). Hypothalamic alteration in neurodegenerative diseases. *Front. Mol. Neurosci.*
13. Beal, M. F., (1994). Neurochemistry and toxin models in Huntington's disease. *Curr. Opin. Neurol., 7*, 542–547.
14. Beal, M. F., (2004). Mitochondrial dysfunction and oxidative damage in Alzheimer's and Parkinson's diseases and coenzyme Q10 as a potential treatment. *J. Bioenerg. Biomembr., 36*, 381–386.
15. Su, K., Bourdette, D., et al., (2013). Mitochondrial dysfunction and neurodegeneration in multiple sclerosis. *Front. Physiol., 4*, 169.
16. Wong, P. C., Pardo, C. A., et al., (1995). An adverse property of a familial ALS-linked SOD1 mutation causes motor neuron disease characterized by vacuolar degeneration of mitochondria. *Neuron., 14*, 1105–1116.
17. Mancuso, M., Coppede, F., et al., (2006). Mitochondrial dysfunction, oxidative stress, and neuro degeneration. *J. Alzheimer's Dis., 10*, 59–73.
18. Rintoul, G. L., &, Reynolds, I. J., (2010). Mitochondrial trafficking and morphology in neuronal injury. *Biochim. Biophys. Acta., 1802*, 143–150.
19. Di Filippo, M., Tozzi, A., et al., (2014). Interferon-β1a protects neurons against mitochondrial toxicity via modulation of STAT1 signaling: Electrophysiological evidence. *Neurobiol. Dis., 62*, 387–393.
20. Qi, X., Lewin, A. S., et al., (2006). Mitochondrial protein nitration primes neurodegeneration in experimental autoimmune encephalomyelitis. *J. Biol. Chem., 281*, 31950–31962.
21. Tang, T. S., Slow, E., et al., (2005). Disturbed Ca^{2+} signaling and apoptosis of medium spiny neurons in Huntington's disease. *Proc. Natl. Acad. Sci. USA., 102*, 2602–2607.

22. Floyd, R. A., Carney, J. M., et al., (1992). Free radical damage to protein and DNA: Mechanisms involved and relevant observations on brain undergoing oxidative stress. *Ann. Neurol., 32*, S22–S27.
23. Gilgun-Sherki, Y., Melamed, E., et al., (2001). Oxidative stress induced-neurodegenerative diseases: The need for antioxidants that penetrate the blood brain barrier. *Neuropharmacology, 40*, 959–975.
24. Zaleska, M. M., &, Floyd, R. A., (1985). Regional lipid peroxidation in rat brain *in-vitro*: Possible role of endogenous iron. *Neurochem. Res., 10*, 397–410.
25. Nunomura, A., Honda, K., et al., (2005). Alzheimer-specific epitopes of tau represent lipid peroxidation-induced conformations. *Free Radic. Biol. Med., 38*, 746–754.
26. Nunomura, A., Perry, G., et al., (1999). RNA oxidation is a prominent feature of vulnerable neurons in Alzheimer's disease. *J. Neurosci., 19*, 1959–1964.
27. Alam, Z. I., Jenner, A., et al., (1997). Oxidative DNA damage in the Parkinsonian brain: An apparent selective increase in 8-hydroxyguanine levels in substantia nigra. *J. Neurochem., 69*, 1196–1203.
28. Taylor, D. M., Gibbs, B. F., et al., (2007). Tryptophan 32 potentiates aggregation and cytotoxicity of a copper/zinc superoxide dismutase mutant associated with familial amyotrophic lateral sclerosis. *J. Biol. Chem., 282*, 16329–16335.
29. Ferrante, R. J., Browne, S. E., et al., (1997). Evidence of increased oxidative damage in both sporadic and familial amyotrophic lateral sclerosis. *J. Neurochem., 69*, 2064–2074.
30. Karg, E., Klivényi, P., et al., (1999). Nonenzymatic antioxidants of blood in multiple sclerosis. *J. Neurol., 246*, 533–539.
31. Wint, D., &, Cummings, J. L., (2016). *Neuropsychiatric Aspects of Cognitive Impairment* (pp. 197–208). Oxford University Press.
32. https://parkinsonsnewstoday.com/2018/02/28/8-common-treatments-parkinsons-disease-2/ (accessed on 23 June 2020).
33. Harvey, A. L., Edrada-Ebel, R., et al., (2015). The re-emergence of natural products for drug discovery in the genomics era. *Nat. Rev. Drug Discov., 14*, 111–129.
34. Upadhyay, S., &, Dixit, M., (2015). Role of polyphenols and other phytochemicals on molecular signaling. *Oxid. Med. Cell. Longev.*, 504253.
35. Ghosal, S., Lal, J., et al., (1989). Immunomodulatory and CNS effects of sitoindosides IX and, X., two new glycowithanolides from *Withania somnifera. Phytother Res., 3*, 201–206.
36. Davis, L., &, Kuttan, G., (2000). Immunomodulatory activity of *Withania somnifera. J. Ethnopharmacol., 71*, 193–200.
37. Mishra, L. C., Singh, B. B., et al., (2000). Scientific basis for the therapeutic use of *Withania somnifera* (ashwagandha): A review. *Altern. Med. Rev., 5*, 334–346.
38. Agarwal, R., Diwanay, S., et al., (1996). Studies on immunomodulatory activity of *Withania somnifera* (ashwagandha) extracts in experimental immune inflammation. *J. Ethanopharmacol., 67*, 27–35.
39. Davis, L., &, Kuttan, G., (2002). Effect of *Withania somnifera* on cell mediated immune response in mice. *J. Exp. Clin. Cancer Res., 21*, 585–590.
40. Chandel, R. S., Kulshrestha, D. K., et al., (1977). Bacogenin A3: A new sapogenin from *Bacopa monniera. Phytochemistry, 16*, 141–143.
41. Karlowicz-Bodalska, K., Han, S., et al., (2017). *Curcuma Longa* as medicinal herb in the treatment of diabetic complications. *Acta Pol. Pharm., 74*, 605–610.

42. Dhanasekaran, M., Tharakan, B., et al., (2007). Neuroprotective mechanisms of ayurvedic antidementia botanical *Bacopa monniera. Phytother. Res., 21*, 965–969.
43. Bhattacharya, S. K., Kumar, A., et al., (1995). Effects of glycowithanolides from *Withania somnifera* on animal model of Alzheimer's disease and perturbed central cholinergic markers of cognition in rats. *Phytother Res., 9*, 110–113.
44. Kuboyama, T., Tohda, C., et al., (2014). Effects of ashwagandha (roots of *Withania somnifera*) on neurodegenerative diseases. *Biol. Pharm. Bull., 37*, 892–897.
45. Sandhir, R., &, Sood, A., (2017). *Neuroprotective Potential of Withania somnifera (ashwagandha) in Neurological Conditions*. Springer International Publishing.
46. Brosseron, F., Krauthausen, M., et al., (2014). Body fluid cytokine levels in mild cognitive impairment and Alzheimer's disease: A comparative overview. *Mol. Neurobiol., 50*, 534–544.
47. Nagats, T., Mogi, M., et al., (2000). Cytokines in Parkinson's disease. *Advances in Research on Neurodegeneration*(pp. 143–151).
48. Swarup, V., Phaneuf, D., et al., (2011). Deregulation of TDP-43 in amyotrophic lateral sclerosis triggers nuclear factor kappaB-mediated pathogenic pathways. *J. Exp. Med., 208*, 2429–2447.
49. Malik, F., Singh, J., et al., (2007). A standardized root extract of *Withania somnifera* and its major constituent withanolide-A elicit humoral and cell-mediated immune responses by up regulation of Th1-dominant polarization in BALB/c mice. *Life Sci., 80*, 1525–1538.
50. Kure, C., Timmer, J., et al., (2017). The immunomodulatory effects of plant extracts and plant secondary metabolites on chronic neuroinflammation and cognitive aging: A mechanistic and empirical review. *Front Pharmacol., 8*, 117.
51. Goswami, S., Saoji, A., et al., (2011). Effect of *Bacopa monnieri* on cognitive functions in Alzheimer's disease patients. *Int. J. Collaborative Res. Intern. Med. Public Health, 3*, 285–293.
52. Rama, B. P., (2018). A review on immunomodulatory effects of plant extracts. *Virology and Immunology J.*, 2.
53. Russo, A., Borrelli, F., et al., (2003). Life nitric oxide-related toxicity in cultured astrocytes: Effect of *Bacopa monniera. Sci., 73*, 1517–1526.
54. Sharma, H. S., Drieu, K., et al., (2000). Role of nitric oxide in blood-brain barrier permeability, brain edema and cell damage following hyperthermic brain injury: An experimental study using EGB-761 and Gingkolide B pretreatment in the rat. *Acta Neurochir., Suppl. 76*, 81–86.
55. Wan, W. B., Cao, L., et al., (2014). EGb761 provides a protective effect against Aβ1-42 oligomer-induced cell damage and blood-brain barrier disruption in an *in vitro*bEnd.3 endothelial model. *PLoS One, 9*, e113126.
56. Jiang, J., Wang, W., et al., (2007). Neuro protective effect of curcumin on focal cerebral ischemic rats by preventing blood-brain barrier damage. *Eur. J. Pharmacol., 561*, 54–62.
57. Pocernich, C. B., Lange, M. L., et al., (2011). Nutritional approaches to modulate oxidative stress in Alzheimer's disease. *Curr. Alzheimer Res., 8*, 452–469.
58. Schmitz, K., Barthelmes, J., et al., (2015). Disease modifying nutricals for multiple sclerosis. *Pharmacol. Ther., 148*, 85–113.
59. Gianforcaro, A., &, Hamadeh, M. J., (2014). Vitamin D as a potential therapy in amyotrophic lateral sclerosis. *CNS Neurosci. Ther., 20*, 101–111.

60. Betti, M., Minelli, A., et al., (2011). Dietary supplementation with α-tocopherol reduces neuro-inflammation and neuronal degeneration in the rat brain after kainic acid-induced status epilepticus. *Free Radic. Res., 45*, 1136–1142.
61. Arreola, R., Alvarez-Herrera, S., et al., (2016). Immunomodulatory effects mediated by dopamine. *J. Immunol Res.,* 3160486.
62. Arreola, R., Becerril-Villanueva, E., et al., (2015). Immunomodulatory effects mediated by serotonin. *J. Immunol Res.,* 354957.

CHAPTER 3

Bio-Implants Derived from Biocompatible and Biodegradable Biopolymeric Materials

NEETHA JOHN

Central Institute of Plastics Engineering and Technology, JNM Campus, Udyogamandal, Kochi – 683501, Kerala, India, E-mail: neethajob@gmail.com

ABSTRACT

It has been very well recognized by current medical field that the scope of biotechnology is very vast. The surgical implant being one among the field of greater importance. Treatment with the help of implants became most viable and can easily adaptable one. This field has grown to forward directions especially in dental implants. Amongst the viable materials in use polymer play a key role as its characteristics supporting the bio implants to a greater extend. Polymers are corrosion resistant, mechanically stronger and biocompatible, hence became one of the top priority material compared to other materials like ceramics and metals.

3.1 INTRODUCTION

An implant is a manufactured medical device to support the available biological systems, replace a missing biological structure, or support a damaged biological structure. These are man-made devices which are different from a transplant, or biomedical tissue. Titanium, silicone, andapatite are used as implants materials which come in contact with body. These materials are biocompatible and also functional materials can be easily used as bio-implants [1]. Implants can accommodate electronics components, e.g., artificial pacemakers and cochlear implants. Implants can be also bioactive like in drug delivery systems in the form of pills or drug-incorporated stents. There are various types of implants that can be classified into groups depending on the application [1–6] shown in Table 3.1.

Materials commonly used in implants are stainless steel (SS316L), aluminum alloys, titanium alloys, polymers, polyetheretherketone, ceramics, and alumina chromium and nickel alloys, etc. If an implant stops to perform the function for which it was inserted it is then said to be failed.

Implant can cause undesirable effects like pain, infection, or toxicity leading to rejection of implant. The causes are deformation of implant, fracture, and loosening of the fixator. It may be due to biocompatibility of implant. High quality stainless steel has been proven successful as human implantation.

TABLE 3.1 Major Types and Functions of Bio-Implants

Sl. No.	Types	Functions	Examples
1.	Cardiovascular	Cardiovascular medical devices are used when the heart valves and the other circulatory system are not functioning properly. Conditions of heart failure, cardiac arrhythmia, ventricular tachycardia, heart valve disease dysfunction, angina, and atherosclerosis are treated with this.	Artificial heart, artificial heart valve, cardiac pacemaker, implantable defibrillator, and coronary stent
2.	Orthopedic	In body, the bones and joints are treated with orthopedic implants which help to solve the issues. Bone fractures, osteoarthritis, scoliosis, spinal stenosis, and chronic pain are treated.	To anchor fractured bones Pins, rods, screws, and plates are used.
3.	Sensory and neurological	They are used for disorders affecting the senses in body the brain and disorders in nerves. They are used mostly in the treatment of conditions such as cataracts, glaucoma, keratoconus, epilepsy, Parkinson's disease, and depression.	Intrastromal corneal ring segment, Intraocular lens, cochlear implant, tympanostomy tube, and neurostimulator

TABLE 3.1 *(Continued)*

Sl. No.	Types	Functions	Examples
4.	Electric	Patients suffering from rheumatoid are treated with electrical implants to relieve pain. Electrical signals send the implant to electrodes in the vagus nerve.	Electric neck implant
5.	Contraception	Contraceptive implants are used to prevent pregnancy and treat conditions of non-pathological menorrhagia.	Copper- and hormone-based intrauterine devices.
6.	Cosmetic	To give body some acceptable aesthetic features.	Implant for nose prosthesis, ocular prosthesis and inject filler to enhance dimensions or fill gaps
7.	Other organs	There are other organ dysfunctions in the body like the gastrointestinal, respiratory, and urological systems. Implants are applied to treat gastroesophageal conditions reflux disease, gastroparesis, respiratory failure, sleep apnea, urinary, and fecal incontinence and erectile dysfunction	Gastric stimulator implantable, nerve stimulator, diaphragmatic/ phrenic, neurostimulator, surgical mesh

Various researches being done in this field as it is diverse and have different directions due to its multidisciplinary nature. This chapter provides ample systems to greatly expand the biomedical applications of polymeric materials and a brief overview of some of the recent developments in the field of polymeric bio-implants.

3.2 POLYMER-BASED BIOMATERIALS

Transportation, building construction, electrical, and electronics, packaging, and aviation industries have attracted considerably with Polymer-based composites [7]. Polymers are used in biomedical applications [8] due to ease of processing, lightweight, flexibility, high strength to weight ratio, greater availability, and higher recyclability is their basic characteristics [9]. Even though metals and ceramics possess higher mechanical properties compared to polymers, polymers still are preferred due to its special characteristics. There can be further improvements by mixing with another polymer, addition of fibers, and nanoparticles [10]. Novel and smart materials are prepared out of polymer matrix with nanoparticle incorporation.

Polymers can be one of the highest among biomaterials which are extensively used in biomedical applications in large multitude. Polymers have appropriate physical, chemical, surface, and biomimetic properties, can be designed and prepared with various structures. Polymers are prepared with relative ease which leads to the versatility attributed by polymers.

Health care industry has great technological advancements over the last two decades particularly in the field of biomaterials. This is considered as the latest revolution in the area of biomaterials. The survival and quality of life and have enhanced because of biomaterials-based technological developments. More efficient and sophisticated medical devices are available in the market today.

Significant changes happened in the lifestyle pattern. The aging population needs more medical attention. There exists a higher social pressure to reduce costs for health care also. There are several factors including increasing awareness among the public, knowing the efficiency and performance of biomaterials as medical devices, etc. Developments in nanotechnology and the delivery of bioactive agents for treating, repairing, and restoring the function of tissues are of great significance today.

Adhesives, coatings, foams, and packaging materials, textiles, high modulus fibers, composites, electronic devices, biomedical devices, optical devices, and high-tech ceramics are major applications of polymers [11]. Polymeric materials are used along with soil, provide mulch, and promote plant growth. Heart valve and blood vessels are made of Dacron, Teflon, and polyurethane (PU) like polymers are now made with biomaterials.

Polymers are attractive bio-materials, they are having excellent machinability, optically transparent used for detection, they are biocompatible with greater thermal and electrical properties, and with high-aspect-ratio for the microstructures. Polymers can be easily used for surface modification and functionalization. This is more applicable in the case of biopolymers including DNA, proteins, and natural polymers. Both synthetic and natural polymer materials are used in soft fabrication techniques.

Engineering plastics like polyetheretherketone is of excellent and notable properties compared to some of the special engineering plastics. There are remarkable advantages like it can take up heat, superior mechanical property, self-lubricating, corrosion-resistant, fire-resistant, irradiation resistant, higher insurability, hydrolytic resistant, and very easily processable. The applications include aerospace, automobile, electronic, electric, medical, and food industry.

Advanced researches are being conducted polymeric biomaterials from various disciplines of polymer chemistry, materials science, biomedical engineering, surface chemistry, biophysics, and biology. In the past few years, polymer-based biomaterial technologies are coming to the commercial applications at a very rapid pace. Polylactic acid (PLA) the most widely used synthetic polymer which introduced by Biscnoff and Walden in 1893 [7]. These are highly biocompatible, with controlled degradation rate and degrade into toxic-free components like CO_2 and water. They are used for biomedical applications, like natural polymers, polysaccharides or proteins, and synthetic polymers.

Other examples are poly(glycolic acid) (PGA), poly(hydroxyl butyrate) (PHB) and poly(ε-caprolactone) (PCL) [8]. Plastics which are good for biomedical applications are polypropylene (PP), PU, and polyethylene (PE) and equally found useful. Some of the polymers are soluble in water. Polyvinyl alcohol (PVA), polyethylene glycol (PEG), polyvinyl acetate, polyacrylic acid (PAA), and guar gum have used for similar applications [12].

Recent developments in the field of biocompatible and biodegradable polymers for the applications in tissue engineering and drug delivery have renewed the quest for more efficient new generation polymers. Gomurashvilli et al. [2] have developed new biodegradable and tissue-resorbable co-poly(ester amides) (PEAs) using a versatile active polycondensation technique.

Amino acids, diols, and dicarboxylic acid components in the backbone have a wide range of mechanical properties and biodegradation rates. These PEAs can be used in drug delivery and tissue development. Chitosan-PEG superabsorbent gel prepared at ambient conditions by Chirachanchai et al. [3] may be very useful for bioactive agents as an injectable carrier. Metal containing polymeric complexes synthesized by Fraser et al. [7] used in biological activities for using as catalysts, act as materials to respond for stimuli centers and structural materials.

Cooper et al. [13] have developed polymers with high molecular weights. It also gives high tensile properties and melts processability similar to synthetic polymers. Hence can be used as similar as radiation sterilizable aromatic polyanhydride. Nicholas et al. [6] have developed a new route for the preparation of fluorescent bioconjugates by living radical polymerization using protein-derived macro initiators. These fluorescent bioconjugates can be easily traceable in biological environments, during biomedical assays. Cycloaddition reactions have been explored by Grayson et al. [14] to prepare macrocyclic poly(hydroxyl styrene). The presence of a phenolic hydroxyl group on each repeat unit in the cyclic polymer gives better scope for attachment of bioactive counterparts. The cyclization technique to have a wider application for preparing a wide range of functionalized macrocycles. Smith and coworkers [15] found poly(N-vinylpyrrolidinone) hydrogels functionalized with drug molecules as promising hydrogels for sustained release of drugs over several days.

Kumar et al. [16] have developed a green route to pegylated amphiphilic polymers with the use of immobilized enzyme, Candida Antarctica lipase B. The ability of these polymers to form the nano-micelles makes them suitable for application in drug delivery systems and in cancer diagnostics. Gong et al. [17] developed a new highly innovative two-photon activated photodynamic therapy (PDT) which has three-fold level of application, (a) a photosensitizer: a porphyrin substituted on the meso positions by chromophores with large two-photon absorption and activated in the near infra-red region in the tissue transparency window and efficiently producing singlet oxygen as the cytotoxic agent; (b) which can target small molecule also target receptor sites on the tumor; and (c) a imaging agent near IR that can correctly give image of the tumor for treatment. Adronov et al. [14] synthesized carborane functionalized dendronized polymers and found them to be useful as potential boron neutron capture therapy (BNCT) agents. Nederberg and coworkers [18] have developed a series of telechelic biodegradable ionomers based on poly(trimethylene carbonate) carrying zwitterionic, anionic, or cationic functional groups for protein drug delivery. Biodegradable ionomers was utilized for protein loading simply by letting the material swell in an aqueous protein solution. The process can be done either directly after loading or after a drying. Protein activity is maintained suggesting that these ionomers may favorably interact with guest proteins and denaturation is suppressed.

3.2.1 CHARACTERIZATION OF POLYMERIC BIOMATERIAL

For any polymer-based medical devices, an in-depth understanding of physical, chemical, biological, and engineering properties is highly relevant to get the performance. Efficient polymer-based medical devices and implants require the development of advanced instrumentation and characterization techniques, along with mathematical models to study the structure-property and functional performance relationships of various polymeric systems. Henderson et al. [19] give some of the concepts of NIST measurement technique can be used for tissue engineering which consists of high-throughput, combinatorial methods to produce test specimens of different material properties and advanced instrumentation procedures used for collecting data with high-resolution, non-invasive, multi-level imaging of cells and 3D visualization. Hassan et al. [20, 21] demonstrated the application of broadband dielectric spectroscopy can be used for the study of degradation polymeric materials.

Polymer characterization can use mass spectrometry (MS) which provides details of the development of ionization techniques like MALDI and ESI combined with the latest developments in mass analyzers such as reflection-TOF and FTICR. This has greatly enhanced the capabilities of MS to better understand the detailed composition of polymeric materials. These techniques are giving better mass resolution and accuracy along with hyphenated techniques, also give better details about the smallest components in polymeric biomaterials to be assessed, especially with regard to the composition of repeat and end groups. Maziarz and coworkers [22, 23] review the applications of MS in the characterization of polymeric biomaterials.

3.2.2 POLYMERIC INTERFACES

This is of very much important to have an in depth understanding of surface/interface characteristics of polymeric biomaterials and precise details of the interaction between their surface and biological entity. This is highly relevant for their successful and safe performance in various applications including biosensing, diagnostics, and medical devices. People are looking for the development of versatile, convenient, and more economical for providing resistant to the surface from fouling by proteins, cells, and microorganisms.

Glasgow et al. [24] review the biocompatibility of surfaces of various hydrophobic, hydrophilic, and heterogenic polymeric biomaterials related to their interactions with proteins and blood platelets. Nandivada et al. [25] give the review in the applications of reactive polymer surfaces created by chemical vapor deposition (CVD) polymerization can work as a strong substrate for biomimetic modifications. Wingkono and co-workers [26] report on the investigation of phase-separated microdomains in a PEG-PCL PU and their effect on osteoblast adhesion. Combinatorial libraries were employed to optimize microphase domain size and shape of PEG-poly(caprolactone) (PCL) PUs in which the effect of these surface structures on osteoblast adhesion was examined using the culturing technique of the cells directly on the libraries.

Tissue engineering that applies the principles of engineering and sciences is considered as an interdisciplinary field in the life development of biological substitutes that can be restored, or improve tissue functions. He et al. [27] reviews the materials requirements for tissue engineering application, and focus on polymer materials including both natural and synthetic polymers. It describes the most widely used and newly developed fabrication technologies for the construction of tissue-engineering scaffolds and surface modification or functionalization techniques of tissue engineering scaffolds. Konno et al. [28] developed a covalently crosslinked hydrogel material from a phospholipid polymer containing 2-methacryloyloxyethyl phosphorylcholine units and p-vinylphenylboronic acid (PMBV) in which PVA act as a cell container.

Hydrogels can be formed in pure water and also in saline water including cell culture medium. The viability of the cells in the entrapped in the hydrogel was not affected. Hydrogel based on the phospholipid polymer can be used as a cell container to preserve and transport the cells in tissue engineering applications. Cooper et al. [29] present the evaluation of bioabsorbable, aliphatic polyester blends comprising PLA and poly(ecaprolactone-co-p-dioxanone) (PCL: PDO) for application as a device for cranial fixation. Duran et al. [30] prepared poly(y-benzyl-Z,-glutamate) (PBLG) polypeptide nanotubes for optical biosensor applications using the synthesis of nanostructures within the pores of a nanoporous membrane as the technique.

3.3 ROLE OF BIOPOLYMERS AS BIO-IMPLANTS

3.3.1 BACTERIAL PLASTICS

Bio-based plastics are starch-based plastics, protein-based plastics, and cellulose-blended plastics. Blends are prepared with conventional plastics such as polyethylene (PE), PP, and polyvinyl alcohol. These are bio-based plastics which are partially biodegradable. Petroleum derived plastics will remain as broken pieces may create additional pollution. These plastics have intrinsic thermal and mechanical weaknesses and they are now discouraged for applications. Bacteria are employed to make the building blocks for biopolymers from renewable sources. To produce bio-based plastics completely resembling conventional plastics, starch, cellulose, fatty acids are used. Bacteria can consume the organic material for growth. Some of the building blocks can be produced microbial for polymerization purposes. There are many structural variations of hydroxyalkanoic acids. They are lactic acid, succinic acid, (R)-3-hydroxypropionic acid, bioethylene produced from dehydration of bioethanol, 1, 3-propanediol and cis-3, 5-cyclohexadiene-1, 2-diols from microbial transformation of benzene and other chemicals. Various bacterial plastics prepared using these materials [31].

Polymerization of hydroxyalkanoates is conducted *in vivo.* All other monomers are polymerized *in vitro* by chemical reactions, leading to the formation of PHA, poly(lactic acid) (PLA), poly(butylene succinate) (PBS), PE, poly(trimethylene terephthalate) (PTT) and poly(p-phenylene) (PPP). These plastics are bio-based. Their properties are identical to those of traditional petroleum-based plastics. PE-based bioethanol leading to bioethylene. They are exactly the same as petroleum-based polyethylene. PHA is available in many varieties based on the structural. This is resulting in variable melting temperature (Tm), glass-transition temperature (Tg) and degradation temperature as reported by Steinbüchel [32], Doi et al., (1995), Wang et al. (2009), Spyros and Marchessault [33], and Galegoa et al. (2000).

3.3.2 BIODEGRADABILITY

Enzymes, bacteria, and fungi like microorganisms are taking part in the degradation of natural and synthetic plastics [34]. The biodegradation of bacterial plastics proceeds actively under different soil conditions according to their different properties. PHA is one of the natural plastics. Microorganisms can synthesis and preserve PHA even if conditions of lower nutrients availability and can undergo degradation and can metabolize it when the carbon or energy source is limited (Williams and Peoples 1996). (R)-3-hydroxybutyric acid is a biodegraded product of poly(3-hydroxybutyrate) (Doi et al., 1992), extracellular degradation of poly [(R)-3-hydroxybutyrate-co-(R)-3-hydroxyvalerate] yields both (R)-3-hydroxybutyrate and (R)-3-hydroxyvalerate [35].

PLA is fully biodegradable under the composting condition in a large-scale operation with temperatures of 60°C and above (Pranamuda and Tokiwa, 1999). PBS can degrade by hydrolysis mechanism and termed as hydro-biodegradable. The polymer molecular weights reduction takes place by hydrolysis at the ester linkages, further degradation may occur by the actions of microorganisms [38]. Polyethylene (PE) can biodegrade by two mechanisms which are hydro-biodegradation and oxo-biodegradation [36].

PBS is biodegradable aliphatic polyesters which is one of the members of biodegradable polymers family. In order to improve the properties number of copolymerization stages are made according to Darwin et al. [37]; Jin et al. (2000, 2001). Phenyl units are introduced into the side chain of PBS, which leads to the better biodegradability of the copolyesters. Jung et al. (1999) successfully synthesized new PBS copolyesters containing alicyclic 1; 4-cyclohexanedimethanol. The applications for PLA are mainly as thermoformed products such as drink cups, take-away food trays, containers, and planter boxes. Polystyrene and PET are partially replaced with PLA material due to the rigidity. Applications include mulch film, packaging film, bags, and 'flushable' hygiene products [38].

Polybutylene succinate (PBS based different blends are tried gave greater toughness. *Poly(3-hydroxybutyrate-co-3-hydroxyvalerate)* (PHBV) and poly(butylene succinate-co-adipate) (PBSA) biodegradable polymers some of them [39]. These polymers were added to a PLA/PBAT blend giving a decrease of thermal properties. There is an observation of an increase in melt flow with PBS, as it is more flexible compared to the other. There is no change in melt flow observed with PBSA or PHBV [40]. PLA degrades into water and carbon dioxide that does not cause any harm; they can be cleared out of human body, thus PLA became the most popular biomedical material in the market.

3.3.3 BIODEGRADABLE ORTHOPEDIC IMPLANTS

The human skeleton consists of separate and fused bones supported and supplemented by ligaments, tendons, muscles, and cartilage. Bones gets arterial blood supply, venous drainage, and nerves. There is a tough fibrous layer with which particular surfaces of bones are covered. The human skeleton is changing always, it changes composition throughout lifespan. In the early stages, a fetus does not have any hard skeleton; bones are formed gradually during nine months in the womb. By birth, all bones are formed but a newborn baby has more bones than an adult. An adult human has 206 bones. A baby is born with around 300 bones. The bones do not have pockets or space left to grow further. The strength of the bones are not the same in all direction, it shows an isotropicity. Bones are not strong and stiff if stressed from side to side.

There are several important factors to be considered in designing biodegradable orthopedic implants. Primarily material should degrade over in time, therefore the scaffolds are functioning as a temporary support and allowing space for newly generated tissue to replace the defect [41, 42]. Second, the initially implanted biomaterials or the degraded materials and the by-products, such as monomers, initiators, and residual solvents, may cause a serious inflammatory or immunogenic response in the body [43]. Finally, the material should possess sufficient strength to sustain the mechanical loads subjected to defective area during the process of healing. The material may be showing a lower mechanical strength as defects are replaced with new tissue designed with scaffolds of orthopedic implants. From the beginning, whatever is the proposed final application must be a considered as the primary concern. Scaffolds can be used as fixation as internal devices to support the defective sites. Tissue repair will be aided by scaffolds implanted by the induced cell migration and proliferation. Scaffolds are used to give out bioactive molecules and cells closers to the organ to enhance defect healing process.

Biomaterials are getting attraction in the usage as implants which help in the regeneration of orthopedic defects [41, 42]. In the United States alone, every year more than 3.1 million orthopedic surgeries are performed [44]. The currently available treatments using no degradable fixation materials have proven effective, even though tissue engineering procedures with biodegradable implants are being considered as more promising future alternative [4, 45]. Biodegradable implants can be processed to give temporary support for bone fractures. They can degrade at the same rate which is used for the natural new tissue formation and their use can be eliminated the need of a second surgery [40]. The scaffold can function as a substrate for seeded cells, to give support for the tissue containing a defect and new tissue formation at the locations of injury [46, 47]. New tissue formations can be made faster by the usage of drugs or molecules which are bioactive. It can be used to treat osteomyelitis like conditions [48, 49].

3.3.4 SYSTEMS WHICH GIVE MECHANICAL SUPPORT

Biodegradable orthopedic are used during the healing process in the form of fixation implants like screws, staples, pins, rods, and suture reinforces the support areas weakened by bone fracture, sports injury, or osteoporosis [50–52]. It is very important that high mechanical strength and stiffness are required for biodegradable devices especially used for orthopedic applications. They are subjected to high loads when the devices are implanted. These biomaterials should show longer biodegradation times [17, 20]. A biodegradable screw made of poly(L-lactide) with a titanium screw in a ligament demonstrated that it could provide a promising alternative in terms of primary fixation strength [53]. A blend of poly(propylene fumarate) (PPF) and poly(propylene fumarate)-diacrylate (PPF-DA) has been molded into a biodegradable fixation plates. It can be used as a bone allograft interbody fusion spacer with acceptable mechanical properties [52]. Scaffolds are providing physical support also have been used to introduce bioactive molecules at the defect site [56]. Scaffolds also can be used to control the release of bioactive molecules and thus accelerating the healing process [57]. In the case of less stable drugs, its effectiveness can be extended by encapsulating them inside a matrix [58]. Nano- or microparticles and hydrogel-based implants systems are some of the delivery systems.

3.4 BIOMATERIALS: PROPERTIES REQUIREMENTS FOR BIOMEDICAL APPLICATION

Biomaterials are inert substance or combination of substances greatly used for implantations. It may be used with a living system to support or replace functions of living tissues or organs. Biomaterials can be even metals, ceramics, natural or synthetic polymers, and composites. Biomaterials [59] can be natural or manmade, which can make the whole part of a living structure. These biomedical devices can perform, augment, or replaces a natural function. This material should have biocompatibility which decides whether the material is suitable for exposure to the body or bodily fluids. It shows its ability to perform and give a proper response in the biological environment. If these materials placed inside the body will allow the body to function normally without creating any complications then these materials are said to be biocompatible. Some of them may cause an allergic reaction to the body once it comes in contact with body fluids. Polymers are the most promising and largest class of biomaterials. It is proved by their widespread use in various medical applications. There is a large number of polymeric biomaterials developed and developments are continuing as of its popularity. The polymer can be synthesized with a wide range of properties and functionality. This becomes the key to the success of polymer-based biomaterials and the ease coupled with low cost.

Components of implants are getting the direct interaction or contact with the human body. Therefore, the safety level of these materials used in biomedical applications is very high. It must be non-toxic, biodegradable, and biocompatible and meet the specifications given in the standards. Three basic properties are required of any biocompatible materials. They are superhydrophobicity, adhesion, and self-healing. These properties make it fit for biomedical applications. To meet all the requirements a large number of research works are going in order to develop materials satisfying all requirements.

3.4.1 SUPER HYDROPHOBICITY

Materials with superhydrophobic surfaces are very difficult to wet. This property exhibited by many plants and insects. It can reduce blood coagulation and unfavorable platelet adhesion if there is superhydrophobicity for these biomaterials. Then they can be making use in biomedical applications [9]. There are many biomaterials showing such properties are produced [60].

3.4.2 ADHESION

Plants and animals are showing this phenomenon of selective adhesion, which is required for those organisms for their survival. These organisms attach temporary or permanently to their surrounding substrates. Adhesion abilities are also important for bacteria, animals, and plants. Polymeric materials with these properties are created and using in biomedical applications by Bassas-Galia et al. [61].

3.4.3 SELF-HEALING

Self-healing is known as the human body repairs the damaged tissues by itself whenever there is an injury. This happens when the portion of the injury is smaller in size. When there is an injury or damage which is beyond self-healing, there is an introduction of alternative material like the use of an implant. Implants may be subjected to various loads; wear and aging which causes failure and needed to replace. Researchers are trying to produce materials that can heal or repair by itself the first types of self-healing biomaterials are composites that irreversibly repaired. Second generation self-healing materials can reversibly restore the damaged matrix. To get improved mechanical and physical properties polymeric self-healing materials, nanoparticles are used. Compared to metals and ceramics, polymers show lower strength and modulus which can be improved by various techniques.

Polymer-based composites have been widely used due to their advantages and ability to handle the loads. Self-healing materials [62, 63] should have the characteristics ability to repair the damaged portion many times. These materials should have the capability to repair the defects in the substrates of any size and also should have reduced maintenance costs. When comparison with the normal substrates, it must exhibit superior quality and should be more economical compared with the conventional materials.

Microcapsules were added to PLA to form a composite material with self-healing property. The microcapsules are filled with additives for healing. It can function when cracked and releasing the self-healing additives to fractured areas. This can also function as nucleating agents to improve the PLA composite's temperature resistance. Self-healing microcapsules can be created by encapsulating the dicyclopentadiene and Grubbs catalyst. It is then released into damage volumes and undergoing polymerization by the chemical reaction of the catalyst. This technique helps in the recovery of the polymer composite's toughness towards facture.

3.4.4 BULK PROPERTIES

The modulus of elasticity of the implant material should be comparable to the bone. It must be showing a uniform distribution of stress at implant so that the relative movement at implant-bone interfaces to be the minimum. To prevent fractures and improve functional stability an implant material should have high tensile and compressive strength. Stress transfer will be improved from the implant to bone if the interfacial shear strength is increased, so that implant is at lower stresses. Yield strength and fatigue strength required for an implant material. Higher yield strength and fatigue strength is essential to prevent sudden fracture of the implant with cyclic loading. Ductility is another bulk property required for making the contours and shapes for an implant. If hardness and toughness increase the wear of implant, material decreases which also prevents the premature fracture of the implants.

3.4.5 SURFACE PROPERTIES

For any implants surface tension and surface energy essentially affects the wettability of implant. This causes wetting of fluid blood and affects the cleanliness of implant surface. Osteoblasts are cells from bone can cause improved adhesion on implant surface. Surface energy can change adsorption of proteins [64]. If the surface roughness of implants increases the surface, area increases and improves the cell attachment to the bone. An implant surface can classify on different criteria, such as roughness, texture orientation, and irregularities [65, 66]. Wennerberg et al. have subdivided implant surfaces as per the surface roughness which can be of three categories: minimally rough (0.5–1 m), intermediately rough (1–2 m), and rough (2–3 m). The implant surface can also be classified according to their surface texture as concave and convex texture and orientation of surface irregularities as anisotropic and isotropic surfaces.

3.4.6 BIOCOMPATIBILITY

The implant material shows the proper response in a biological environment is referred as biocompatible. It also referred as the corrosion resistance and cytotoxicity of the products. Corrosion resistance [67–69] basically means the release of metallic ions from metal surface to the surrounding environment. There are many types of corrosion crevice corrosion, pitting corrosion, galvanic corrosion, electrochemical corrosion, etc. Clinical significance of corrosion is that the implant made up of bio-material should have the corrosion resistance. Corrosion can results into rough surface, weak restoration, release of elements from the metal or alloy, toxic reactions, etc. There can be allergic reactions in patients due to corrosion.

3.5 SPECIALIZED BIO-IMPLANTS

3.5.1 TOTAL AND PARTIAL KNEE REPLACEMENT

Among the knee replacement procedures, about 90% is the total knee replacement. During TKR procedure, the repair of the knee joint is by covering the thighbone with metal and encasing the shinbone with a plastic frame. The procedure makes replacement of the rough and irregular surface of the worn bone with a smooth surface. The undersurface of the kneecap will be replaced with a plastic surface to decrease the pain and provide a smooth functional joint. The process is to remove some of the parts of bones and cartilage. In case of a partial knee replacement, only the part of the knee that's damaged or arthritic will be replaced. It requires only smaller operation procedure and involves less bone and blood loss and also produces less pain. These are the advantages to this approach. Partial knee replacement patients will get a faster recovery time compared to TKR procedures. The disadvantage of the process is that there can be arthritis developed which needs another surgery in the parts of the knee that are not replaced [70, 71].

3.5.2 HIP JOINT REPLACEMENT

Joints in any parts of the body are important components of the skeletal system. It is positioned at bone joints for the transmission of loads from bone to bone by muscular action; also, there can be some relative motion of the component bones. Tissue of a bone is complex in nature and the composite consisting of soft and strong protein collagen and brittle hydroxyapatite. Bone is an anisotropic material with mechanical properties that differ in the longitudinal (axial) and transverse (radial) directions. The cartilage is a coating on each connecting surface, which consists of body fluids that lubricate and provide an interface with a very low coefficient of friction that provides the bone sliding movement. The human hip joint occurs at the junction between the pelvis and the upper leg (thigh) bone, or femur. Large rotary motion is allowable at the hip by a ball-and-socket type of joint. The top of the femur terminates in a ball-shaped head that fits into a cuplike cavity within the pelvis.

There are national and international standards on which the orthopedic community is taking guidelines. For better mechanical properties of medical-grade UHMWPE is preferred, medical-grade UHMWPE is as per the standard ASTM F-648 and ISO 5834. It contains the specifications of the unconsolidated resin powder and consolidated stock material. Processing to be done in various stages to produce highly crosslinked polyethylene for hip and knee bearings. These steps are to promote crosslinking by an irradiation process. Residual stresses are removed by thermal irradiation after the processing step which will increase the level of crosslinking. There will be also a sterilization step given to those implants. The properties of UHMWPE are less influenced by processes of irradiation, thermal processing, and sterilization [72, 73].

3.5.3 POLYMERS IN DENTAL IMPLANTS

There are methods to replace a missing tooth by the use of many materials as an implant. Great advancements occurred in the field of science and technology related to the materials for dental implants [74]. There are many types of polymers like ultrahigh molecular weight PU, polyamide, polymethylmethacrylate, polytetrafluoroethylene, and PU used as materials for dental implants. There is a great amount of researches and advancements in the field of biomaterials available for dental implants. Newer materials came up like zirconia, roxolid, and surface-modified titanium implants. These materials have the satisfactory functional requirements and also esthetically pleasing. The earlier material, methyl methacrylate resin implants became failures in many cases [74–76]. In 1969, Hodosh reported that polymers were biologically useful substances [77]. Polymethacrylate based tooth-replica implants was the polymer dental implant developed by Milton Hodosh. When natural tooth replacing polymers are found to be the ideal material for the restoration of function and appearance [78].

Polymers were selected for many reasons [78]. The physical properties of the polymers can be easily altered and compositions may be changed easily depending upon the applications. Polymers can be changed into porous or softer form, polymers can be processed can have larger productivity. They do not generate microwaves or electrolytic currents as in the case of metals. It shows fibrous connective tissue attachment, it can be easily analyzed compared with metals and give better esthetics. Disadvantages are like inferior mechanical properties and poor adhesion to living tissues.

3.5.4 NEURAL IMPLANTS

Electrical circuitry is implanted into the nerve cells to activate the parts and structures of the nervous system are called neural implants. Many experiments in the 1960s, material sciences, and the progress in medical and neuroscience lead to advancements in therapies of neurological diseases. This will lead to repair and rehabilitation of lost functions of human systems [80, 81]. Neuromodulation is the process of stimulation of the central nervous system (CNS). These structures which will be modulating the nerve excitability and neurotransmitters release [80]. Patients suffering from Parkinson's disease (PD) will be cured by suppressing tremor and movement disorders by deep brain stimulation. Similarly, treatment for epilepsy and other psychiatric diseases like depression and obsessive-compulsive disorder are done with the help of vagal nerve stimulation [82, 83]. These treatments are now expanded to psychiatric diseases and many more applications. Some of them are in the development stages in preclinical and clinical trials.

Motor implants to restore grasping [85] stance and gait [86, 87] as well as ventilation [88] by the electrical stimulation of the diaphragm have been developed and introduced into preclinical studies or even as commercial products to the market. Lower number of patients benefit from this system due to some technical issues. There is a limited number of studies about the performance of the implants in patients due lack of availability. Genitally deaf children and adults who have lost their hearing at a later point in life [88] have been implanted and were able to hear and to communicate with these implants. Developed implants can connect to the brain stem [89] and midbrain auditory structures [91] when tumors have destroyed the pathways from the ear to the cortex, will restore sound perception.

The retinal process to restore vision using implantation of complex electrical stimulators into the eyes of blind people is one of the modern technological progresses. This gives as a miniaturization technology helps the development such implants [80, 92]. Benefits and demerits have to be studied clearly and carefully in any medical and surgical treatment. It is also important that the patient should give the final consent for implantation. All neural implants approved as a medical device have to fulfill general requirements. They should make any harm to the body and should stay stable and functional over a certain period of time as decades. Proper design of any neurotechnical interface is the major key challenge for the creation of any neural implant. Many electrical sites have to give proper close contact with neural tissue to activate the nerve cells. Nerves are very delicate and soft tissue gets damaged by hard materials when forces due to movements of the implants. Polymers are found as the optimal material class. Because there is no response to implantation, gives long-term stability in any hostile environment, low material stiffness, and good electrical insulation compared to metallic conductors [93–98].

3.5.5 IMPLANTS HELPS IN CONTROLLED DRUG RELEASE

Biodegradable polymers can work for shorter times and slowly degrade if there are desirable conditions under the controlled mechanism, into products which are easily eliminated in the body's metabolic pathways [97]. Biodegradable polymers are more popular than no degradable delivery system, as they are eventually absorbed or metabolized and removed from the body by excretion. This method totally eliminates the need for surgery for the removal of the implant after the completion of the therapy.

The major advantage of this system include administration of a therapeutic drug in a controlled manner at the required delivery rate, maintaining the concentration of drug within the optimum therapeutic range for a prolonged treatment duration, minimum side effects and needs of frequent dosage is minimized. Controlled drug delivery systems are effectively used to control of hypercholesterolemia [100]. LDL cholesterol biosynthesis can be controlled by slow release of Simvastatin for a prolonged period of time which also gives prevention of coronary heart disease.

Anticancer drugs have poor performance against solid tumors in which drug penetration into the tumor is prevented. Toxicity is a limiting factor for the dose to be increased. Polymeric biodegradable poly-L-lactic acid (PLA) and poly(L-lactic acid-co-glycolic acid) (PLGA) copolymer can be administrated locally as an implant carrying an anticancer drug may a suitable method of concentrating the drug near the tumor site. S. O. Adeosun et al. studied that PLA can degrade hydrolytically and suitable fillers can be incorporated in the implants to reduce the cost [101]. There is a large difference in elastic modulus between the metal fixture and the bone. This can cause residual stress in the screw holes and also give adverse effect of rigid plates on bones and get the stress on the implants. PLA like biopolymers can solve these issues and can replace metals.

3.6 METHODS FOR MAKING BIO-IMPLANTS

3.6.1 3D PRINTING

Shortage of donor organs remains as a major concern in the medical [102] field and researchers are continuously searching for new methods to mimic or even replicate organs [103]. In such circumstances one of the best solution found out by researchers are the use of Bio-implants. A bio-implant is an implant with a biological component that is placed in a cavity of the human body for 30 days or more [104]. It can restore, support, or enhance the functions of the human tissues. It can also maintain the compatibility and conformity with the tissues along with the acceptability by the body. It gives the strength to the materials and the intactness of the implant. Biological materials such as cells, protein, etc., using bioprinting are prepared from biological implants. 3D printing of implants can be considered as organ printing. It uses ideally autologous cells to reduce the chances of rejection by the body. This can also reduce the waiting time for the replacement organ.

Two components are needed for making biological implants. A bioprinter containing materials such as living cells which predetermine the 3D form for creating the organ and biochemical reactor in which the manufactured organ can mature *in vitro*. Organ printing is defined as a computer-aided processing which cells or cell-laden biomaterials are placed in the form of aggregates, which then serve as building blocks and are further assembled into a 3D functional organ. Solid objects with complex shapes are manufactured by additive manufacturing methods and referred as 3D printing. This is a method used to manufacture objects by making layers of material arranged one over the other to get the finished article. Fused deposition modeling (FDM) is an additive manufacturing method. Thin strands of molten thermoplastics materials are laid down on each other using a print-head controlled by a computer-aided design (CAD) software. The printed object will form when the material gets solidified over the print surface.

Acrylonitrile-butadiene-styrene (ABS) resin can be used in FDM printers and personal desktop printers with ABS but now shifted to PLA due to its bio-compostability, pleasant smell as well as low shrinkage and good printability. PLA based materials used in FDM printers are not perfect. Printing fine details could be very much challenging due to melting and flowing is affected by temperature and viscosity of the melt. PLA based materials has the drawback as it weak in temperature resistance, which can cause deformation of printed objects under elevated temperatures may be during storage and shipping or usage.

3D bone structure requires porosity, for the flow of nutrients, blood, oxygen, and mineral. Production of such structure remains a problem using conventional methods. Blends made up of PLA and poly-ε-caprolactone (PCL) is a suitable material that gives properties required. Bone grafting method is used for the repair of bones that are severely damaged or lost completely. Arthritis, traumatic injury, and surgery for bone tumor are very common in the senior people. New researches for the design of new materials for the wide application are very much necessary. There can be permanent or temporary bone replacement depending on the properties of the material. A permanent bone replacement can use when a bone is missing due to some conditions. A temporary implant is used when the implant could be removed when the treatment is completed [101]. The selection will be depending on various factors, the purpose clinical application, defect area size, mechanical, and strength properties, material availability, and required bioactivity, material handling, cost aspects, and ethical concepts [105, 106].

3.6.2 ELECTROSPINNING

Electrospinning is one of the nanotechnologies with the largest diversity of applications. This can be used for the processing of a wide range of basic materials and the production of a diversity of electrospun nanofibers assembly organization. It helps in making of 1D to 3D products and diversity of forms with different arrangements [107–111]. The nanofibers obtained by electrospinning can be of polymeric, ceramic, or composite nature. Each of these groups has a wide portfolio of applications (Table 3.2) [113–115]. Biomedical applications of the polymeric nanofibers obtained through electrospinning are multiple [116, 117].

TABLE 3.2 Classification of Electro Spun Products

Electrospun Nanofibers	**Polymeric**	**Biomedical**	Regenerative medicine, tooth implants, bone, blood vessel implants, neural tissue engineering, structural tissues, such as cartilages, muscles, ligaments, dressings, meshes, medial prostheses.
			Structures for controlled drug release, cosmetics, drug delivery, and nanofibrous drug delivery system, polymer-drug blend fiber system, hemostatic devices.
	Polymeric	**Industrial**	Filtering mediums, separating membranes, nanosensors
	Ceramics	**Biomedical/ Industrial**	Biomechanical devices, biosensors, bone reconstruction, catalysts, biosensors, membranes, storage batteries.
			Aerospatiale products, finishing treatments by chemical, storage devices for information
	Composites	**Regenerative Medicine**	Implants

The most common technique of electrospinning is by using the electrostatic production method of nanofibers. Electric power is used to make polymer fibers with diameters 2 nm to several micrometers. Preparation is done from polymer solutions or melts. It is a versatile method to produce continuous fibers on a scale of nanometers. It is difficult to achieve using any other standard spinning techniques. It is a very simple way of preparing nanofibers. There are several parameters that will influence the formation and structure of nanofibers prepared using electrospinning [118].

3.7 GLOBAL IMPLANT SURVEY

The highest growing market as per the analysis done for each segment is the neurostimulation market is about the bio-implants. This is because of consumer awareness and technological advancements. Another fact is that neurological disorders increasing in the aging population. Because of that, there is drastic development of advanced implantable neurostimulation devices. These can be cost-effective products with FDA approvals and passed all clinical trials [119].

The largest market is in the orthopedic implants according to the total shares in the world implants market till 2015. This is due to the increasing obesity and an increase in osteoporosis-related fractures in all age groups. Patients that opt for orthopedic implant surgeries have increased from the age group of fewer than 55 to over 80 years. New technologies developed like the creation of patient-specific implants, knee arthroplasty, 3D technology, and sculptural based CAD are being developed which have numerous benefits over implants. Figure 3.1 shows an overview of the implant usage statics [120].

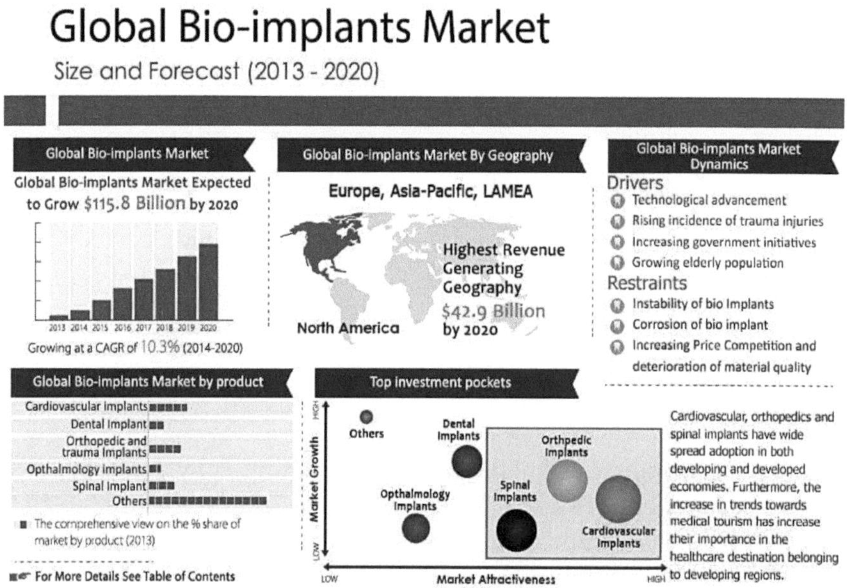

FIGURE 3.1 Global bio-implants market.

Figure 3.1 shows an idea about the implant market and its future scope. The growth rate is calculated as 10.3%, and the maximum market share of implants for the cardiovascular and next is dental and spinal implants. Most attracted implants are orthopedic, cardiovascular, and spinal as they are the important organs of the body. The world can create a greater sources for income from bio-implants which is found steadily growing across the globe. There are pros and cons for any substances as it is for bio-implants also. The price for the bio-implants and the deterioration of bio-implants are some of the major concerns. The world is looking for better opportunities as researches on bio-implants are going at a faster pace.

3.8 CONCLUSIONS

It is a saying that when you are evaluating the present situations and predicting the future, one must also reconsider the past. The modern medical field has begun to understand, realize, and utilize the benefits of biotechnology in health care similar to all other sectors. Design and material concepts for surgical implants are a study of material sciences and biomechanical sciences. Implants are being considered as a first treatment option by patients as it is so popular amongst over the other treatment modalities. It became o the mainstream for a dental practice. It has grown greater still a long way to achieve in the field of bio-implants. The process and investigation of biological component show that polymer material can be the best material which equally acts and reacts with better characteristics of biomaterials. It has good mechanical properties and can be used for the feature implants.

Polymers are highly corrosion resistant and wear-resistant materials. This material is highly biocompatible with the human body and suitable for the bioimplantation. The bone properties depend on the age and vary with factors like torsion, compression, tension, and bending. These factors would be easily matched with polymer materials as it has the best mechanical properties for better implants. Polymers are the highest percentage of materials with the properties also with the better biocompatibility can act as the reference material in place of ceramics and metals.

Biomaterials have a lot of advantages and can be used for many applications. Biomedical science and materials engineering can be combined as they are promising and largely falls in biomaterials and tissue engineering. Medical technologies are changing the way of lives and treatments as many advances being made in this discipline. The average lifespan of the human has increased as general medicine had improved all over the globe. On the contrary, the average life expectancy of over 80 years in many developed countries has created the problem of age-related illness. It is also a social concern to take care of the growing population. The aging process can be a happier one. But for many people, growing old give negative effects due to pains and stress. Bio-implants can give a great support to the aged people.

KEYWORDS

- **acrylonitrile-butadiene-styrene**
- **boron neutron capture therapy**
- **chemical vapor deposition**
- **computer-aided design**
- **fused deposition modeling**
- **mass spectrometry**

REFERENCES

1. Wong, J. Y., Bronzing, J. D., & Peterson, D. R., (2012). *Boca Raton* (p. 281). Florida: CRC Press.
2. McLatchie, G., Borley, N., & Chikwe, J., (2013). *Oxford Handbook of Clinical Surgery*(p. 794). Oxford, UK: OUP Oxford.
3. Ibrahim, H., Esfahani, S. N., Poorganji, B., Dean, D., & Elahinia, M., (2017). *Materials Science and Engineering: C., 70*(1), 870–888.
4. Nowosielski, R., Cesarz-Andraczke, K., Sakiewicz, P., Maciej, A., Jakóbik-Kolon, A., & Babilas, R., (2016). *Arch. Metall. Mater., 61*(2), 807–810.
5. Ritabh, K., Richard, A. L., Stephen, G., Benedict, A. C., & Rami, J. A., (2006). *Journal of Orthopedic Research, 24*(9), 1799–1802.
6. David, X. F., & Joseph, P. M., (2015). *British Journal of Radiology, 88*, 1056.
7. Ibrahim, I. D., Jamiru, T., Sadiku, E. R., Kupolati, W. K., Agwuncha, S. C., et al., (2016). *Composite Interfaces, 23*(1), 15–36.
8. Fuchs, E. R., Field, F. R., Roth, R., & Kirchain, R. E., (2008). *Composites Science and Technology, 68*, 1989–2002.
9. Victor, S. P., & Muthu, J., (2014). *Mater Sci. Eng. C Mater. Biol. Appl., 39*, 150–160.
10. Ma, Z., Mao, Z., & Gao, C., (2007). *Colloids. Surf. B Biointerfaces, 60*, 137–157.
11. Scognamiglio, F., Travan, A., Rustighi, I., Tarchi, P., Palmisano, S., et al., (2016). *J. Biomed. Mater. Res. B Appl. Biomater, 104*(3), 626–639.
12. Pertici, G., (2016). *From Fundamentals to Translational Medicine* (p. 1).
13. Lin, J. H., Lu, C. T., He, C. H., Huang, C. C., Lou, C. W., et al., (2011). *Journal of Composite Materials, 45*(19), 1945–1951.
14. Zare, Y., & Shabani, I., (2016). *Mater. Sci. Eng. C Mater. Biol. Appl., 60*, 195–203.
15. Bassas-Galia, M., Follonier, S., Pusnik, M., & Zinn, M., (2016). *Perale G Hilborn, J Eds*, pp. 31–64.

16. Mao, C., Liang, C., Luo, W., Bao, J., Shen, J., et al., (2009). *Journal of Materials Chemistry, 19*(47), 9025–9029.
17. Thakur, V. K., & Kessler, M. R., (2015). *Polymer, 69*, 369–383.
18. Nederberg, F., Atthoff, B., Bowden, T., Hilborn, J. et al., (2008). Mahapatro and Kulshrestha; Polymers for Biomedical Applications, ACS Symposium Series; American Chemical Society: Washington, DC. *ACS Symp. Ser.* Chapter 15.
19. Henderson, L. A., Kipper, M. J., & Chiang, M. Y. M., (2008). Mahapatro and Kulshrestha; *Polymers for Biomedical Applications*, ACS Symposium Series; American Chemical Society: Washington, DC. *ACS Symp. Ser.,* Chapter 8.
20. Hassan, M. K., Wiggins, J. S., Storey, R. S., et al., (2008). Mahapatro and Kulshrestha; *Polymers for Biomedical Applications*, ACS Symposium Series; American Chemical Society: Washington, DC. *ACS Symp. Ser.,* Chapter 9.
21. Angelova, N., & Hunkeler, D., (1999). *Tibtech, 17*, 409.
22. Maziarz, E. P., Liu, X. M., & Wood, T. D., (2008). Mahapatro and Kulshrestha; Polymers for Biomedical Applications, ACS Symposium Series; American Chemical Society: Washington, DC. *ACS Symp. Ser.,* Chapter 10.
23. Wertz, J. T., Mauldin, T. C., & Boday, D. J., (2014). *CS Appl. Mater. Interfaces, 6*(21), 18511–18516.
24. Glasgow, K., Dhara, D., Mahapatro, & Kulshrestha, (2008). *Polymers for Biomedical Applications.* American Chemical Society: Washington, DC. *ACS Symp. Ser.,* Chapter 16.
25. Nandivada, H., Chen, H. Y., Elkasabi, Y., Lahann, J., Mahapatro, & Kulshrestha, (2008). *Polymers for Biomedical Applications*. American Chemical Society: Washington, DC. *ACS Symp. Ser.,* Chapter 17.
26. Wingkono, G., Meredith, C., Mahapatro, & Kulshrestha, (2008). *Polymers for Biomedical Applications.* American Chemical Society: Washington, DC. *ACS Symp. Ser.,* Chapter 18.
27. He, W., Feng, Y., Ramakrishna, S., Mahapatro, & Kulshrestha, (2008). *Polymers for Biomedical Applications*. American Chemical Society: Washington, DC. *ACS Symp. Ser.,* Chapter 19.
28. Konno, T., Ishihara, K., Mahapatro, & Kulshrestha, (2008). *Polymers for Biomedical Application.* American Chemical Society: Washington, DC. *ACS Symp. Ser.,* Chapter 20.
29. Cooper, K., Li, Y., Lowenhaupt, B., Yuan, J., Zimmerman, M., Mahapatro, & Kulshrestha, (2008). *Polymers for Biomedical Applications*. American Chemical Society: Washington, DC. *ACS Symp. Ser.,* Chapter 21.
30. Duran, H., Lau, K. H. A., Lübbert, A., Jonas, U., Steinhart, M., Knoll, W., Mahapatro, & Kulshrestha, (2008). *Polymers for Biomedical Applications*. American Chemical Society: Washington, DC. *ACS Symp. Ser.,* Chapter 22.
31. Guo-Qiang, C., (2010). *Introduction of Bacterial Plastics PHA, PLA, PBS, PE, PTT, and PPP*. Springer-Verlag Berlin Heidelberg.
32. Steinbüchel, A., (1991). *Biomaterials Macmillan* (pp. 125–213). Basingstoke.
33. Spyros, A., & Marchessault, R. H., (1996). *Macromolecules, 29*, 2479–2486.
34. Gu, J. D., Ford, T. E., Mitton, D. B., & Mitchell, R., (2000). *The Uhlig Corrosion Handbook* (2nd edn., pp. 915–927). Wiley, New York.
35. Luzier, W. D., (1992). *Proc National Academic Sci. U.S.A, 89*, 839–842.
36. Bonhomme, S., Cuer, A., Delort, A. M., Lemaire, J., Sancelme, M., & Scott, C., (2003). *Polym. Degrad. Stab., 81*, 441–452.

37. Darwin, P. R. K., Abdelilah, A., Elise, D., Josefina, L. C., & Sebastian, M. G., (2003). *Polymer., 44*, 1321–1330.
38. Aamer, A. S., Fariha, H., Abdul, H., & Safia, A., (2008). *Biotechnol. Adv., 26*, 246–265.
39. Avérous, L., & Pollet, E., (2012). Chapter 2. In: Avérous, L., & Pollet, E., (eds.), *Green Energy and Technology Series* (pp. 13–40). Springer: London, UK.
40. Pivsa-Art, W., Chaiyasat, A., Pivsa-Art, S., Yamane, H., & Ohara, H., (2013). *Energy Procedia, 34*, 549–554.
41. Jackson, D. W., & Simon, T. M., (1999). *Clin. Orthop., 367S*, 31–45.
42. Temenoff, J. S., & Mikos, A. G., (2000). *Biomaterials, 21*, 2405–2412.
43. Fleming, J. E., Muschler, G. F., Boehm, C., Lieberman, I. H., & McLain, R. F., (2004). In: Goldberg, V., & Caplan, A., (eds.), *Orthopedic Tissue Engineering: Basic Science and Practice* (pp. 51–65). Marcel Decker, New York.
44. Athanasiou, K. A., Agrawal, C. M., Barber, F. A., & Burkhart, S. S., (1998). *Arthroscopy, 14*, 726–737.
45. Heath, C. A., (2000). Cells for tissue engineering. *Trends Biotechnol., 18*, 17–19.
46. Vacanti, J. P., Langer, R., Upton, J., & Marler, J. J., (1998). *Adv. Drug Delivery Rev., 33*, 165–182.
47. Barbensee, J. E., Mc Intire, L. V., & Mikos, A. G., (2000). *Pharm. Res., 17*, 497–504.
48. Ambrose, C. G., Clyburn, T. A., Louden, K., Joseph, J., Wright, J., Gulati, P., et al., (2004). *Clinical Orthop., 421*, 293–299.
49. Hansoo, P., Johnna, S. T., & Antonios, G. M., (2007). *Biodegradable Orthopedic Implants* (p. 55). Springer, Chapter 4.
50. Caborn, D. N., Urban, W. P. J., Johnson, D. L., Nyland, J., & Pienkowski, D., (1997). *Arthroscopy, 13*, 229–232.
51. Higashi, S., Yamamuro, T., Nakamura, T., Ikada, Y., Hyon, S. H., & Jamshidi, K., (1986). *Biomaterials, 7*, 183–187.
52. Timmer, M. D., Carter, C., Ambrose, C. G., & Mikos, A. G., (2003). *Biomaterials, 24*, 4707–4714.
53. Claes, L. E., Ignatius, A. A., Rehm, K. E., & Scholz, C., (1996). *Biomaterials, 17*, 1621–1626.
54. Disegi, J. A., & Wyss, H., (1989). *Orthopedics, 12*, 75–79.
55. Rupp, S., Krauss, P. W., & Fritsch, E. W., (1997). *Arthroscopy, 13*, 61–65.
56. Holland, T. A., & Mikos, A. G., (2003). *J. Control Release, 86*, 1–14.
57. Luginbuehl, V., Meinel, L., Merkle, H. P., & Gander, B., (2004). *Eur. J. Pharm. Biopharm., 58*, 197–208.
58. Holland, T. A., Tessmar, J. K. V., Tabata, Y., & Mikos, A. G., (2003). *J. Control Release, 94*, 101–114.
59. Jeong, B., Kim, S. W., & Bae, Y. H., (2002). *Adv. Drug Deliv. Rev., 54*, 37–51.
60. Ibrahim, I. D., Sadiku, E. R., Jamiru, T., Hamam, A., & Kupolati, W. K., (2017). *Current Trends Biomedical Eng. and Biosci., 4*(4).
61. Zare, Y., & Shabani, I., (2016). Polymer/metal nano composites for biomedical applications. *Mater. Sci. Eng. C Mater. Biol. Appl., 60*, 195–203.
62. Bassas-Galia, M., Follonier, S., Pusnik, M., & Zinn, M., (2016). Natural Polymers: A source of inspiration. In: Perale, G., & Hilborn, J., (eds.), *Bioresorbable Polymers for Biomedical Applications* (pp. 31–64).

63. Mao, C., Liang, C., Luo, W., Bao, J., Shen, J., et al., (2009). Preparation of lotus-leaf-like polystyrene micro- and nanostructure films and its blood compatibility. *Journal of Materials Chemistry, 19*(47), 9025–9029.
64. Thakur, V. K., & Kessler, M. R., (2015). Self-healing polymer nano-composite materials: A review. *Polymer, 69*, 369–383.
65. Jozsa, L., & Reffy, A., (2012). Folia histochem. *Cytochem, 18*, 195–200.
66. Lee, J. M., Salvati, A. E., & Detts, F., (1992). *J. Bone Jt. Surgery*, (2012), 74b, 380–384.
67. Hansen, D. C., (2008). Metal Corrosion in the Human Body: The Ultimate Bio-Corrosion Scenario. *The ElectrochemicalSociety Interface, 17*, 31–34.
68. Bakker, D., Grote, J. J., Vrouenraets, C. M. F., Hesseling, S. C., De Wijn, J. R., Van, B. C. A., Heimke, G., et al., (1990). Elsevier Science Publication, Amsterdam (pp. 99–104).
69. Judee, G. E., Nemeno, S. L., Wojong, Y., Kyung, M. L., & JeongIk, L., (2014). *BioMed. Research International* (Vol. 2014).
70. Naveen, K. A., & Gangadhara, S. B., (2015). *International Journal of Engineering Research and Technology (IJERT), 4*(05).
71. Rajabhushan, D., & Vishnuvardhan, R. R., (2017). *International Journal of Advance Engineering and Research Development (IJAERD), 4*(8).
72. Nino, K., & Ana, P., (2016). Fixture for automated 3D scanning. *International Conference on Manufacturing Engineering and Materials, ICMEM 2016*, Nový Smokovec, Slovakia.
73. Revell, P. A., (ed) (2008). Joint replacement technology. Woodhead Publishing Limited, Cambridge, UK, pp. 83–84.
74. Mäkelä, K., (2010). *Academic Dissertation*. Helsinki.
75. Hulbert, S. F., & Bennett, J. T., (1975). *Journal of Dent. Res., 54,* Spec No B, B153–B157.
76. Misch, C. E., (1999). *Implant Dentistry, 8*, 90.
77. Sykaras, N., Iacopino, A. M., Marker, V. A., Triplett, R. G., & Woody, R. D., (2000). *Int. J. Oral Maxillofac Implants, 15*, 675–690.
78. Lemons, J. E., (1990). *J. Am. Dent. Assoc., 121*, 716–719.
79. Carvalho, T. L., Araújo, C. A., Teófilo, J. M., & Brentegani, L. G., (1997). *Int. J. Oral. Maxillofac. Surg., 26*, 149–152.
80. Kawahara, H., (1983). *Int. Dent. J., 33*, 350–375.
81. Stieglitz, T., & Meyer, J. U., (2006). In: Urban, G. A., (ed.), *BIOMEMS* (Vol. 4, pp. 71–137). Springer-Verlag: Dordrecht, The Netherlands.
82. Stieglitz, T., & Meyer, J. U., (2006). In: Urban, G. A., (ed.), *BIOMEMS* (Vol. 3, pp. 41–70). Springer: Dordrecht, The Netherlands.
83. Albert, G. C., Cook, C. M., Prato, F. S., & Thomas, A. W., (2009). *Neurosci. Biobehav. R., 33*, 1042–1060.
84. Groves, D. A., & Brown, V. J., (2005). *Neurosci. Biobehav. R., 29*, 493–500.
85. Rijkhoff, N. J. M., (2004). *Child. Nerv. Syst., 20*, 75–86.
86. Brindley, G. S., (1972). *J. Physiol., 222*, 135–136.
87. Brindley, G. S., Polkey, C. E., Rushton, D. N., & Cardozo, L., (1986). *J. Neurol. Neurochir. Psychiatr., 49*, 1104–1114.
88. Rupp, R., & Gerner, H., (2004). *J. Biomed. Tech., 49*(4), 93–98.
89. Agarwal, S., Triolo, R. J., Kobetic, R., Miller, M., Bieri, C., Kukke, S., Rohde, L., & Davis, J. A. Jr., (2003). *J. Rehabil. Res. Dev., 40*, 241–252.

90. Guiraud, D., Stieglitz, T., Koch, K. P., Divoux, J. L., & Rabischong, P., (2006). *J. Neural. Eng., 3*, 268–275.
91. Glenn, W. W., Holcomb, W. G., Shaw, R. K., Hogan, J. F., & Holschuh, K. R., (1976). *Ann. Surg., 183*, 566–577.
92. Loizou, P. C., (1999). *IEEE Eng. Med. Biol., 18*, 32–42.
93. Clark, G. M., (2008). *J. Rehabil. Res. Dev., 45*, 651–693.
94. Dettman, S. J., Pinder, D., Briggs, R. J., Dowell, R. C., & Leigh, J. R., (2007). *Ear Hearing, 28*, 11S–18S.
95. Colletti, V., Shannon, R. V., Carner, M., Veronese, M., & Colletti, L., (2009). *Prog. Brain. Res., 175*, 333–345.
96. Lim, H. H. L., Joseph, T., Battmer, G., Samii, R. D., Samii, M. A., Patrick, J. F., & Lenarz, M., (2007). *J. Neurosci., 27*, 13541–13551.
97. Stieglitz, T., (2009). *J. Neural. Eng., 6*, 1–11.
98. Alekha, K. D., & Greggrey, C., (1998). *Cud Worth, Journal of Pharmacological and Toxicological Methods, 40*(1), 1–12.
99. Kamala, K., Jhansi, R., Vijay, T., & Baruah, D. K. M., (2015). *Journal of Pharmacy and Biological Sciences, 10*(3), 49–52.
100. Hakata, T., Sato, H., Watanabe, Y., & Matsumoto, M., (1994). *Chemical and Pharmaceutical Bulletin, 42*, 1138–1142.
101. Salsa, T., Pina, M. E., & Teixeira-Dias, J. J. C., (1996). *Applied Spectroscopy, 50*(10), 1314–1318.
102. Benjamin, H., (2015). *International Journal of Precision Engineering and Manufacturing.*
103. Lavine, M., Roberts, L., & Smith, (2002). *Science, 295*(5557), 995.
104. U.S. Food and Drug Administration and Guidance, (2015). IDE.
105. Khatiwala, C., Law, R., Shepherd, B., Dorfman, S., & Csete, M., (2012). *Gene Therapy and Regulation, 7*(1).
106. Meng, B. H., Sum, H. N., & Yong-Jin, Y., (2015). *Inte. J. of Precision Engineering and Manufacturing, 16*(5), 1035–1046.
107. Danu, C. M., Nechita, E., & Manea, L. R., (2015). *Studies and Scientific Researches Economics* (21st edn., p. 14).
108. Manea, L. R., Nechita, E., Danu, M. C., & Agop, M., (2015). *J. Comput. Theor. Nanosci., 12*(11), 4693.
109. Manea, L. R., Stanescu, I., Nechita, E., & Agop, M., (2015). *J. Comput. Theor. Nanosci., 12*(11), 4373.
110. Calin, M. A., Manea, L. R., Schacher, L., Adolphe, D., Leon, A. L., Potop, G. L., & Agop, M., (2015). *Journal of Nanomaterials.*
111. Leon, A. L., & Manea, L., (2008). *4th International Textile Clothing and Design Conference, Book of Proceedings* (pp. 803–806).
112. Scarlet, R., Manea, L. R., Sandu, I., Martinova, L., Cramariuc, O., & Sandu, I. G., (2012). *Revista De Chimie, 63*(7) 688.
113. Scarlet, R., Manea, L. R., Sandu, I., Cramariuc, B., & Sandu, A. V., (2012). *Revista De Chimie, 63*(8)777.
114. Vasilica, P., Liliana-Rozemarie, M., & Gabriel, P., (2009). In: Das, D. B., Nassehi, V., & Deka, L., (eds.), *ISC 2009 Industrial Simulation Conference 2009* (pp. 352–355). Loughborough, UK, Eurosis-ETI.
115. Manea, L. R., & Scarlet, R., (2015). *Advanced Fibers and Yarns, 2*, (Bacau Alma Mater).

116. Nedjari, S., Hebraud, A., & Schlatter, G., (2015). Electro spinning. *Principles Practice and Possibilities* (p. 173).
117. Manea, L. R., et al., (2016). *IOP Conf. Ser. Mater. Sci. Eng., 145*, 03, 2006, IOP Conference Series: Materials Science and Engineering.
118. Salma, A., Puspita, D., & Swarnali, I., (2016). *IOSR Journal of Pharmacy and Biological Sciences, 11*(3) Ver. IV. pp. 123–132.
119. Article on ResearchandMarkets.com, April 5, 2018, (2018). *Region: Global,* p. 133.
120. Medical Implants Market By Type (Orthopedic Implants, cardiac implants, Stents, Spinal Implants, Neurostimulators, dental implants, Breast Implants, facial Implants) and By Materials (Metallic, Ceramic, Polymers, Natural)-Global Opportunity Analysis and Industry Forecast, 2014–2022, Article Published in Allied Market Research, July 2019.
121. Doi, Y., Kitamura, S., Abe, H., (1995). Microbial synthesis and characterization of poly(hydroxybutyrateco – hydroxyhexanoate). *Macromolecules 23*, 4822–4828.
122. Wang, H. H., Li, X. T., Chen, G. Q., (2009). Production and characterization of homopolymer polyhydroxyheptanoate (P3HHp) by a fadBA knockout mutant Pseudomonas putida KTOY06 derived from P. putida KT2442. *Process Biochem 44*, 106–111.
123. Galegoa, N., Rozsaa, C., Sa'nchez, R., et al., (2000) Characterization and application of poly(b-hydroxyalkanoates) family as composite biomaterials. *Polym. Test, 19*, 485–492.
124. Williams, S. F., & Peoples, O. P, (1996). Biodegradable plastics from plants. *Chem. Tech. 38*, 38–44.
125. Doi, Y., Kumagai, Y., Tanahashi, N., & Mukai, K., (1992). Structural effects on biodegradation of microbial and synthetic poly(hydroxyalkanoate). In: Vert, M., Feijen, J., Albertsson, A., Scott, G., Chiellini, E. (eds.). *Iodegradable Polymers and Plastics*. Royal Society of Chemistry, Cambridge.
126. Pranamuda, H., & Tokiwa, Y., (1999). Degradation of poly (L-lactide) by strains belonging to genus Amycolatopsis. *Biotechnol Lett 21*, 901–905.
127. Aamer, A. S., Fariha, H., Abdul, H., et al., (2008). Biological degradation of plastics: a comprehensive review. *Biotechnol. Adv. 26*, 246–265.
128. Jin, H. J., Lee, B. Y., Kim, M. N., et al., (2000a). Properties and biodegradation of poly(ethylene adipate) and poly(butylene succinate) containing styrene glycol units. *Eur. Polym. J. 36*, 2693–2698.
129. Jin, H. J., Kim, D. S., Kim, M. N., et al., (2001). Synthesis and properties of poly(butylene succinate) with N-hexenyl side branches. *J. Appl. Polym. Sci. 81*, 2219–2226.
130. Jung, I. K., Lee, K. H., Chin, I. J., et al., (1999). Properties of biodegradable copolyesters of succinic acid-1, 4-butanediol/succinic acid-1, 4-cyclohexanedi- methanol. *J. Appl. Polym. Sci. 72*, 553–561.

CHAPTER 4

Versatile Nature of Poly(Vinylpyrrolidone) in Clinical Medicine

K. R. DHANYA,[1] P. MEREENA LUKE,[1,2] SABU THOMAS,[1] DIDIER ROUXEL,[3] and NANDAKUMAR KALARIKKAL[1]

[1]*International and Inter-University Center for Nanoscience and Nanotechnology, Mahatma Gandhi University, Kottayam – 686560, Kerala, India, E-mail: k_r_dhanya@yahoo.co.in (K. R. Dhanya)*

[2]*Chemical Faculty, Polymers Technology Department, Gdansk University of Technology, Gdansk – 80233, Poland*

[3]*Institute Jean Lamour, Université de Lorraine, Nancy Cedex – 50840–54011, France*

ABSTRACT

Unique polymer, poly(vinylpyrrolidone has highly vast and interesting medical applications. Diverse properties in every field, makes it as a prominent polymer. On combining with substrates, PVP shows better performance, biocompatibility, and dimensional stability. When it is taking orally, the emitting is really easier and faster and completely through the kidney. One of the superior properties of PVP is its compatibility and its important application is the substitute for blood plasma. This chapter solely summarizes the utilization of PVP in the therapeutic area.

4.1 INTRODUCTION

The nontoxic behavior of PVP imparts special attention in the last decades and its strong binding capacity and better solubility make faster execution in the field of medical health care. PVP is also used in the non-medical sector due to its excellent chemical as well as physical properties. PVP forms stable complexes and yields further recognition in the pharmaceutical field and medicine [9–14, 20, 48]. Another important area of PVP is the use in ophthalmic solutions and in drug tablets. Industrial and commercial applications leads to paper, cosmetics, fibers, textiles, adhesives [21–30], coatings, inks, industrial, environmental, optical, electrical [31–33, 72], photography [73–81], food, and household. Quantitative analysis of PVP in accordance with its use leads to medicine and its versatile nature such as lack of toxicity, easier film formation and adhesive capacity, PVP can readily utilize for the synthesis of nanogels [82–86]. PVP shows an amorphous nature and due to this character, the scattering loss must be very low and can be used for various purposes.

4.2 STRUCTURE AND SYNTHESIS

Poly(vinylpyrrolidone) is known to be polyvidone or povidone and it is depicted in the section of polymers of water-soluble as shown in Figure 4.1. In dry conditions, it can absorb moisture and in the solution state, exhibit fine wetting characteristics and can easily form films.

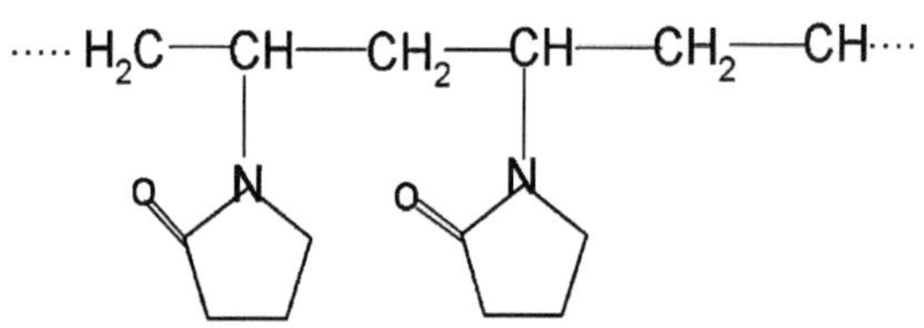

FIGURE 4.1 Structure of poly(vinyl pyrrolidone).

It is synthesized by free radical polymerization reaction using free radical initiator azobisisobutyronitrile. PVP is prepared from N-vinylpyrrolidone (NVP) and radical polymerization of NVP in bulk, solution, or in suspension, yields poly(vinylpyrrolidone) (Figure 4.2).

PVP varies from molecular weight and by this variation; it can be applicable to various fields like cosmetic area, industrial field, and pharmaceuticals.

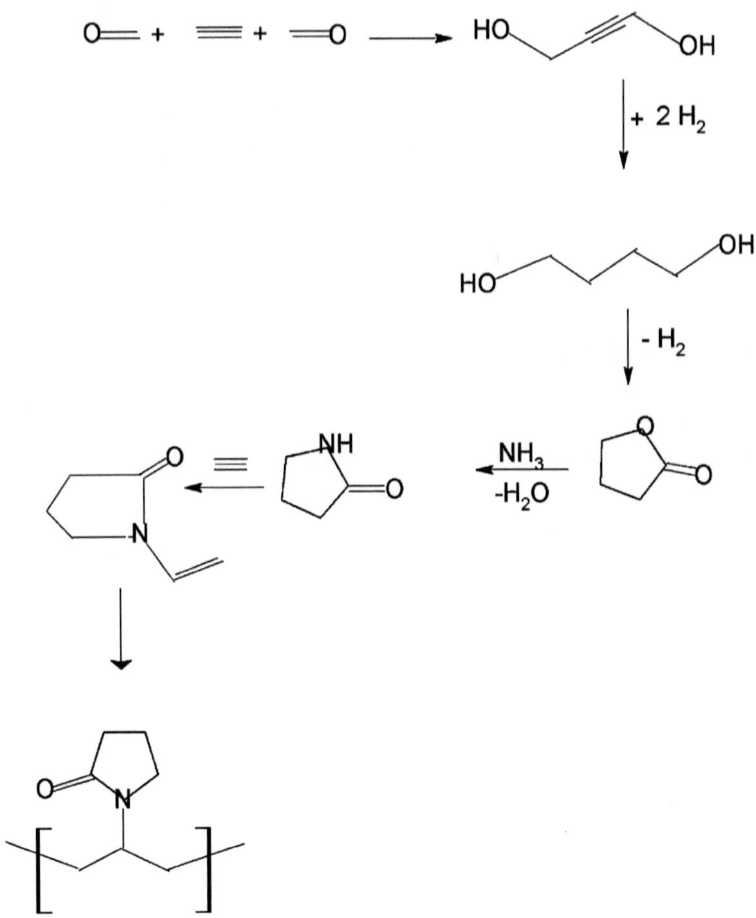

FIGURE 4.2 Synthesis of poly(vinyl pyrrolidone) from formaldehyde and acetylene.

4.3 PROPERTIES

PVP is soluble in all types of organic solvents and under normal conditions and is quite stable. The reason behind this remarked solubility is that they contain both hydrophilic and hydrophobic groups in the chain structure. The inert, hygroscopic, colorless, and temperature resistant behavior of PVP makes it a unique polymer [87, 88]. The glass transition temperature is related with its molecular weight. PVP films are fairly hard and transparent. Binding with natural as well as synthetic resins, PVP forms clear films. Chemically modified PVP are extremely inert in nature. In powder state, it is comparatively stable and if applying temperature on PVP, it becomes yellow in color. PVP is in contact with light, the formation of gelation happens.

When PVP is used as a drug, easily released, and penetrate through the kidney [89]. The dosage and absorption of PVP into body is very less and is inversely linked with its molecular weight [89, 90]. Molecular weight is an important factor in such kinds of analysis. When the molecular weight is higher, it is difficult to go through body membranes and generate some reactions which must be somewhat disturbing to our mechanical functions [91].

4.4 MEDICAL APPLICATIONS OF PVP

PVP is available in various forms such as tablets, granules, pellets [92], several kinds of hydrogels [93–100], powders, syrups, oral solutions, coatings in medical applications [101–103], etc. PVP has a wide range of properties and hence can be used in biomaterial applications. PVP polymer is extensively used in medical sectors. It is used as a blood plasma expander and binder substance in tablets. When use it orally, or the substance abusers injected, autopsies shown that crospovidone can contribute much larger in pulmonary vascular injury. Besides all these characteristics, the povidone-iodine complex has disinfectant properties and can be used widely in the ointment, pessaries, liquid soaps, and surgical scrubs.

Poly(vinylpyrrolidone) has better film formation ability and affinity towards amphiphilic surfaces. It has excellent sorption behavior and is generally used as a base substrate against toxins and viruses.

PVP is frequently used in drugs and chewable tablets due to the small shape and size of particles [104, 105].

This polymer exhibits a unique holding capacity to grasp all components together and finally rapid releasing the medicinal component. On transdermal deliveries, PVP is considered as an additive which reduces the crystallization behavior in drugs [105, 107]. Patel et al. gave explanations in the case of solid drugs, PVP act as a membrane coat and a drug capsule with controlled release, but when PVP coated in antidepressant drug system, discharge of venlafaxine is normally delayed [108].

PVP is also used for ophthalmic applications also and the major function is to reduce severe eye irritation. It generally acts as a pore former in drug release controlling membranes and layers [109].

PVP-based bio-adhesive films can be extensively used in buccal release of fentanyl [110] and preferred surface area was only 1–2 cm^2 equivalent to a 10 cm^2 transdermal patch afford a healing effect [111]. Nanofibrous PVP is used for wound healing applications, developed by electrospinning method. Other types of nanofibrous membranes for potential applications such as enhanced wound healing and drug release include PVP blended emodin [112]. An optimum combination of PVP-Alginate with nanoparticles of silver plays a vital role in wound dressings [113], vascular, nervous system, cartilage, bone, and ligament regeneration [114–116].

Silver nanoparticles bounded PVP polymers shows superior effect on angiogenesis [117]. This study was proposed by Kang et al. Silver nanoparticle dopped PVP-Chitosan have remarkable property of reducing serious infections in connection with medical tools. Cancer treatments are widely carried out using PVP grafted carbon nanotubes; it can also used as dentistry, orthopedics, as well as controlled drug delivery. PVP-Hypericin complex [118–121] is widely used for anti-cancer photodynamic analysis and treatments. Due to its superior properties such as nontoxic, nonantigenic, and biocompatible nature, it is widely used in the delivery of genes/drugs.

Graphene-based materials are used in various fields such as biomedical, drug, and vaccine delivery shows exciting properties. Nanocomposite of graphene, PVP, and polyaniline can be used for measuring cholesterol levels in medical diagnosis. Both graphene oxides coated as well as non-coated PVP on human immune cells describes outstanding immune enhancement properties and such kinds of studies were carried out by Zhi et al. PVP coated nanostructured cobalt ferrite shows extremely different properties such as biocompatibility, magnetic, and antioxidant property and this excellent antioxidant property help them to use in cancer treatments. Co-polymers of PVP have extensive use in various fields.

Veeran et al. explained polycaprolacone-PVP co-polymers are used in antituberculosis drug carriers and antitumor drug encapsulation. Kodaira et al. and Kamada et al. proposed PVP copolymerized with dimethyl maleic anhydride nowadays used for renal and various kinds of other drug delivery systems. PVP-Iodine complex shows disinfectant properties and it is used during surgery, antibacterial, and antimicrobial agent in medical devices. Apt et al. [14]; Benevento et al. [16]; Ciulla, Yanai, et al. proved PVP-Iodine can be used as an antiseptic agent in intraocular surgery and can have a capacity to diminish the effect of conjunctival bacterial flora. Jones et al. suggested the use in urinary catheters, ureteral stents.

These types of devices can be used in cardiovascular diseases which will help a free motion through the vasculature.

PVP is considered as a water-soluble matrix and have excellent applications in drug delivery [122, 123]. It has been observed that the rate of cytotoxicity of pharmaceutical ingredients reduces generally due to the solid diffusion structure of polymer [124] and shows defensive properties towards macrophage action [125, 126]. Due to the nontoxic nature of PVP, it has superior application in plasma replacement and describes a clear solution [127–129].

PVP has the ability to bind chemically with dyes, insulin, and all types of other drugs [130]. The rate of aggregation of erythrocytes [131] normally increases on adding PVP by intravenously and also diminishes the count of platelet adhesiveness in human blood by both *in vivo* and *in vitro* types of patterns [132] and even in high concentrations, PVP functions in human erythrocytes in frozen storage. At this stage, PVP induce plasma proteins [133], fibrinogen, and antihemophilic factor VIII and coagulation factors [134]. The excellent nature of PVP functioning on embryonic development in humans is the other most significant area to be concerned. It is firstly considered as the substitute for blood plasma and other extensive areas of use such as cosmetics, pharma, and industry, PVP is selected as an apt polymer.

PVP is the main part of therapeutic systems; namely wound dressing and prostaglandin release devices. PVP shows distinctive sorption behavior and acts as a matrix for agents removing viruses from blood. When it is used as a tablet, polymers remain in the body, it did not metabolize and very little amount of high molecular fractions will exist within the body. Generally accessible in an extensive range of molecular weights, i.e., 2500–3,000,000 g mol^{-1}, hence it is used for pharmaceutical applications. Universal solubility in water and other organic solvents are the essential property of PVP [135].

The presence of polar lactam group makes it highly hydrophilic in nature and nonpolar methylene group attributed the lipophilic character. These excellent features have led to the extensive application in pharmaceutical field. Wet granulation method is applicable in the production of tablets, syrups, drops, injectables, and in film coatings [136]. PVP exhibits film forming capacity and are applicable in tablets, transdermal systems, and therapeutic sprays. PVP and its copolymers are also widely useful in biomedical areas [137–142]. N-vinyl pyrrolidone and its copolymers based hydrogels used in wound healing applications. PVP modified materials have been widely used in medical devices, in polyurethane (PU) central venous catheters have superior properties when covered with PVP polymer [143].

PVP bounded PU catheters were highly hydrophilic in nature and can absorb less amounts. PVP exhibits the film-forming capability thus it can be used in tablets, transdermal systems, and therapeutic sprays of fibrinogen and fibronectin. Minimum amount of protein adsorption is related to a reduction of *Staphylococcus aureus* and *Staphylococcus epidermidis*. Blending of PVP with suitable materials led to outstanding applications. If PVP is considered as an additive, BSA adsorption capacity is comparatively less and the blood compatibility of PES membrane becomes better. Wang et al. reported the significance of this work [144] in their research on PVP. PVP polymer has numerous applications and it can be tested in artificial cartilage [145], artificial skin, drug vehicle, and cardiovascular devices [146, 147]. PVP hydrogel systems are used in nasal release of acyclovir (ACV) and these are completely harmless and safe materials. PVP-ethylcellulose composite was used for coating pellets for topiramate delivery [148]. The role of PVP extended in wide way, it can also act as a pore-former and increases the rate of drug delivery from 23% to 29%. PVP-poloxamer 407 combined to form a thermally reversible in situ gel as drug release matrix for human growth hormone and can be applied in pediatric hypopituitary dwarfism [149]. PVP hydrogels are also used for skin regeneration and wound dressing uses and the membrane prepared has excellent characteristic properties such as transparency, fast pain-relieving property, and compatibility [150–162]. PVP bounded paracetamol tablets release the drug very rapidly compared to others such as tablets with gelatine or hydroxypropyl cellulose. The function of PVP as a binder and the tablets containing povidone were usually very harder [163]. PVP shows unique stability to the conditions of Phenytoin tablet formulations [164] and when combined with triesters of citric acid, get a soft gelatine capsule that contains insoluble drugs [165] and the pharma based applications of PVP is shown in Table 4.1.

TABLE 4.1 Functions and Medicinal Grades of Poly(Vinylpyrrolidone)

Functions of Poly (Vinyl Pyrrolidone)	Medicinal Forms
Adhesive	Adhesive gels, transdermal systems, etc.
Lyophilizing agent	Oral lyophilisates and injectables
Stabilizer	Suspensions, dry syrups, etc.
Toxicity reducer	Injectables and oral preparations
Hydrophiliser	Suspensions
Solubilizing agent	Oral solutions and topical solutions
Taste masking	Oral solutions and chewing tablets
Film-forming agent	Tablets, ophthalmic solutions
Better bioavailability	Tablets, pellets, and trans-dermal systems
Binder	Granules and tablets

For cartilage replacement, Shi, and Xiong performed research based on PVP-PVA hyrogel and these manufactured by F/T method, F/T cycles, and also 60Co gamma-irradiation [166, 167]. For the treatment of oral squamous cell carcinoma, the buccoadhesive PVA patches of 5-fluorouracil and its general nature as well as properties by the effects of PVP were studied by Ghareeb and Mohammad [168] and the patches were analyzed by the mucoadhesive mechanical properties. For steroid release, PVP coatings were employed for beclomethasone dipropionate microparticles within a hydrofluoroalkane propellant [169] from high pressure metered inhaler. The addition of PVP into a system decreases the nature of aggregation of microparticles and hence obtains somewhat stable suspensions.

The substitution of opacified natural lens by a polymeric intraocular one in Cataract surgery is also carried out using PVP like polymeric materials. Conventional methods used PMMA, silicon, and acrylic-based polymers but due to the high stiffness, they cannot perform in normal way. PVP-PVA hydrogel overcomes this difficulty and is used widely in ophthalmology related applications [170] and on further studies as well as research shows, the nature of hydrogel is completely related to human lens. The wide use of PVP copolymers in various kinds of drugs such as indomethacin, tolbutamide, nifedipine improves its water solubility and bioavailability [171]. These polymers with drugs are normally applicable to hot melt extrusion process and the extrudates have excellent dissolution capacity. Topiramate drug carries PVP in the form of a binder and has very high efficiency in drug layering technique and it also changes the drug delivery, solubility, etc. PVP based films are used widely for the buccal release of fentanyl and these systems are much more beneficial in therapeutic treatments.

Most significant role of precipitation agents is its use as drug excipient. The unfavorable consequence of most of the drugs is its low solubility in water, reduced bioavailability, etc. When on adding PVP composites with these drugs, the rate of dissolution and drug solubility is remarkably increases. If we are considering the insoluble drugs, PVP is widely applicable as a precipitating agent. The remarkable property of this polymer is that, used for extension of agonist and the drug release method in capsule shell as well as membranes and disintegrating agents in solid drugs. PVP coated Membrane with the capsule shell in a dry condition is not easily release. Insoluble Polyvinylpyrrolidone is known as Crospovidone and prepared by the physical crosslinking reaction of PVP with a monomer containing bi-functional unit in presence of alkali at 100°C. This is also used as a perfect disintegrant tablet formulation and forms complexes with other substrates. In ophthalmic solutions, role of PVP termed as a demulcent or moisturizer and on blended with polyethylene glycol (PEG) 400 and dextran 70, it is useful for slight irritations, eye dryness, wind, and sun [172]. Discussing the nature of PVP, it is physiologically inert but hard, transparent, oxygen-permeable material and on crosslinking reactions, PVP forms various crosslinked networks [173].

This polymer is comparatively stable at normal conditions and can form complexes with various other compounds such as dyes, organic substartes, etc. Complexes of PVP can remove by filtration process.

Based on the application of PVP as a binder, it can form hard granules with high flow property but very low friability [174]. PVP when combined with germicidal compounds such as bisphenols and chlorinated phenols shows low toxicity and side effects of skin reactions. One of the most important points to be remembered is that chemicals like nicotine, formamide, and potassium cyanide show little oral toxic nature in PVP solutions. Copolymers of PVP are used to enhance the bioavailability nature of less soluble tolbutamide, indomethacin, and nifedipine [175–178] drugs.

PVP-heparins composite have superior properties compared to other molecules and this complex have anticoagulant activities due to heparin nature and mostly soluble in all types of organic solvents. This polymer has widely used in oral liquid dosage forms and the role of PVP in such solutions is as dispersing agents. Crosslinked PVP can be used to treat diarrhea and based on research, an excellent capacity to treat burns and completely removes toxins. When PVP combines Phenobarbitals and secobarbitals, the renal emission of barbityrates and the toxic effects of ameliorate. Hence from all the experimental records, surely state that polyvinylpyrrolidone polymer has higher effect on reducing the pharmacological actions of drugs include sulfathiazole, sodium salicylate, cholramphenicol insulin, Phenobarbital, procaines, and p-amino benzoicacid. PVP based hydrogels are widely used in controlled drug delivery systems and sustainable release is the main benefit for drug release investigations and the medicinal constituent emitted has been used for extended time intervals. This results in adding high concentrations of medicinal components to a desired location for a longer period of time.

4.5 OTHER APPLICATIONS

PVP is used for other kinds of applications as given below:

1. Optical and electrical applications;
2. Membranes;
3. Adhesives;
4. Ceramics;
5. Paper, coatings, and inks;
6. Household, industrial;
7. Photographic and lithographic applications;

8. Textile;
9. Environmental use;
10. Fibers.

4.5.1 OPTICAL AND ELECTRICAL APPLICATIONS

PVP based materials are used in screens, PCB (printed circuit boards), Cathode ray tubes, storage devices, and solar cells. The function of PVP is like a dispersant in PCB. PVP based metal nanoparticle confirms the use in flexible photonics and electronics, medical imaging, optical biosensors. Gold and Silver nanoparticles incorporated PVP composites are used for different optical devices. Nanoparticles of silver covered PVP are used in interconnections, microelectronics, bio-MEMS implants, and microfluidic devices. Conducting inks based on polypyrrole-Ag nanoparticles incorporated by PVP is used as a protection agent. For the use in printed electronics, ceramic capacitors PVP bounded metallic nanoparticles found extensive applications. PVP is also used as a surface modifier in many applications and PVP coated Nickel nanocomposites were used in MLCC electrodes and printed electronics.

Various kinds of nanoparticles, nanorods, and nanocubes included PVP is used to control particle size as well as particle aggregation. In many composites, PVP is mainly used as a stabilizer and filler. It can be used in satellite devices, energy dissipation, and low thermal expansions [179].

Vanadium oxide-PVP composites are generally used in lithium-ion batteries. These products show high mechanical strength, dimensional stability, and high electrical conductivity. Solar cells and dye-sensitized cells based PVP materials have found remarkable properties. In Nickel-Cd and NiH batteries), the function of PVP is as a coating agent, binder, and an adhesive. Thus, the use of PVP in energy storage devices has considerably higher effect comparing to other applications.

4.5.2 MEMBRANES

PVP is considered as an efficient additive for membrane fabrication, nano, micro, and macro filtration process in hemodialysis, wastewater treatment, gas separation and many others. Its excellent compatibility with variety of compounds and strong polar nature makes it a successive polymer for wide range of applications. Poly(acrylonitrile copolymerized with butadiene-styrene-PVP membrane is used in water treatment, pharmaceutical, and biotechnology applications. PES coated PVP membrane is usually used in water purification studies. PVP is a versatile polymer used as a template to create well-organized mesoporous silica for enzyme immobilization reactions. The extensive use of PVP as hydrophilizer in membranes and PPO-PVP blends are used in carbon molecular sieving (CMS) membranes.

4.5.3 ADHESIVES

The application of PVP in this area strikes the non-toxic nature and hence can be used in medical adhesives for special care such as wound dressing and medical electrodes. Adhesive films made up of PVP have superior adherence character to metals, plastics, PET, PU polymers, minerals, and textile. It shows superior property as a dispersant in many polymers and pigments.

4.5.4 CERAMICS

PVP function as stress-relieving agent in ceramic films and a binder in high-temperature fire prepared products. PVP is when added with clay minerals; it is used in fields such as soil science and geochemistry. The wide application of PVP as a binder at high temperature fire prepared products. The manufacture of yellow ceramic pigment with Ag NPs was identified by Mastre et al. PVP is also used as a stabilizer and adsorbent for colloidal clay minerals.

4.5.5 PAPER, COATINGS, AND INKS

PVP has a wide variety of applications such as coatings, inks, dispersions, waxes, etc. The function of this polymer in paints depicted as a protective colloid like substance and leveling agent. Another major role tends its superior dye adsorption capacity.

4.5.6 HOUSEHOLD, INDUSTRIAL, PHOTOGRAPHIC, AND LITHOGRAPHIC APPLICATIONS

The extensive applicational areas include cosmetics and detergents, and products from industry level provide excellent compatible nature and dimensional stability. The use of PVP in lithographic studies is the formation of patterns in films [180–182].

4.5.7 TEXTILE, FIBERS, AND ENVIRONMENTAL

For the synthesis of glass, plastics, and fibers, PVP is a commonly used material. It is generally used in the dying process in textiles and printing domains. These polymer-based fabrics are commonly used in surgical cloths, wound cloths, and sports wear. Waste-water purification, oil, and dye removal, etc., are other major roles of PVP. Filters incorporated by PVP (TiO_2-PVP) are used in self-cleaning and water filtration purposes. Studies relating that these compositions have a greater effect on gastronomical diseases.

4.6 CONCLUSION

The world of clinical and experimental medicine is enormous, vibrant, and diverse. The present chapter describes the vast properties of the polyvinylpyrrolidone polymer. Polyvinylpyrrolidone has numerous applications which include medical and nonmedical. The diverse nature of PVP is useful to research in all direction and this review is absolutely benefitted to readers. PVP has incredible opportunities in future research and developments and this particular polymer has extensive properties and uses in other areas such as optical and electrical, fibers, ceramics, and paper, coatings, and inks, household applications, photographic, and lithographic studies. But the pharmaceutical industry and medicine is the most recognizing field for PVP polymers. The chapter reviews the wide sectors and detailed areas of PVP polymer.

KEYWORDS

- **adhesives**
- **carbon molecular sieving**
- **ceramics**
- **lithographic applications**
- **N-vinylpyrrolidone**
- **printed circuit boards**

REFERENCES

1. Zhi, X., Fang, H., Bao, C., Shen, G., Zhang, J., Wang, K., Guo, S., et al., (2013). The immunotoxicity of graphene oxides and the effect of PVP-coating. *Biomaterials, 34*, 5254–5261.
2. Kariduraganavar, M. Y., Kittur, A. A., & Kamble, R. R., (2014). Chapter 1-Polymer synthesis and processing. In: Sangamesh, K., Cato, L., & Meng, D., (eds.), *Nat. Synth. Biomed. Polym.* (1st edn., pp. 1–31). Elsevier Inc.
3. Haaf, F., Sanner, A., & Straub, F., (1985). Polymers of N-vinylpyrrolidone: Synthesis, characterization and uses. *Polym. J., 17*, 143–152.
4. Buhler, V., (2008). *Kollidon Poly(Vinylpyrrolidone) Excipients for the Pharmaceutical Industry* (9th edn.) BASF-The Chemical Company, Ludwigshafen, Germany.
5. Sneader, W., (2005). *Drug Discovery*. Chichester, UK: John Wiley & Sons Ltd.
6. Raimi-Abraham, B. T., Mahalingam, S., Edirisinghe, M., & Craig, D. Q. M., (2014). Generation of poly(N-vinylpyrrolidone) nanofibers using pressurized gyration. *Mater. Sci. Eng. C, 39*, 168–176.
7. Folttmann, B. H., & Quadir, A., (2008). Excipients in pharmaceuticals: An overview. *Drug Deliv. Technol., 8*, 22–27.
8. Halake, K., Birajdar, M., Kim, B. S., Bae, H., Lee, C., Kim, Y. J., Kim, S., et al., (2014). Recent application developments of water-soluble synthetic polymers. *J. Ind. Eng. Chem., 1843*, 1–6.
9. Schwartz, J. A., Contescu, C. I., & Putyera, K., (2004). *Dekker Encyclopedia of Nanoscience and Nanotechnology* (Vol. 2). Boca Raton, FL; CRC Press.
10. Oechsner, M., & Keipert, S., (1999). Polyacrylic acid/polyvinylpyrrolidone bipolymeric systems. I. Rheological and mucoadhesive properties of formulations potentially useful for the treatment of dry-eye-syndrome. *Eur. J. Pharm. Biopharm., 47*, 113–118.
11. Abd El-Rehim, H. A., Hegazy, E. S. A., Hamed, A. A., & Swilem, A. E., (2013). Controlling the size and swell ability of stimuli-responsive polyvinylpyrrolidone-poly(acrylic acid) nanogels synthesized by gamma radiation induced template polymerization. *Eur. Polym. J., 49*, 601–612.
12. Kadłubowski, S., Henke, A., Ulański, P., Rosiak, J. M., Bromberg, L., & Hatton, T. A., (2007). Hydrogels of polyvinylpyrrolidone (PVP) and poly(acrylic acid) (PAA) synthesized by photo induced cross linking of homopolymers. *Polymer (Guildf), 48*, 4974–4981.
13. Chun, M. K., Cho, C. S., & Choi, H. K., (2002). Mucoadhesive drug carrier based on interpolymer complex of poly(vinyl pyrrolidone) and poly(acrylic acid) prepared by template polymerization. *J. Contr. Rel., 81*, 327–334.
14. Apt, L., Isenberg, S., Yoshimori, R., & Paez, J. H., (1984). Chemical preparation of the eye in ophthalmic surgery. III. Effect of povidone-iodine on the conjunctiva. *Arch. Ophthalmol., 102*, 728–729.
15. Apt, L., Isenberg, S. J., Yoshimori, R., & Spierer, A., (1989). Outpatient topical use of povidone-iodine in preparing the eye for surgery. *Ophthalmology, 96*, 289–292.
16. Benevento, W. J., Murray, P., Reed, C. A., & Pepose, J. S., (1990). The sensitivity of *Neisseria gonorrhoeae*, Chlamydia trachomatis, and herpes simplex type II to disinfection with povidone-iodine. *Am. J. Ophthalmol., 109*, 329–333.
17. Loftsson, T., Frikdriksdóttir, H., Sigurkdardóttir, A. M., & Ueda, H., (1994). The effect of water-soluble polymers on drug-cyclodextrin complexation. *Int. J. Pharm., 110*, 169–177.

18. Foster, A., & Klauss, V., (1995). *Ophthalmia neonatorum* in developing countries. *N. Engl. J. Med., 332*, 600–601.
19. Burks, R. I., (1998). Povidone-iodine solution in wound treatment. *Phys. Ther., 78*, 212–218.
20. Chaudhary, U., Nagpal, R. C., Malik, A. K., & Kumar, A., (1998). Comparative evaluation of antimicrobial activity of polyvinylpyrrolidone (PVP)-iodine versus topical antibiotics in cataract surgery. *J. Ind. Med. Assoc., 96*, 202–204.
21. Chun, M. K., Cho, C. S., & Choi, H. K., (2002). Mucoadhesive drug carrier based on interpolymer complex of poly(vinyl pyrrolidone) and poly(acrylic acid) prepared by template polymerization. *J. Contr. Rel., 81*, 327–334.
22. Fechine, G. J. M., Barros, J. G., & Catalani, L. H., (2004). Poly(N-vinyl-2-pyrrolidone) hydrogel production by ultraviolet radiation: New methodologies to accelerate cross linking. *Polymer (Guildf), 45*, 4705–4709.
23. Diaz, D. C. I., Falson, F., Guy, R. H., & Jacques, Y., (2007). *Ex vivo* evaluation of bioadhesive films for buccal delivery of fentanyl. *J. Contr. Rel., 122*, 135–140.
24. Kamal, H. M., Alaul, A. M., Sarwaruddin, C. A. M., Dafader, N. C., Haque, M. E., & Akter, F., (2008). Characterization of poly(vinyl alcohol) and poly(vinyl pyrrolidone) co-polymer blend hydrogel prepared by application of gamma radiation. *Polym. Plast. Technol. Eng., 47*, 662–665.
25. Jain, P., & Banga, A. K., (2010). Inhibition of crystallization in drug-in-adhesive type transdermal patches. *Int. J. Pharm., 394*, 68–74.
26. Banerjee, S., Chattopadhyay, P., Ghosh, A., Datta, P., & Veer, V., (2014). Aspect of adhesives in transdermal drug delivery systems. *Int. J. Adhes. Adhes., 50*, 70–84.
27. Kadlubowski, S., (2014). Radiation-induced synthesis of nanogels based on poly (N-vinyl-2-pyrrolidone): A review. *Radiat. Phys. Chem., 102*, 29–39.
28. Schulz, M., Fussnegger, B., & Bodmeier, R., (2010). Drug release and adhesive properties of crospovidone-containing matrix patches based on polyisobutene and acrylic adhesives. *Eur. J. Pharm. Sci., 41*, 675–684.
29. Vakalopoulos, K. A., Daams, F., Wu, Z., Timmermans, L., Jeekel, J. J., Kleinrensink, G. J., Van, D. M. A., & Lange, J. F., (2013). Tissue adhesives in gastrointestinal anastomosis: A systematic review. *J. Surg. Res., 180*, 290–300.
30. Wang, G., Zhang, T., Ahmad, S., Cheng, J., & Guo, M., (2013). Physicochemical and adhesive properties, microstructure and storage stability of whey protein-based paper glue. *Int. J. Adhes. Adhes., 41*, 198–205.
31. Slistan-Grijalva, A., Herrera-Urbina, R., Rivas-Silva, J. F., Ávalos-Borja, M., Castillón-Barraza, F. F., & Posada-Amarillas, A., (2008). Synthesis of silver nanoparticles in a polyvinylpyrrolidone (PVP) paste, and their optical properties in a film and in ethylene glycol. *Mater. Res. Bull., 43*, 90–96.
32. Rodríguez, J., Navarrete, E., Dalchiele, E. A., Sánchez, L., Ramos-Barrado, J. R., & Martín, F., (2013). Polyvinylpyrrolidone-LiClO4 solid polymer electrolyte and its application in transparent thin film super capacitors. *J. Power Sour., 237*, 270–276.
33. Ruecha, N., Rangkupan, R., Rodthongkum, N., & Chailapakul, O., (2014). Novel paper-based cholesterol biosensor using graphene/polyvinylpyrrolidone/polyaniline nanocomposite. *Biosens Bioelectron, 52*, 13–19.
34. Rao, V. V. R. N., & Kalpalatha, A., (1987). Electrical conduction mechanism in poly(vinyl pyrrolidone) films. *Polymer (Guildf), 28*, 648–650.

35. Lu, X., Zhao, Y., Wang, C., & Wei, Y., (2005). Fabrication of CdS nanorods in PVP fiber matrices by electro spinning. *Macromol. Rapid Commun., 26*, 1325–1329.
36. Hida, Y., & Kozuka, H., (2005). Photoanodic properties of sol-gel-derived iron oxide thin films with embedded gold nanoparticles: Effects of polyvinylpyrrolidone in coating solutions. *Thin Solid Films, 476*, 264–271.
37. Yan, E., Huang, Z., Xin, Y., Zhao, Q., & Zhang, W., (2006). Polyvinylpyrrolidone/tris(8-quinolinolato) aluminum hybrid polymer fibers by electro spinning. *Mater. Lett., 60*, 2969–2973.
38. Huang, J., & Gao, L., (2006). Anisotropic growth of In(OH)3 nano cubes to nanorods and nano sheets via a solution-based seed method. *Cryst. Growth Des., 6*, 1528–1532.
39. Subba, R. C. V., Han, X., Zhu, Q. Y., Mai, L. Q., & Chen, W., (2006). Dielectric spectroscopy studies on (PVPþPVA) polyblend film. *Microelectron. Eng., 83*, 281–285.
40. Jing, C., Hou, J., Zhang, Y., & Xu, X., (2007). Preparation of thick, crack-free germanosilicate glass films by polyvinylpyrrolidone and study of the UV bleachable absorption band. *J. Non Cryst. Solids, 353*, 4128–4136.
41. Couto, G. G., Klein, J. J., Schreiner, W. H., Mosca, D. H., De Oliveira, A. J. A., & Zarbin, A. J. G., (2007). Nickel nanoparticles obtained by a modified polyol process: Synthesis, characterization, and magnetic properties. *J. Coll. Interf. Sci., 311*, 461–468.
42. Wang, S., & Shi, G., (2007). Uniform silver/polypyrrole core-shell nano particles synthesized by hydrothermal reaction. *Mater. Chem. Phys., 102*, 255–259.
43. Zhang, L., Xu, Y., Wu, D., Sun, Y., Jiang, X., & Wei, X., (2008). Effect of polyvinylpyrrolidone on the structure and laser damage resistance of sol-gel silica anti-reflective films. *Opt. Laser Technol., 40*, 282–288.
44. Tseng, T. T., & Tseng, W. J., (2009). Effect of polyvinylpyrrolidone on morphology and structure of In_2O_3 nanorods by hydrothermal synthesis. *Ceram. Int., 35*, 2837–2844.
45. Wei, S. F., Lian, J. S., & Jiang, Q., (2009). Controlling growth of ZnO rods by polyvinylpyrrolidone (PVP) and their optical properties. *Appl. Surf. Sci., 255*, 6978–6984.
46. He, T., Wang, C., Pan, X., & Wang, Y., (2009). Nonlinear optical response of Au and Ag nanoparticles doped polyvinylpyrrolidone thin films. *Phys. Lett. A, 373*, 592–595.
47. Narayanan, M., Ma, B., & Balachandran, U., (2010). Improved dielectric properties of lead lanthanum zirconatetitanate thin films on copper substrates. *Mater. Lett., 64*, 22–24.
48. Feng, X., (2010). Synthesis of Ag/polypyrrole core-shell nanospheres by a seeding method. *Chin. J. Chem., 28*, 1359–1362.
49. Shi, Z., Wang, H., Dai, T., & Lu, Y., (2010). Room temperature synthesis of Ag/polypyrrole core-shell nanoparticles and hollow composite capsules. *Synth. Met., 160*, 2121–2127.
50. Qiao, J., Fu, J., Lin, R., Ma, J., & Liu, J., (2010). Alkaline solid polymer electrolyte membranes based on structurally modified PVA/PVP with improved alkali stability. *Polymer (Guildf), 51*, 4850–4859.
51. Laforgue, A., (2011). All-textile flexible supercapacitors using electrospun poly(3, 4-ethylenedioxythiophene) nanofibers. *J. Power Sourc., 196*, 559–564.
52. Kim, J. W., Hong, S. J., & Kwak, M. G., (2011). Characteristics of eco-friendly synthesized SiO_2 dielectric nanoparticles printed on Si substrate. *Microelectron. Eng., 88*, 797–801.

53. Jung, Y. J., Govindaiah, P., Choi, S. W., Cheong, I. W., & Kim, J. H., (2011). Morphology and conducting property of Ag/poly(pyrrole) composite nanoparticles: Effect of polymeric stabilizers. *Synth. Met., 161*, 1991–1995.
54. Cheng, Y. T., Uang, R. H., & Chiou, K. C., (2011). Effect of PVP-coated silver nanoparticles using laser direct patterning process by photothermal effect. *Microelectron. Eng., 88*, 929–934.
55. Ravi, M., Pavani, Y., Kiran, K. K., Bhavani, S., Sharma, A. K., & Narasimha, R. V. V. R., (2011). Studies on electrical and dielectric properties of PVP:KBrO4 complexed polymer electrolyte films. *Mater. Chem. Phys., 130*, 442–448.
56. Rajeswari, N., Selvasekarapandian, S., Karthikeyan, S., Prabu, M., Hirankumar, G., Nithya, H., & Sanjeeviraja, C., (2011). Conductivity and dielectric properties of polyvinyl alcohol-polyvinylpyrrolidone poly blend film using non-aqueous medium. *J. Non Cryst. Solids, 357*, 3751–3756.
57. Liu, C., Nan, J., Zuo, X., Xiao, X., & Shu, D., (2012). Synthesis and electrochemical characteristics of an orthorhombic LiMnO2 cathode material modified with poly(vinyl-pyrrolidone) for lithium ion batteries. *Int. J. Electrochem. Sci., 7*(8), 7152–7164.
58. Park, J. G., Akhtar, M. S., Li, Z. Y., Cho, D. S., Lee, W., & Yang, O. B., (2012). Application of single walled carbon nanotubes as counter electrode for dye sensitized solar cells. *Electrochim. Acta, 85*, 600–604.
59. Chao, S., Ma, B., Liu, S., Narayanan, M., & Balachandran, U., (2012). Effects of pyrolysis conditions on dielectric properties of PLZT films derived from a polyvinylpyrrolidone-modified sol-gel process. *Mater. Res. Bull., 47*, 907–911.
60. Ummartyotin, S., Bunnak, N., Juntaro, J., Sain, M., & Manuspiya, H., (2012). Hybrid organic-inorganic of ZnS embedded PVP nano composite film for photoluminescent application. *Compt. Rend. Phys., 13*, 994–1000.
61. Khan, W. S., Asmatulu, R., Ahmed, I., & Ravigururajan, T. S., (2013). Thermal conductivities of electrospun PAN and PVP nanocomposite fibers incorporated with MWCNTs and NiZn ferrite nanoparticles. *Int. J. Therm. Sci., 71*, 74–79.
62. Choi, S., Kim, Y., Yun, J. H., Kim, I., & Shim, S. E., (2013). PVP-assisted synthesis of dense silica-coated graphite with electrically insulating property. *Mater. Lett., 90*, 878–879.
63. Lin, H., Yan, X., & Wang, X., (2013). Controllable synthesis and down-conversion properties of flower-like NaY(MoO4)2 micro crystals via polyvinylpyrrolidone-mediated. *J. Solid State Chem., 204*, 266–271.
64. Howe, K. S., Clark, E. R., Bowen, J., & Kendall, K., (2013). A novel water-based cathode ink formulation. *Int. J. Hydro. Energy, 38*, 1731–1736.
65. Zhai, D., Zhang, T., Guo, J., Fang, X., & Wei, J., (2013). Water-based ultraviolet curable conductive inkjet ink containing silver nano-colloids for flexible electronics. *Coll. Surf. A Physicochem. Eng. Asp., 424*, 1–9.
66. Altecor, A., Li, Q., Lozano, K., & Mao, Y., (2014). Mixed-valent VOx/polymer nanohybrid fibers for flexible energy storage materials. *Ceram. Int., 40*, 5073–5077.
67. Onlaor, K., Thiwawong, T., & Tunhoo, B., (2014). Electrical switching and conduction mechanisms of nonvolatile write-once-read-many-times memory devices with ZnO nanoparticles embedded in polyvinylpyrrolidone. *Org. Electron., 15*, 1–9.
68. Kubota, S., Morioka, T., Takesue, M., Hayashi, H., Watanabe, M., & Smith, R. L., (2014). Continuous supercritical hydrothermal synthesis of dispersible zero-valent copper nanoparticles for ink applications in printed electronics. *J. Supercrit. Fluids, 86*, 33–40.

69. Park, S. H., & Kim, H. S., (2014). Flash light sintering of nickel nanoparticles for printed electronics. *Thin Solid Films, 550*, 575–581.
70. Neiva, E. G. C., Bergamini, M. F., Oliveira, M. M., Marcolino, L. H., & Zarbin, A. J. G., (2014). PVP-capped nickel nanoparticles: Synthesis, characterization and utilization as a glycerol electrosensor. *Sens. Actuat. B Chem., 196*, 574–581.
71. Omastová, M., Bober, P., Morávková, Z., Peřinka, N., Kaplanová, M., Syrový, T., Hromádková, J., et al., (2014). Towards conducting inks: Polypyrrole-silver colloids. *Electrochim. Acta, 122*, 296–302.
72. Sheng, L., Ren, J., Miao, Y., Wang, J., & Wang, E., (2011). PVP-coated grapheme oxide for selective determination of ochratoxin A via quenching fluorescence of free aptamer. *Biosens. Bioelectron., 26*, 3494–3499.
73. Svanholm, E., (2007). *Printability and Ink-Coating Interactions in Inkjet Printing.* Dissertation. Karlstad University, Karlstad, Sweden. ISBN: 91-7063-104-2.
74. Park, O., (2003). *The Function of Fountain Solution in Lithography.* Fuji Hunt Photographic Chemicals, Inc., Orange Park, Florida.
75. Schmidt, T., Mönch, J. I., & Arndt, K. F., (2006). Temperature-sensitive hydrogel pattern by electron-beam lithography. *Macromol. Mater. Eng., 291*, 755–761.
76. Burkert, S., Schmidt, T., Gohs, U., Mönch, I., & Arndt, K. F., (2007). Patterning of thin poly(N-vinyl pyrrolidone) films on silicon substrates by electron beam lithography. *J. Appl. Polym. Sci., 106*, 534–539.
77. Lu, C., Cheng, M. M. C., Benatar, A., & Lee, L. J., (2007). Embossing of high aspect-ratio-microstructures using sacrificial templates and fast surface heating. *Polym. Eng. Sci., 47*, 830–840.
78. Färm, E., Kemell, M., Santala, E., Ritala, M., & Leskelä, M., (2010). Selective-area atomic layer deposition using poly(vinyl pyrrolidone) as a passivation layer. *J. Electrochem. Soc., 157*, K10–K14.
79. Lee, B. M., Kang, D. W., Jung, C. H., Choi, J. H., Hwang, I. T., Hong, S. K., & Lee, J. S., (2011). Patterning of polymer nanocomposite resists containing metal nanoparticles by electron beam lithography. *J. Nanosci. Nanotechnol., 11*, 7390–7393.
80. Misra, N., Biswal, J., Dhamgaye, V. P., Lodha, G. S., & Sabharwal, S., (2013). A comparative study of gamma, electron beam, and synchrotron X-ray irradiation method for synthesis of silver nanoparticles in PVP. *Adv. Mater. Lett., 4*, 458–463.
81. Moga, K. A., Bickford, L. R., Geil, R. D., Dunn, S. S., Pandya, A. A., Wang, Y., Fain, J. H., et al., (2013). Rapidly-dissolvable micro needle patches via a highly scalable and reproducible soft lithography approach. *Adv. Mater., 25*, 5060–5066.
82. Kadlubowski, S., (2014). Radiation-induced synthesis of nano gels based on poly(N-vinyl-2-pyrrolidone): A review. *Radiat. Phys. Chem., 102*, 29–39.
83. An, J. C., Weaver, A., Kim, B., Barkatt, A., Poster, D., Vreeland, W. N., Silverman, J., & Al-Sheikhly, M., (2011). Radiation-induced synthesis of poly(vinylpyrrolidone) nanogel. *Polymer, 52*, 5746–5755.
84. Teodorescu, M., Morariu, S., Bercea, M., & Săcărescu, L., (2016). Viscoelastic and structural properties of poly(vinyl alcohol)/poly(vinylpyrrolidone) hydrogels. *RSC Adv., 6*, 39718–39727.
85. Nho, Y. C., & Park, K. R., (2002). Preparation and properties of PVA/PVP hydrogels containing chitosan by radiation. *J. Appl. Polym. Sci., 85*, 1787–1794.

86. Yu, H., Xu, X., Chen, X., Hao, J., & Jing, X., (2006). Medicated wound dressings based on poly(vinyl alcohol)/poly(*N*-vinyl pyrrolidone)/chitosan hydrogels. *J. Appl. Polym. Sci., 101*, 2453–2463.
87. Folttmann, B. H., & Quadir, A., (2008). Excipients in pharmaceuticals: An overview. *Drug Deliv. Technol., 8*, 22–27.
88. Halake, K., Birajdar, M., Kim, B. S., Bae, H., Lee, C., Kim, Y. J., Kim, S., et al., (2014). Recent application developments of water-soluble synthetic polymers. *J. Ind. Eng. Chem., 1843*, 1–6.
89. Schwarz, W., (1990). *PVP: A Critical Review of the Kinetics and Toxicology of Polyvinylpyrrolidone (Povidone)*. CRC Press: Boca Raton, FL.
90. Sneader, W., (1990). *Drug Discovery*. Chichester, UK: John Wiley & Sons Ltd., 2005. Polyvinylpyrrolidone (Povidone), CRC Press: Boca Raton, FL.
91. NPCS Board of Consultants & Engineers, (2009). *The Complete Book on Water Soluble Polymers (Google eBook)*. Asia Pacific Business Press Inc., Delhi, India.
92. Yang, M., Xie, S., Li, Q., Wang, Y., Chang, X., Shan, L., Sun, L., Huang, X., & Gao, C., (2014). Effects of polyvinylpyrrolidone both as a binder and pore-former on the release of sparingly water-soluble topiramate from ethylcellulose coated pellets. *Int. J. Pharm., 465*, 187–196.
93. Abd El-Rehim, H. A., Hegazy, E. S. A., Hamed, A. A., & Swilem, A. E., (2013). Controlling the size and swell ability of stimuli-responsive polyvinylpyrrolidone-poly(acrylic acid) nanogels synthesized by gamma radiation induced template polymerization. *Eur. Polym. J., 49*, 601–612.
94. Kadłubowski, S., Henke, A., Ulański, P., Rosiak, J. M., Bromberg, L., & Hatton, T. A., (2007). Hydrogels of polyvinylpyrrolidone (PVP) and poly(acrylicacid) (PAA) synthesized by photo induced cross linking of homopolymers. *Polymer (Guildf), 48*, 4974–4981.
95. Fechine, G. J. M., Barros, J. G., & Catalani, L. H., (2004). Poly(N-vinyl-2-pyrrolidone) hydrogel production by ultraviolet radiation: New methodologies to accelerate cross linking. *Polymer (Guildf), 45*, 4705–4709.
96. Kamal, H. M., Alaul, A. M., Sarwaruddin, C. A. M., Dafader, N. C., Haque, M. E., & Akter, F., (2008). Characterization of poly(vinylalcohol) and poly(vinyl pyrrolidone) co-polymer blend hydrogel prepared by application of gamma radiation. *Polym. Plast. Technol. Eng., 47*, 662–665.
97. Ma, R., Xiong, D., Miao, F., Zhang, J., & Peng, Y., (2009). Novel PVP/PVa hydrogels for articular cartilage replacement. *Mater. Sci. Eng. C, 29*, 1979–1983.
98. Abbaszadeh, F., Moradi, O., Norouzi, M., & Sabzevari, O., (2013). Improvement single-wall carbon nanotubes (SWCNTs) based on functionalizing with monomers 2-hydroxyethylmethacryate (HEMA) and N-vinylpyrrolidone (NVP) for pharmaceutical applications as cancer therapy. *J. Ind. Eng. Chem., 20*(5), 2895–2900.
99. Dafader, N. C., Haque, M. E., & Akhtar, F., (2005). Effect of Kappa-Carrageenan on the properties of poly(vinylpyrrolidone) hydrogel prepared by the application of radiation. *Polym. Plast. Technol. Eng., 44*, 1339–1346.
100. Dafader, N. C., Haque, M. E., & Akhtar, F., (2005). Synthesis of hydrogel from aqueous solution of poly(vinyl pyrrolidone) with agar by gamma-rays irradiation. *Polym. Plast. Technol. Eng., 44*, 243–251.
101. Jones, D. S., Djokic, J., & Gorman, S. P., (2005). The resistance of polyvinylpyrrolidone-iodine-poly(e-caprolactone) blends to adherence of *Escherichia coli*. *Biomaterials, 26*, 2013–2020.

102. Wang, B. L., Liu, X. S., Ji, Y., Ren, K. F., & Ji, J., (2012). Fast and long-acting antibacterial properties of chitosan-Ag/polyvinylpyrrolidone nanocomposite films. *Carbohydr. Polym., 90*, 8–15.
103. Babcock, D. E., Hergenrother, R. W., Craig, D. A., Kolodgie, F. D., & Virmani, R., (2013). *In vivo* distribution of particulate matter from coated angioplasty balloon catheters. *Biomaterials, 34*, 3196–3205.
104. Haaf, F., Sanner, A., & Straub, F., (1985). Polymers of N-vinylpyrrolidone: Synthesis, characterization and uses. *Polym. J., 17*, 143–152.
105. Schwartz, J. A., Contescu, C. I., & Putyera, K., (2004). *Dekker Encyclopedia of Nanoscience and Nanotechnology* (Vol. 2). Boca Raton, FL; CRC Press.
106. Jain, P., & Banga, A. K., (2010). Inhibition of crystallization in drug-in-adhesive type transdermal patches. *Int. J. Pharm., 394*, 68–74.
107. Banerjee, S., Chattopadhyay, P., Ghosh, A., Datta, P., & Veer, V., (2014). Aspect of adhesives in transdermal drug delivery systems. *Int. J. Adhes. Adhes., 50*, 70–84.
108. Patel, H. A., Shah, S., Shah, D. O., & Joshi, P. A., (2011). Sustained release of venlafaxine from venlafaxine-montmorillonite-polyvinylpyrrolidone composites. *Appl. Clay Sci., 51*, 126–130.
109. Verma, R., (2003). Development and evaluation of extended release formulations of isosorbide mononitrate based on osmotic technology. *Int. J. Pharm., 263*, 9–24.
110. Diaz, D., Consuelo, I., Falson, F., Guy, R. H., & Jacques, Y., (2007). *Ex vivo* evaluation of bioadhesive films for buccal delivery of fentanyl. *J. Contr. Rel., 122*, 135–140.
111. Diaz, D. C. I., Falson, F., Guy, R. H., & Jacques, Y., (2007). *Ex vivo* evaluation of bioadhesive films for buccal delivery of fentanyl. *J. Contr. Rel., 122*, 135–140.
112. Dai, X. Y., Nie, W., Wang, Y. C., Shen, Y., Li, Y., & Gan, S. J., (2012). Electrospun emodin polyvinylpyrrolidone blended nanofibrous membrane: A novel medicated biomaterial for drug delivery and accelerated wound healing. *J. Mater. Sci. Mater. Med., 23*, 2709–2716.
113. Singh, R., & Singh, D., (2012). Radiation synthesis of PVP/alginate hydrogel containing nanosilver as wound dressing. *J. Mater. Sci. Mater. Med., 23*, 2649–2658.
114. Raimi-Abraham, B. T., Mahalingam, S., Edirisinghe, M., & Craig, D. Q. M., (2014). Generation of poly(N-vinylpyrrolidone) nanofibers using pressurized gyration. *Mater. Sci. Eng. C, 39*, 168–176.
115. Sebe, I., Szabó, B., Nagy, Z. K., Szabó, D., Zsidai, L., Kocsis, B., & Zelkó, R., (2013). Polymer structure and antimicrobial activity of polyvinylpyrrolidone-based iodine nanofibers prepared with high-speed rotary spinning technique. *Int. J. Pharm., 458*, 99–103.
116. Mogoşanu, G. D., & Grumezescu, A. M., (2014). Natural and synthetic polymers for wounds and burns dressing. *Int. J. Pharm., 463*, 127–136.
117. Kang, K., Lim, D. H., Choi, I. H., Kang, T., Lee, K., Moon, E. Y., Yang, Y., et al., (2011). Vascular tube formation and angiogenesis induced by polyvinylpyrrolidone-coated silver nanoparticles. *Toxicol. Lett., 205*, 227–234.
118. Kubin, A., Loew, H. G., Burner, U., Jessner, G., Kolbabek, H., & Wierrani, F., (2008). How to make hypericin water-soluble. *Pharmazie, 63*, 263–269.
119. Huang, Z., Xu, H., Meyers, A. D., Musani, A. I., Wang, L., Tagg, R., Barqawi, A. B., & Chen, Y. K., (2008). Photodynamic therapy for treatment of solid tumors-Potential and technical challenges. *Technol. Cancer Res. Treat, 7*, 309–320.

120. Penjweini, R., Loew, H. G., Eisenbauer, M., & Kratky, K. W., (2013). Modifying excitation light dose of novel photosensitizer PVP-Hypericin for photodynamic diagnosis and therapy. *J. Photochem. Photobiol. B, 120*, 120–129.
121. Penjweini, R., Smisdom, N., Deville, S., & Ameloot, M., (2014). *Transport and Accumulation of Pvp-Hypericin in Cancer and Normal Cells Characterized.*
122. Buhler, V., (2005). *Polyvinylpyrrolidone Excipients for Pharmaceuticals Povidone, Crospovidone and Copovidone*. Springer-Verlag, Berlin, Germany.
123. Kadajji, V. G., & Betageri, G. V., (2011). Water soluble polymers for pharmaceutical applications. *Polymer, 3*(4), 1972–2009.
124. Ng, C. L., Lee, S. E., Lee, J. K., et al., (2016). Solubilization and formulation of chrysosplenol C in solid dispersion with hydrophilic carriers. *International Journal of Pharmaceutics, 512*(1), 314–321.
125. Zhi, X., Fang, H., Bao, C., et al., (2013). The immunotoxicity of grapheme oxides and the effect of PVP-coating. *Biomaterials, 34*(21), 5254–5261.
126. Escamilla-Rivera, V., Uribe-Ramırez, M., Gonzalez-Pozos, S., Lozano, O., Lucas, S., & De Vizcaya-Ruiz, A., (2016). Protein corona acts as a protective shield against Fe_3O_4-PEG inflammation and ROS-induced toxicity in human macrophages. *Toxicology Letters, 240*(1), 172–184.
127. Singleton, A. J., (1953). The use of polyvinylpyrrolidone as a plasma volume expander in preventing or combating shock. *Texas Reports on Biology and Medicine, 11*, 138–143.
128. Robinson, B. V., Sullivan, F. M., Borzelleca, J. F., & Schwartz, S. L., (1990). *PVP–a Critical Review of the Kinetics and Toxicology of Polyvinylpyrrolidone (Povidone).* Lewis Publishers, Chelsea.
129. Kumar, R. P., & Abraham, A., (2016). PVP-coated naringenin nanoparticles for biomedical applications-*In vivo* toxicological evaluations. *Chemico-Biological Interactions, 257*, 110–118.
130. Gropper, A. L., Raisz, L. G., & Amspacher, W. H., (1952). Plasma expanders. *Int. Abstr. Surg., 95*, 521.
131. Kort, J., & Heinlein, H., (1943). Functional and morphological investigations of the action of colloidal blood substitutes with special attention to the action of periston. *Arch. Klin. Chir., 205*, 230.
132. Sanbar, S. S., Zweifler, A. J., & Smet, G., (1967). *In-vivo* and *in-vitro* effects of polyvinylpyrrolidone on platelet adhesiveness in human blood. *Lancet, 2*, 917.
133. Burstein, M. J., (1961). On the separation of, 8-lipoproteins of the serum after flocculation by polyvinylpyrrolidone. *J. Physiol. (Paris) 53*, 519.
134. Perkins, H. A., Rolfs, M. R., Thacher, C., & Richards, V., (1966). Effect of polyvinylpyrrolidone on plasma coagulation factors. *Proc. Soc. Exp. Biol. Med., 123*, 667.
135. Buhler, V., (2005). *Polyvinylpyrrolidone Excipients for Pharmaceuticals*. Springer, Berlin Heidelberg.
136. Florence, A. T., & Attwood, D., (2006). *Physicochemical Principles of Pharmacy.* Pharmaceutical Press, London.
137. Li, L., Yin, Z., Li, F., Xiang, T., Chen, Y., & Zhao, C., (2010). *J. Membr. Sci., 349*, 56.
138. Carr, D. A., & Peppas, N. A., (2009). *Macromol. Biosci., 9*, 497.
139. Gao, B., Hu, H., Guo, J., & Li, Y., (2010). *Colloids Surf., B, 77*, 206.
140. Wang, D., Hill, D. J. T., Rasoul, F., & Whittaker, A. K., (2011). *Radiat. Phys. Chem., 80*, 207.
141. Liu, S., Jones, L., & Gu, F. X., (2012). *Macromol. Biosci., 12*, 608.

142. Ran, F., Nie, S., Li, J., Su, B., Sun, S., & Zhao, C., (2012). *Macromol. Biosci., 12*, 116.
143. Francois, P., Vaudaux, P., Nurdin, N., Mathieu, H. J., Descouts, P., & Lew, D. P., (1996). *Biomaterials, 17*, 667.
144. Wang, H., Yu, T., Zhao, C., & Du, Q., (2009). *Fibers Polym., 10*, 1.
145. Seal, B. L., Otero, T., & Panitch, A., (2001). *Mater. Sci. Eng., 4*, 147.
146. Cascone, M. G., Sim, B., & Downes, S., (1995). *Biomaterials, 16*, 569.
147. Giusti, P., Lazzeri, L., & Barbani, N., (1993). *J. Mater. Sci. Mater. Med., 4*, 538.
148. Yang, M., Xie, S., Li, Q., Wang, Y., Chang, X., Shan, L., Sun, L., et al., (2014). Effects of polyvinylpyrrolidone both as a binder and pore-former on the release of sparingly water-soluble topiramate from ethylcellulose coated pellets. *Int. J. Pharm., 465*, 187–196.
149. Taheri, A., Atyabi, F., & Dinarvnd, R., (2011). Temperature-responsive and biodegradable PVA:PVP K30: poloxamer 407 hydrogel for controlled delivery of human growth hormone (hGH). *J. Pediatr. Endocrinol. Metab., 24*, 175–179.
150. Dai, X. Y., Nie, W., Wang, Y. C., Shen, Y., Li, Y., & Gan, S. J., (2012). Electrospun emodin polyvinylpyrrolidone blended nanofibrous membrane: A novel medicated biomaterial for drug delivery and accelerated wound healing. *J. Mater. Sci. Mater. Med., 23*, 2709–2716.
151. Darwis, D., Hilmy, N., Hardiningsih, L., & Erlinda, T., (1993). Poly(N-vinylpyrrolidone) hydrogels: 1. Radiation polymerization and crosslinking of N-vinylpyrrolidone. *Radiat. Phys. Chem., 42*, 907–910.
152. Fogaça, R., & Catalani, L. H., (2013). PVP hydrogel membranes produced by electrospinning for protein release devices. *Soft Mater., 11*, 61–68.
153. Himly, N., Darwis, D., & Hardiningsih, L., (1993). Poly(n-vinylpyrrolidone) hydrogels, 2. Hydrogel composites as wound dressing for tropical environment. *Radiat. Phys. Chem., 42*, 911–914.
154. Hu, Y., Yan, P., Fu, L., Zheng, Y., Kong, W., Wu, H., & Yu, X., (2017). Polyvinyl alcohol/polyvinyl pyrrolidone cross linked hydrogels induce wound healing through cell proliferation and migration. *J. Biomater. Tissue Eng., 7*, 310–318.
155. Ignatova, M., Manolova, N., & Rashkov, I., (2007). Novel antibacterial fibers of quaternized chitosan and poly(vinyl pyrrolidone) prepared by electro spinning. *Eur. Polym. J., 43*, 1112–1122.
156. Kadłubowski, S., Henke, A., Ulański, P., Rosiak, J. M., Bromberg, L., & Hatton, T. A., (2007). Hydrogels of polyvinylpyrrolidone (PVP) and poly(acrylic acid) (PAA) synthesized by photo induced cross linking of homopolymers. *Polymer (Guildf), 48*, 4974–4981.
157. Park, K. R., & Nho, Y. C., (2004). Preparation and characterization by radiation of hydrogels of PVA and PVP containing aloe vera. *J. Appl. Polym. Sci., 91*, 1612–1618.
158. Razzak, M. T., Darwis, D., & Zainuddin, S., (2001). Irradiation of polyvinyl alcohol and polyvinyl pyrrolidone blended hydrogel for wound dressing. *Radiat. Phys. Chem., 62*, 107–113.
159. Silva, D. A., De Hettiarachchi, B. U., Nayanajith, L. D. C., Milani, M. D. Y., & Motha, J. T. S., (2011). Development of a PVP/kappa-carrageenan/PEG hydrogel dressing for wound healing applications in Sri Lanka. *J. Natl. Sci. Found. Sri Lanka, 39*, 25–33.
160. Singh, B., & Pal, L., (2011). Radiation cross linking polymerization of sterculia polysaccharide PVA-PVP for making hydrogel wound dressings. *Int. J. Biol. Macromol., 48*, 501–510.

161. Darwis, D., Hilmy, N., Hardiningsih, L., & Erlinda, T., (1993). Poly(N-vinylpyrrolidone) hydrogels: 1. Radiation polymerization and cross linking of N-vinylpyrrolidone. *Radiat. Phys. Chem., 42*, 907–910.
162. Himly, N., Darwis, D., & Hardiningsih, L., (1993). Poly(n-vinylpyrrolidone) hydrogels: 2. Hydrogel composites as wound dressing for tropical environment. *Radiat. Phys. Chem., 42*, 911–914.
163. Jun, Y. B., Min, B. H., Kim, S. I., & Kim, Y. I. J., (1989). Preparation and evaluation of acetaminophen tablets. *Kor. Pharm. Sci., 19*, 123–128.
164. Jachowicz, R., (1987). Dissolution rates of partially water soluble drugs from solid dispersion systems. II. Phenytoin. *Int. J. Pharm., 35*, 7–12.
165. White, R. K., (1994). *Pharmaceutical Compositions Containing Polyvinylpyrrolidone and a Tri-Ester and a Process of Manufacture Thereof.* Internat. Patent WO/1994/025008.
166. Shi, Y., Xiong, D., & Zhang, J., (2014). Effect of irradiation dose on mechanical and biotribological properties of PVA/PVP hydrogels as articular cartilage. *Tribol. Int., 78*, 60–67.
167. Shi, Y., & Xiong, D., (2013). Microstructure and friction properties of PVA/PVP hydrogels for articular cartilage repair as function of polymerization degree and polymer concentration. *Wear, 305*, 280–285.
168. Ghareeb, M. M., & Mohammad, H. A., (2013). Study the effects of secondary polymers on the properties of buccoadhesive polyvinyl alcohol patches of 5-Flourouracil. *Int. J. Pharm. Pharm. Sci., 5*, 484–488.
169. Jones, S. A., Martin, G. P., & Brown, M. B., (2006). Manipulation of beclomethasone hydrofluoroalkane interactions using biocompatible macromolecules. *J. Pharm. Sci., 95*, 1060–1074.
170. Leone, G., Consumi, M., Greco, G., Bonechi, C., Lamponi, S., Rossi, C., & Magnani, A., (2011). A PVA/PVP hydrogel for human lens substitution: Synthesis, rheological characterization, and in vitro biocompatibility. *J Biomed Mater Res B Appl Biomater, 97*(2), 278-88..
171. Forster, A., Hempenstall, J., & Rades, T., (2001). Characterization of glass solutions of poorly water-soluble drugs produced by melt extrusion with hydrophilic amorphous polymers. *J. Pharm. Pharmacol., 53*, 303–315.
172. Urban, J., Snow, N. H., Tavlarakis, P., Urban, J. J., & Snow, N., (2011). Determination of total polyvinylpyrrolidone (PVP) in ophthalmic solutions by size exclusion chromatography with ultraviolet-visible detection. *J. Chromatogr Sci., 49*(6), 457–462.
173. Burnette, L. W., (1962). A review of the physiological properties of PVP. *Proc. Sci. Sect. TGA, 38*, 1–4.
174. Chowhan, Z. T., Amaro, A. A., & Ong, J. T. H., (1992). Punch geometry and formulation considerations in reducing tablet friability and their effect on *in vitro* dissolution. *J. Pharm. Sci., 81*, 290–294.
175. Jijun, F., Lishuang, X., Xiaoli, W., Shu, Z., Xiaoguang, T., Xingna, Z., et al., (2011). Nimodipine (NM) tablets with high dissolution containing nm solid dispersions prepared by hot melt extrusion. *Drug Dev. Ind. Pharm., 37*, 934–944.
176. Chokshi, R. J., Sandhu, H. K., Iyer, R. M., Shah, N. H., Malick, W. A., & Zia, H., (2005). Characterization of physico-mechanical properties of indomethacin and polymers to assess their suitability for hot-melt extrusion process as a means to manufacture solid dispersion/solution. *J. Pharm. Sci., 94*, 2463–2474.

177. Dua, K., Pabreja, K., & Ramana, M. V., (2010). Preparation, characterization, and *in vitro* evaluation of aceclofenac solid dispersions. *ARS Pharm., 51*, 57–76.
178. Dua, K., Pabreja, K., Ramana, M. V., & Bukhari, N. I., (2011). Preparation, characterization, and *in vitro* evaluation of aceclofenac PVP-solid dispersions. *J. Dispers. Sci. Technol., 32*, 1151–1157.
179. Khan, W. S., Asmatulu, R., Ahmed, I., & Ravigururajan, T. S., (2013). Thermal conductivities of electrospun PAN and PVP nanocomposite fibers incorporated with MWCNTs and NiZn ferrite nanoparticles. *Int. J. Therm. Sci., 71*, 74–79.
180. Schmidt, T., Mönch, J. I., & Arndt, K. F., (2006). Temperature-sensitive hydrogel pattern by electron-beam lithography. *Macromol. Mater. Eng., 291*, 755–761.
181. Burkert, S., Schmidt, T., Gohs, U., Mönch, I., & Arndt, K. F., (2007). Patterning of thin poly(N-vinyl pyrrolidone) films on silicon substrates by electron beam lithography. *J. Appl. Polym. Sci., 106*, 534–539.
182. Frank, Y. X., & Sidlgata, V. S., (2011). *Nanoimprint Lithography Processes for Forming Nanoparticles*. US Patent WO2011094672 A2.
183. Ciulla, T. A., Starr, M. B., & Masket, S., (2002). Bacterial endophthalmitis prophylaxis for cataract surgery: an evidence-based update. *Ophthalmology, 109*, 13–24.

CHAPTER 5

Heat Shock Protein 90 (Hsp90) Inhibitory Potentials of Some Chalcone Compounds as Novel Anti-Proliferative Candidates

DEBARSHI KAR MAHAPATRA,[1] SAYAN DUTTA GUPTA,[2] SANJAY KUMAR BHARTI,[3] TOMY MURINGAYIL JOSEPH,[4] JÓZEF T. HAPONIUK,[4] and SABU THOMAS[5]

[1]*Department of Pharmaceutical Chemistry, Dadasaheb Balpande College of Pharmacy, Nagpur, Maharashtra – 440037, India, E-mail: mahapatradebarshi@gmail.com*

[2]*Department of Pharmaceutical Chemistry, Gokaraju Rangaraju College of Pharmacy, Hyderabad, Telangana – 500090, India, E-mail: sayandg@rediffmail.com*

[3]*Institute of Pharmaceutical Sciences, Guru Ghasidas Vishwavidyalaya (A Central University), Bilaspur, Chhattisgarh – 495009, India, E-mail: skbharti.ggu@gmail.com*

[4]*Chemical Faculty, Polymers Technology Department, Gdansk University of Technology, Gdansk – 80233, Poland, E-mails: say2tomy@gmail.com (T. M. Joseph), jozeph.haponiuk@pg.edu.pl (J. T. Haponiuk)*

[5]*School of Chemical Sciences and International and Inter-University Center for Nanoscience and Nanotechnology, Mahatma Gandhi University, Kottayam – 686560, Kerala, India*

ABSTRACT

Cancer is one of the most deadly diseases that have affected mankind. After cardiac diseases, it is known to be the next demon that caused mortality among millions of people. Heat shock protein 90 (Hsp90) is a homodimer polypeptide, where each monomer's N, M, and C domain is conserved. Hsp90 is over-expressed in cancer cells (more than 90%) in association with the co-chaperones along with associated client proteins (Met, EGFR, Akt, Her2, and c-Raf) which are responsible towards the chemotherapeutic resistance. Globally, scientists tried to find strategies for the management of cancer by exploring hidden natural products and rationally developing drugs by systematically thoughtful insights into the biological targets. This chapter highlights the plausible functions of a new biological target, Hsp90 in the proliferation of cancer and its management by chalcone based small ligands. The encouraging results of these small molecules at preclinical stages in modulating the biological target in multiple oncogenic pathways have certainly attracted pharmaceutical exerts for better optimization, the establishment of an accurate structure-activity-relationships, and superior chemotherapeutic applications.

5.1 INTRODUCTION

Cancer is one of the most deadly diseases that have affected mankind. After cardiac diseases, it is known to be the next demon that caused mortality among millions of people [1]. The disease is fast progressing in this modern era and is expected to affect millions more by the year 2050 [2]. World Health Organization (WHO), several National Cancer Institutes (NCIs) of respective nations, and eminent bodies involved in cancer research have put forward numerous demographic data and predictions which has forced intellectuals and scientists in developing new strategies to combat the upcoming giant [3]. Every continent faces the incoming problems related to cancer of diverse origins such as liver, esophageal, breast, lungs, brain, testicular, ovarian, etc. [4]. Various agencies have laid down guidelines for the prevention of cancer and have blamed critical issues like food habits; where low dietary fiber intake, minimum green vegetables, and fruit consumption, frequent intake of red meat and saturated fats, and recurrent intake of recreational compounds (alcohol, smoking, weeds, etc.), for plausible augmentation in the number of cases [5].

For the complete eradication of the disease across the globe, numerous efforts took place towards the development of novel therapeutically active leads for chemotherapy where the majority of them did not show promising results [6]. The understanding of diverse biochemical pathways related to cellular proliferation, apoptosis, and metastasis represents a defensible approach. Globally, scientists tried to find strategies for the management of cancer by exploring hidden natural products and rationally developing drugs by thoroughly understanding the biological targets [7]. These apoptotic-inducers and cell-cycle regulators from diversity are the present-day weapons available to mankind [8].

Every person on this planet has a unique identity which is due to the difference in genomic outlook, clinical features, and environmental conditions [9]. The discovery of biomarkers of genetic, genomic, and clinical origin has hopefully revolutionized the growing area of pharmacotherapeutics by allowing quick prediction of susceptibility towards any particular disease, assessment of therapy progress reports, enabling targeting facilities, drastically reducing the side-effects, toxicity, and adverse effects, permit rapid shift from one approach to the other, allowing better clinical decisions by the medical practitioners, developing inhibitors through structure-based drug design (SBDD) approach and ultimately amplifying the clinical response of the patient [10]. With the advancements in medical technology, diagnostics, and instruments, this area of pharmacogenomics and personalized medicine are regarded as the future pillars in modern chemotherapeutics [11].

In an analogous way, identifying the plausible functions of a new biological target, heat shock protein 90 (Hsp90) in the management of cancer by chalcone based small ligands reflects the central theme of this chapter.

5.2 HSP90

5.2.1 STRUCTURE

Hsp90 is a homodimer polypeptide, where each monomer's N, M, and C domain is conserved [12] (Figure 5.1). It was experimentally proven that these three regions are highly flexible, i.e., a disturbance in any of the segments affects the other significantly [13]. The N domain (molecular weight ≈ 35 kDa, 1–216 amino acids) contains the ATP binding sites (Bergerat cleft), where the inhibitor molecules also attach itself while suppressing the chaperoning function of Hsp90 [14]. The M-domain (molecular weight ≈ 35 kD, 262–524 amino acids) in the middle segment of the chaperone that is proteolytically resistant [15]. The amino acids of this segment can bind to client proteins, few co-chaperones, and γ-phosphate of ATP [16]. The C-domain (molecular weight ≈ 12 kD, 525–709 amino acids) is the site of protein dimerization, which has a short pentapeptide motif (Met-Glu-Glu-Val-Asp or MEEVD) which serves as an acceptor for tetratricopeptide repeat (TPR) containing co-chaperones [17]. A connector loop or chain between the 'N' and 'M' domain is present which comprises 30–70 charged basic and acidic amino acid residues [18]. The structure of this domain is yet to be resolved at the atomic level.

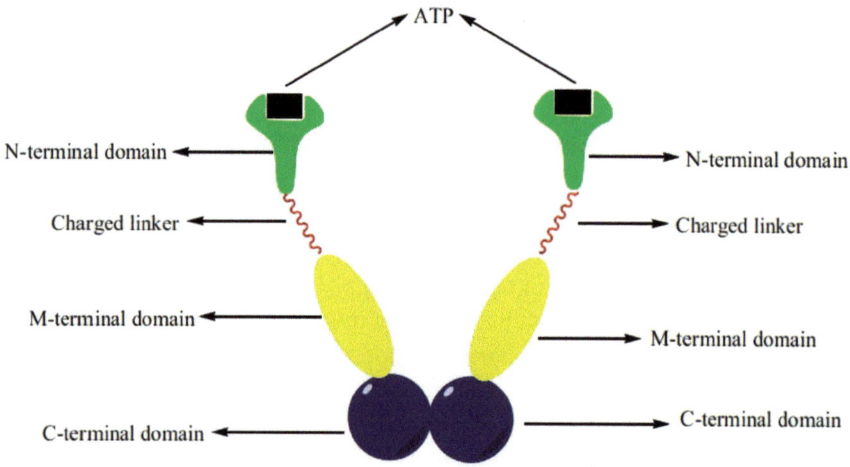

FIGURE 5.1 Schematic representation of Hsp90's structure. The protein exists as a symmetrical dimer with three domains per monomer. The N terminal (colored green) region binds the ATP which is essential for carrying out the chaperoning function. The M domain is separated from the N domain by a flexible charged linker. It contains a loop that is essential for ATP hydrolysis. The C terminal provides an interface between the two monomers.

5.2.2 FUNCTION

Hsp90 carries out its chaperoning function with the aid of ATP and co-chaperones (Figure 5.2). ATP provides the required energy for the entire chaperoning process by hydrolyzing to ADP and phosphate [19]. The process starts with the binding of client protein to Hsp90's middle domain with the help of co-chaperones [20]. Thereafter, the ATP binds to the N domain of the chaperone, resulting in the dimerization of the N and M segments [21]. The N and M domain then comes in contact with one another, which leads to the generation of Hsp90's "closed form" configuration [22]. The ATP hydrolysis and client protein repairing take place in this closed configuration of Hsp90. Thereafter, ADP, phosphate, and the repaired client protein are released from the protein complex [23]. It was also observed that sometimes ATP first attaches itself to the chaperone's N-terminal Bergerat fold and thereafter the client polypeptides bind to Hsp90 with co-chaperones [24]. This is followed by the usual chaperoning cycle, i.e., attainment of the closed-state, hydrolysis of ATP, repair of proteins, and release of ATP and matured proteomes [25] (Figure 5.3).

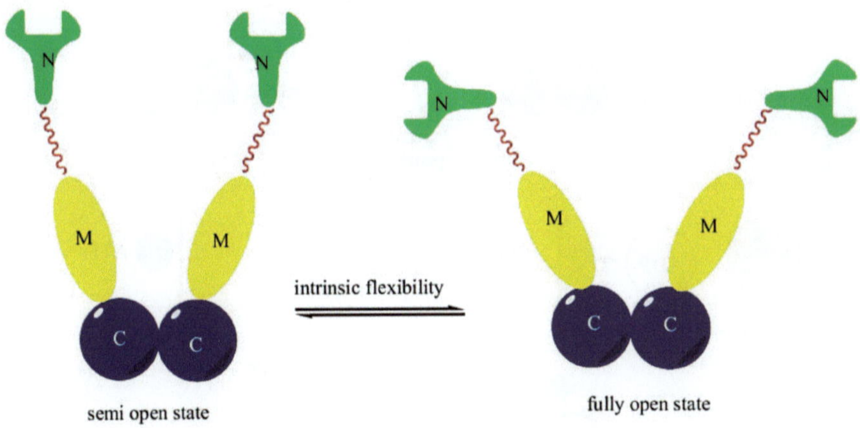

FIGURE 5.2 Resting phase conformations of Hsp90. The intrinsic flexibility of the protein results in two forms (semi-open and fully-open) during its apo-state.

5.2.3 HSP90 AND CANCER

Molecular chaperones are a group of proteins, which convert a newly synthesized immature polypeptide into a mature one [26]. Additionally, they also assist in the refolding of denatured proteins in the cell, i.e., they aid in the repair of damaged proteins (often referred to as client proteins) [27]. Hsp90 is one such ATP dependent chaperone which is synthesized during cellular stress like heat, hypoxia, DNA damage, UV radiation exposure, etc. It is the most abundant chaperone found within a damaged cell [28]. This chaperone has turned out to be an encouraging target for cancer chemotherapy because of the following observations:

1. Several oncogenic proteins are clients of Hsp90, i.e., they are dependent on Hsp90 for maintaining their structure and function. Hence, inhibiting Hsp90's chaperoning function will lead to the suppression of several oncogenic proteins at one shot (Figure 5.4) [29].

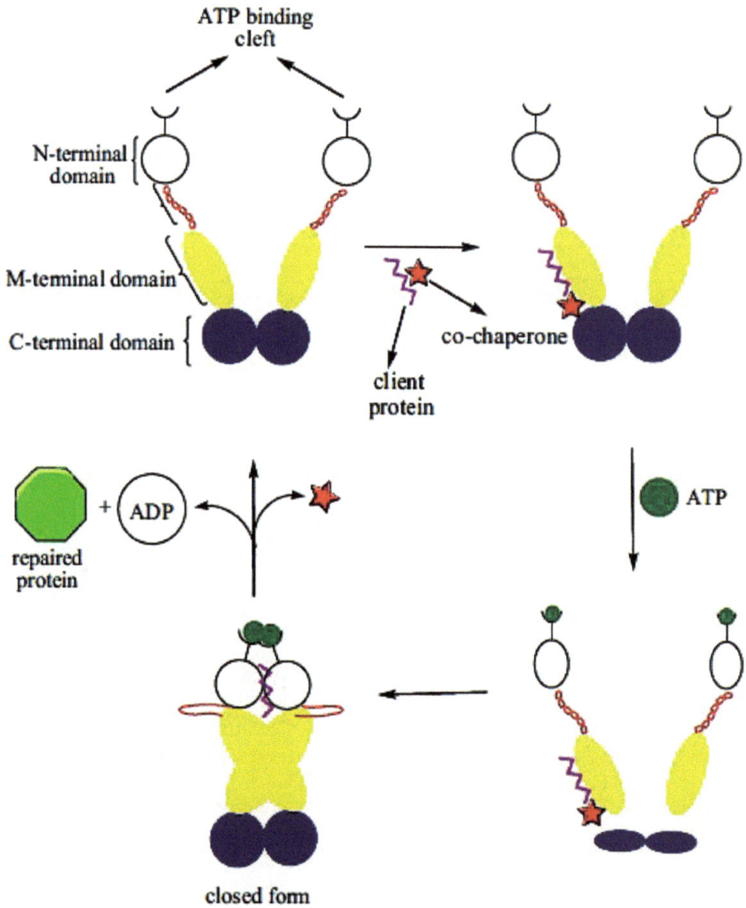

FIGURE 5.3 Schematic illustrations for the chaperoning cycle of Hsp90 protein. First, the client proteins attach to the M domain with the aid of co-chaperones. This is followed by ATP binding, which triggers the transformation of Hsp90 from its "open" to "closed" conformation. The client proteins after being repaired (inside the closed form of Hsp90) are released along with ADP and phosphate ion.

2. Proteins responsible for the development of resistance to chemotherapeutic agents are also clients of Hsp90. Therefore, the chances of developing resistance to Hsp90 inhibitors are reduced to a considerable extent [30].
3. Hsp90 is over-expressed in cancer cells (more than 90%) in association with co-chaperones whereas Hsp90 of normal cells comprises 1–2% of cellular proteins that reside in a free and un-complexed

state. Hence, compounds antagonizing Hsp90's function will exhibit fewer side effects [31].

4. The morphology (amino acid sequence) of the ATP binding domain present in the Hsp90 varies considerably from cellular proteins of topoisomerases, biotin carboxylase, hexokinase, etc., origin which uses ATP for performing the normal functions.

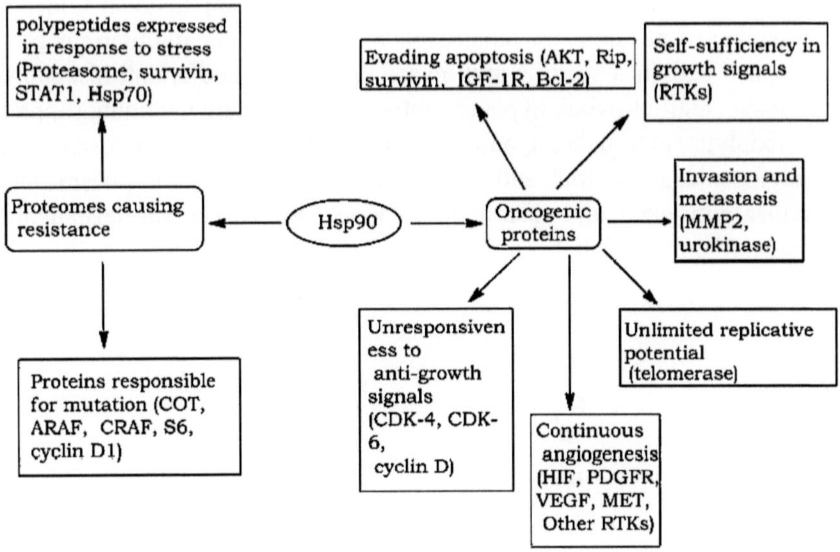

FIGURE 5.4 Association of Hsp90 with cancer.

Therefore, Hsp90 inhibitor never suppresses the cellular proteins and is therefore considered as less-toxic [32]. The above facts provided the impetus for the design and discovery of novel Hsp90 inhibitors for the treatment of cancer.

5.3 CHALCONES

Chalcones (Figure 5.5) are the abundantly present natural products class in nature either in free or in diverse complexed forms. The term chalcone was coined by Kostanecki and Tambor in the 19th century who first synthesized these chromophoric compounds. The term originated from the word of Greek origin "*chalcos*" which means "bronze," named after the bronze-like color of the natural chalcones [33]. The prop-2-ene-1-one scaffold comprising of two aromatic rings attached together with a three carbon α, β unsaturated carbonyl bridge. It has received attention among the research community owing to a large number of merits such as easy steps for laboratory-oriented synthesis via diverse procedures such as direct crossed-coupling reaction, Julia-Kocienski reaction, carbonylated Heck coupling reaction, Suzuki-Miyaura reaction, Sonogashira isomerization coupling, microwave-assisted reactions, solid acid catalyst mediated reactions, Friedel-Crafts reaction, one-pot reactions, solvent-free reactions, etc., [34].

The low-molecular-weight feature facilitate swift computational studies, a large number of positions for replaceable hydrogen atoms, multifarious pharmacological activities such as anti-invasive, anti-spasmodic, anti-histaminic, anti-oxidant, anti-retroviral, hypnotic, anti-gout, antibacterial, immunosuppressant, anxiolytic, anti-malarial, anti-nociceptive, anti-hypertensive, osteogenic, anti-angiogenic, anti-filarial, anti-arrhythmic, anti-protozoal, anti-obesity, anti-tubercular, antifungal, anti-ulcer, hypolipidemic, anti-cancer, anti-leishmanial, anti-inflammatory, anti-diabetic, anti-steroidal, anti-platelet, etc., [35, 36].

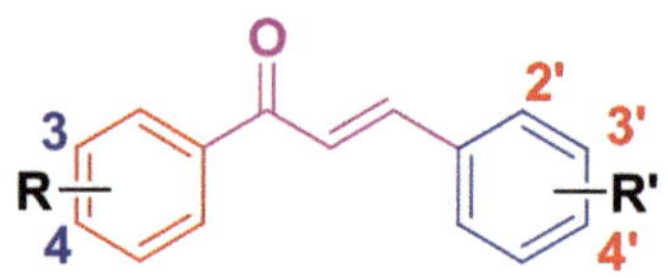

FIGURE 5.5 Structure of chalcone.

In addition to the above stated pharmacological significance, the pharmacophore expresses potent inhibition of cystic fibrosis transmembrane conductance regulator (CFTR), P-glycoprotein (P-gp), fundamental metabolic enzymes, and antagonism of various factors. Non-pharmacological applications include insecticide, scintillator, polymerization catalyst, artificial sweetener, fluorescent whitening agent, organic brightening agent, fluorescent polymers, and analytical receptor for Fe(III) determination [37].

The scaffold serves as an intermediate for the synthesis of several heterocyclic compounds such as pyrimidine, pyrrole, thiazole, thiophene, indole, isoxazole, oxazole, benzodiazepine, benzoxazepine, benzothiazepine, etc. In the last few years, hybridization has been perceived to be an emerging technique to develop analogs with significantly high biological activity. The fusion of naphthoquinone, chromene, naphthalene, coumarin, sulfonamide, imidazolone, pyrazoline, bifendate, piperazine, etc., has led to several folds enhancement in the activity [38].

Various natural chalcones like liquiritigenin, xanthohumol, flavokawain B, xanthoangelol, isoliquiritigenin, cardamonin, echinatin, licochalcone, kuwanon, 4-hydroxyderricin, broussochalcone, macdentichalcone, etc., are the eminent candidates with wide pharmacological perspectives in cancer, diabetes, infection, inflammation, and many other activities [39]. The benzylideneacetophenone scaffold is the open chain intermediates in aurones synthesis of flavones and acts as precursors of flavonoids and isoflavonoids. Chalcones and flavonones are isomeric in nature and in the presence of acid or base, they readily undergo interconversion. The edible chalcones have been found to play a pivotal role as Michael acceptor in Michael addition reaction. This small template aids in the structure elucidation of flavanone, flavonoid, chromanochromane, and tannins [40]. Chalcone or chalconoids exist in two forms; *cis* and *trans*, where the *trans* isomeric form is thermodynamically more stable than the *cis* form. The conjugated double bond system with a delocalized π-electron system facilitates several electron transfer reactions and produces a decrease in redox potential which promotes numerous interactions with the biological targets [41].

5.4 CHALCONES AS HSP90 INHIBITORS

In a computational study, the anti-breast cancer potential of a novel 3-phenylquinolinylchalcone derivative, (*E*)-3-(3-(4-methoxyphenyl) quinolin-2-yl)-1-phenylprop-2-en-1-one (Figure 5.6A) was studied by establishing the possible interaction with the Hsp90 (PDB ID: 1UYE) through molecular docking approach. The study presented significant inhibition of the molecular target (Gscore-10.0 kcal/mol) via interaction with the amino acid Phe138 residue through hydrogen bonding [42].

A novel chalcone molecule (Figure 5.6B) finds future application in treating triple-negative breast cancer, one of the most belligerent form of neoplasm by selectively impairing the growth of MDA-MB-231 breast cancer cells through ubiquitin-proteasome pathway where degradation of Hsp90 client proteins such as EGFR, Met, Her2, c-Raf, Akt, and Cdk4 remained the hallmark characteristic of Hsp90 inhibition. However, the mRNA levels after expression of Met and Akt were not associated with the transcriptional regulation, suggesting an alternative pathway, known as "ubiquitin-proteasome pathway." The treatment of ligand in escalating concentration presented cleavage of PARP, caspase 3/8 and downregulation of anti-apoptotic protein Bcl-2. The study opened avenues for treating this complex form of cancer by preventing metastasis to bone, brain, lung, and liver and enhancing the five-year survival rate [43].

In an effort towards drug discovery, the dose-dependent concentration of 2,'4'-dimethoxychalcone (Figure 5.6C) has been found to considerably inhibit the growth of iressa-resistant non-small cell lung cancer (NSCLC, H1975) by the disruption of Hsp90 chaperoning function. This exemplified a pioneering step for overcoming the possible drug resistance induced by Met amplification and EGFR mutation [44].

The anti-prostate cancer role of methoxylated chalcone (Figure 5.6D) has been explored by Kim and co-workers where the low-molecular-weight-ligands (LMWLs) prevent the translocation of Hsp90-androgen receptor complex in the cytoplasm under the androgen-non-responsive state. The compound paved the pathway towards anti-androgen therapy with its unique mechanism of action and finds application in incurable castrate-resistant prostate cancer conditions [45].

Based on the principles of rational drug design, a novel series of chalcone molecules were rationally designed by linking the resorcinol and the trimethoxyphenyl ring. The prop-2-ene-1-one molecule (Figure 5.6E), designed through SBDD way expressed inhibition of Hsp90 which results in complete suppression of several oncogenic molecules such as EGFR, Her2, Met, and Akt proteins, prevents H1975 cell proliferation (GI_{50} of 48 μM), and overcoming gefitinib-resistance. The molecular docking expressed a noteworthy interaction with Hsp90 target along with several interesting results which clearly supported the disruption of Hsp90 chaperone machinery. The resorcinol ring binds with the hydrophilic portion whereas the trimethoxyphenyl ring dominates in the hydrophobic part (–7.91 kcal/mol) while interacting with the Thr184, Asp93, Leu48, and Val186 residues through hydrogen-bonding and Van der Waals contact [46].

Similar to the above context, licochalcone A (Figure 5.6F), and a natural product targeting Hsp90 has been seen to effectively overcome gefitinib-resistant in NSCLC. Screening against *in vitro* model H1975 cancer cells at 70 μM showed the profound concentration-dependent cytotoxic effect by downregulating the expression levels of EGFR, Her2, Hsp70, Met, Akt, and Hsp90. The results have been found to be analogous to the positive control, geldanamycin, where protein folding machinery was disrupted, with the consequent degradation of Hsp90 in the cytoplasm, and transcriptional upregulation of cochaperone Hsp70. Further evidence from *in silico* studies confirmed inhibition of Hsp90 by effectual binding (–8.84 kcal/mol) with the amino acid residues of the hydrophilic region (Asp93, Asp54, Thr184) as well as hydrophobic region (Phe138, Trp162, Leu107, Val150, Tyr139) through both hydrogen-bonding and Van der Waals contact [47].

The chalcone bioisosterics (Figure 5.6G) developed from resorcinol-based N-benzyl benzamide derivatives displayed tremendous growth inhibition (GI_{50} of 0.42 μM) against NSCLC (H1975) by apoptotic pathway (PARP and Caspase cleavage). The chalcone-like compound inhibited the Hsp90 (IC_{50} 5.3 nM), the ubiquitous molecular chaperone along with its protein clients Her2, EGFR, Met, Akt, and c-Raf as confirmed by the immunostaining and western immunoblot analysis. Inhibition of this imperative molecular target will lead to stabilization and maturation of many oncogenic proteins. Furthermore, the novel inhibitor showed weak inhibition of P_{450} isoforms (1A2, 2C9, 2C19, 2D6, and 3A) with IC_{50} values of <5 μM. Moreover, the compound inhibited the tumor growth in a mouse xenograft model bearing subcutaneous H1975 [48].

The secondary metabolites (flavokawains A, B, and C) obtained from the kava plant (*Piper methysticum*) have been found to circumvent the gefitinib-resistant in NSCLC in a similar context. The metabolite B (Figure 5.6H) and synthetic flavokawain derivative (Figure 5.6I) gained attention among the scientific community owing to its trait in disordering Hsp90 chaperoning utility along with impairing the growth of H1975 cancer cells (IC_{50} value of 33.5 μM). In addition to it, the natural product drastically reduces the expression levels of EGFR, Met, Her2, Akt, Cdk4, Hsp70, and Hsp90 in a concentration-dependent manner. The studies indicated towards potential suppression of multiple oncogenic signaling pathways (PI3K-Akt-mTOR and Ras-Raf-Mek-Erk) simultaneously, thereby lessening the prospects of molecular feedback loops and mutations, which lead to the plausible overcoming of tumor resistance in NSCLC [49, 50].

Flavokawain B (Figure 5.6H) has also been reported to demonstrate antiproliferative activity against human oral carcinoma HSC-3 cells and hepatotoxicity against HepG2 cells by inducing cell-cycle arrest, transcriptional responses, and apoptosis through disruption of Hsp90 chaperone machinery [51, 52].

The role of chalcones was perceived to be universal and not restricted to anticancer applications. Seo et al. identified the anti-infective perspective Hsp90 inhibitory potential of 2,'4'-dihydroxychalcone (2,'4'-DHC) (Figure 5.6J) in *Aspergillus fumigatus*. The chalcone analog significantly decreased the growth of fungus by selectively suppressing the Hsp90 in the Hsp90-calcinurin pathway by binding with the ATPase domain (Figure 5.6) [53].

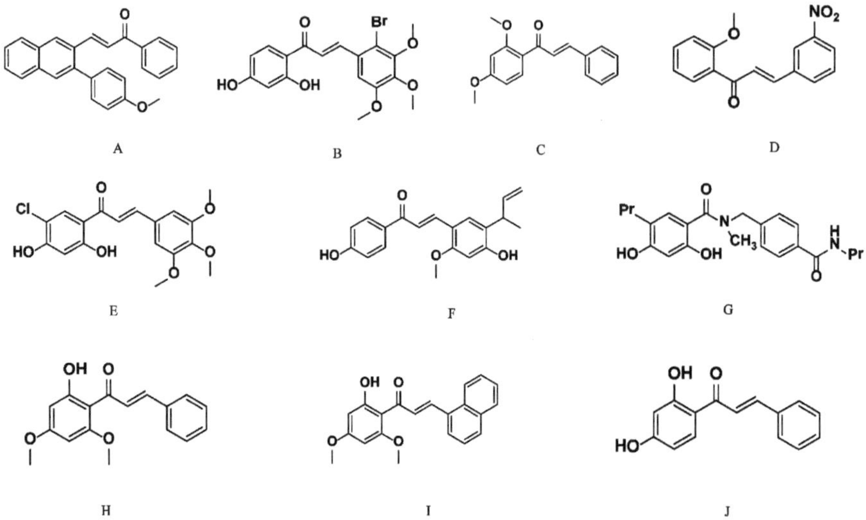

FIGURE 5.6 Chalcone based Hsp90 inhibitors.

5.5 CONCLUSION

The chapter emphasized on the displayed promises laid by the chalcone scaffold bearing molecules originated from the nature and laboratory in the complete inhibition of the critical anti-cancer molecular target Hsp90 as well as downregulating the client proteins such as EGFR, Her2, Met, Akt, and c-Raf. As the first journey started 25 years ago with discovery programs headed by academia, it has now gained attention among the scientists of all fields. About 20 small molecules developed in academia-industry collaborations as competitive Hsp90 inhibitors have reached the clinical trial stage either alone or with another anti-cancer drug combination, which certainly indicated towards the future avenues of pharmacotherapeutics for cancer of multiple origins. In the present scenario, these clinically active candidates are still miles away from the regulatory approval, although, it is quite motivating for the medicinal chemists and researchers for further studies. The encouraging results of these small molecules at preclinical stages in modulating the biological target in multiple oncogenic pathways have certainly attracted pharmaceutical exerts for better optimization, establishment of an accurate structure-activity-relationships, development towards utility in multiple diseases (inflammation, AIDS, different kinds of infections, metabolic diseases, and many more) and superior chemotherapeutic applications.

KEYWORDS

- **cystic fibrosis transmembrane conductance regulator**
- **heat shock protein 90**
- **low-molecular-weight-ligands**
- **p-glycoprotein**
- **structure-based drug design**
- **tetratroicopeptide repeat**

REFERENCES

1. Asati, V., Mahapatra, D. K., & Bharti, S. K., (2016). PI3K/Akt/mTOR and Ras/Raf/MEK/ERK signaling pathways inhibitors as anticancer agents: Structural and pharmacological perspectives. *Eur. J. Med. Chem.*, *109*, 314–341.
2. Asati, V., Mahapatra, D. K., & Bharti, S. K., (2014). Thiazolidine-2, 4-diones as multi-targeted scaffold in medicinal chemistry: Potential anticancer agents. *Eur. J. Med. Chem.*, *87*, 814–833.
3. Asati, V., Mahapatra, D. K., & Bharti, S. K., (2017). K-Ras and its inhibitors towards personalized cancer treatment: Pharmacological and structural perspectives. *Eur. J. Med. Chem.*, *125*, 299–314.
4. Mahapatra, D. K., Asati, V., & Bharti, S. K., (2017). MEK inhibitors in oncology: A patent review (2015-Present). *Exp. Opin. Ther. Pat.*, *27*(8), 887–906.
5. Asati, V., Bharti, S. K., Mahapatra, D. K., Asati, V., & Budhwani, A. K., (2016). Triggering PIK3CA mutations in PI3K/Akt/mTOR axis: Exploration of newer inhibitors and rational preventive strategies. *Curr. Pharm. Design*, *22*(39), 6039–6054.
6. Asati, V., Bharti, S. K., & Mahapatra, D. K., (2016). Mutant B-Raf kinase inhibitors as anticancer agents. *Anti-Cancer Agents Med. Chem.*, *16*(12), 1558–1575.
7. Mahapatra, D. K., & Bharti, S. K., (2017). *Handbook of Research on Medicinal Chemistry: Innovations and Methodologies*. New Jersey: Apple Academic Press.
8. Mahapatra, D. K., & Bharti, S. K., (2016). *Drug Design*. New Delhi: Tara Publications Private Limited.
9. Mahapatra, D. K., & Bharti, S. K., (2019). *Medicinal Chemistry with Pharmaceutical Product Development*. New Jersey: Apple Academic Press.
10. Bharti, S. K., & Mahapatra, D. K., (2015). Promises of personalized medicine in 21st century. In: Pearce, E. M., Howell, B. A., Pethrick, R. A., & Zaikov, G. E., (eds.), *Physical Chemistry Research for Engineering and Applied Sciences*. New Jersey: Apple Academic Press.
11. Bharti, S. K., & Mahapatra, D. K., (2015). Biopharmaceuticals: An introduction to biotechnological aspects and practices. In: Joswik, R., Zaikov, G. E., & Haghi, A. K., (eds.), *Life Chemistry Research: Biological Systems*. New Jersey: Apple Academic Press.

12. Whitesell, L., & Lindquist, S. L., (2005). HSP90 and the chaperoning of cancer. *Nature Rev. Cancer*, *5*(10), 761.
13. Rutherford, S. L., & Lindquist, S., (1998). Hsp90 as a capacitor for morphological evolution. *Nature*, *396*(6709), 336.
14. Queitsch, C., Sangster, T. A., & Lindquist, S., (2002). Hsp90 as a capacitor of phenotypic variation. *Nature*, *417*(6889), 618.
15. Wiech, H., Buchner, J., Zimmermann, R., & Jakob, U., (1992). Hsp90 chaperones protein folding in vitro. *Nature*, *358*(6382), 169.
16. Richter, K., & Buchner, J., (2001). Hsp90: Chaperoning signal transduction. *J. Cell Physiol.*, *188*(3), 281–290.
17. Trepel, J., Mollapour, M., Giaccone, G., & Neckers, L., (2010). Targeting the dynamic HSP90 complex in cancer. *Nature Rev Cancer*, *10*(8), 537.
18. Wandinger, S. K., Richter, K., & Buchner, J., (2008). The Hsp90 chaperone machinery. *J. Biol. Chem.*, *283*(27), 18473–18477.
19. Buchner, J., (1999). Hsp90 & Co.–a holding for folding. *Trend Biochem. Sci.*, *24*(4), 136–141.
20. Young, J. C., Moarefi, I., & Hartl, F. U., (2001). Hsp90: A specialized but essential protein-folding tool. *J. Cell Biol.*, *154*(2), 267.
21. Pearl, L. H., & Prodromou, C., (2006). Structure and mechanism of the Hsp90 molecular chaperone machinery. *Annu. Rev. Biochem.*, *75*, 271–294.
22. Taipale, M., Jarosz, D. F., & Lindquist, S., (2010). HSP90 at the hub of protein homeostasis: Emerging mechanistic insights. *Nature Rev. Mol. Cell Biol.*, *11*(7), 515.
23. Neckers, L., & Workman, P., (2012). Hsp90 molecular chaperone inhibitors: Are we there yet?. *Clin. Cancer Res.*, *18*(1), 64–76.
24. Bose, S., Weikl, T., Bügl, H., & Buchner, J., (1996). Chaperone function of Hsp90-associated proteins. *Science*, *274*(5293), 1715–1717.
25. Subbarao, S. A., Kalmár, É., Csermely, P., & Shen, Y. F., (2004). Hsp90 isoforms: Functions, expression and clinical importance. *FEBS Lett.*, *562*(1–3), 11–15.
26. Chiosis, G., Vilenchik, M., Kim, J., & Solit, D., (2004). Hsp90: The vulnerable chaperone. *Drug Discov. Today*, *9*(20), 881–888.
27. Pearl, L. H., & Prodromou, C., (2000). Structure and *in vivo* function of Hsp90. *Curr. Opin. Struct. Biol.*, *10*(1), 46–51.
28. Solit, D. B., & Chiosis, G., (2008). Development and application of Hsp90 inhibitors. *Drug Discov. Today*, *13*(1/2), 38–43.
29. Maloney, A., & Workman, P., (2002). HSP90 as a new therapeutic target for cancer therapy: The story unfolds. *Exp. Opin. Biol. Ther.*, *2*(1), 3–24.
30. Mahalingam, D., Swords, R., Carew, J. S., Nawrocki, S. T., Bhalla, K., & Giles, F. J., (2009). Targeting HSP90 for cancer therapy. *British J. Cancer*, *100*(10), 1523.
31. Workman, P., (2003). Overview: Translating Hsp90 biology into Hsp90 drugs. *Curr. Cancer Drug Targ.*, *3*(5), 297–300.
32. Bagatell, R., & Whitesell, L., (2004). Altered Hsp90 function in cancer: A unique therapeutic opportunity. *Mol. Cancer Ther.*, *3*(8), 1021–1030.
33. Mahapatra, D. K., Bharti, S. K., & Asati, V., (2015). Anti-cancer chalcones: Structural and molecular target perspectives. *Eur. J. Med. Chem.*, *98*, 69–114.
34. Mahapatra, D. K., Bharti, S. K., & Asati, V., (2015). Chalcone scaffolds as anti-infective agents: Structural and molecular target perspectives. *Eur. J. Med. Chem.*, *101*, 496–524.

35. Mahapatra, D. K., Asati, V., & Bharti, S. K., (2015). Chalcones and their therapeutic targets for the management of diabetes: Structural and pharmacological perspectives. *Eur. J. Med. Chem.*, *92*, 839–865.
36. Mahapatra, D. K., & Bharti, S. K., (2016). Therapeutic potential of chalcones as cardiovascular agents. *Life Sci.*, *148*, 154–172.
37. Mahapatra, D. K., Bharti, S. K., & Asati, V., (2017). Chalcone derivatives: Anti-inflammatory potential and molecular targets perspectives. *Curr. Topic Med. Chem.*, *17*(28), 3146–3169.
38. Mahapatra, D. K., Asati, V., & Bharti, S. K., (2019). Recent therapeutic progress of chalcone scaffold bearing compounds as prospective anti-gout candidates. *J. Crit. Rev.*, *6*(1), 1–5.
39. Mahapatra, D. K., Asati, V., & Bharti, S. K., (2019). Anti-inflammatory perspectives of chalcone based NF-κB inhibitors. In: Mahapatra, D. K., & Bharti, S. K., (eds.), *Pharmacological Perspectives of Low Molecular Weight Ligands*. New Jersey: Apple Academic Press.
40. Mahapatra, D. K., Asati, V., & Bharti, S. K., (2019). Natural and synthetic prop-2-ene-1-one scaffold bearing compounds as molecular enzymatic targets inhibitors against various filarial species. In: Torrens, F., Mahapatra, D. K., & Haghi, A. K., (eds.), *Biochemistry, Biophysics, and Molecular Chemistry: Applied Research and Interactions*. New Jersey: Apple Academic Press.
41. Mahapatra, D. K., Bharti, S. K., & Asati, V., (2019). Recent perspectives of chalcone based molecules as protein tyrosine phosphatase 1B (PTP-1B) inhibitors. In: Mahapatra, D. K., & Bharti, S. K., (eds.), *Medicinal Chemistry with Pharmaceutical Product Development*. New Jersey: Apple Academic Press.
42. Mahto, M. K., Yadav, S., Ram, K. S., Gangulia, S., & Bhaskar, M., (2014). 3-phenylquinolinylchalcone derivatives: Pharmacophore modeling, 3d-qsar analysis and docking studies as anti-cancer agents. *Int. J. Bioassay*, *2*, 2–99.
43. Oh, Y. J., & Seo, Y. H., (2017). A novel chalcone-based molecule, BDP inhibits MDA-MB-231 triple-negative breast cancer cell growth by suppressing Hsp90 function. *Oncol. Rep.*, *38*(4), 2343–2350.
44. Seo, Y. H., (2015). Discovery of 2′, 4′-dimethoxychalcone as a Hsp90 inhibitor and its effect on iressa-resistant non-small cell lung cancer (NSCLC). *Arch Pharma. Res.*, *38*(10), 1783–1788.
45. Kim, Y. S., Kumar, V., Lee, S., Iwai, A., Neckers, L., Malhotra, S. V., & Trepel, J. B., (2012). Methoxychalcone inhibitors of androgen receptor translocation and function. *Bioorg. Med. Chem. Lett.*, *22*(5), 2105–2109.
46. Jeong, C. H., Park, H. B., Jang, W. J., Jung, S. H., & Seo, Y. H., (2014). Discovery of hybrid Hsp90 inhibitors and their anti-neoplastic effects against gefitinib-resistant non-small cell lung cancer (NSCLC). *Bioorg. Med. Chem. Lett.*, *24*(1), 224–227.
47. Seo, Y. H., (2013). Discovery of licochalcone A as a natural product inhibitor of Hsp90 and its effect on gefitinib resistance in non-small cell lung cancer (NSCLC). *Bull Korean Chem. Soc.*, *34*(6), 1917–1920.
48. Park, S. Y., Oh, Y. J., Lho, Y., Jeong, J. H., Liu, K. H., Song, J., Kim, S. H., Ha, E., & Seo, Y. H., (2018). Design, synthesis, and biological evaluation of a series of resorcinol-based N-benzyl benzamide derivatives as potent Hsp90 inhibitors. *Eur. J. Med. Chem.*, *143*, 390–401.

49. Seo, Y. H., & Oh, Y. J., (2013). Synthesis of flavokawain B and its anti-proliferative activity against gefitinib-resistant non-small cell lung cancer (NSCLC). *Bull Korean Chem. Soc.*, *34*(12), 3782–3786.
50. Seo, Y. H., & Park, S. Y., (2014). Synthesis of flavokawain analogues and their anti-neoplastic effects on drug-resistant cancer cells through Hsp90 inhibition. *Bull. Korean Chem. Soc.*, *35*(4), 1154–1158.
51. Pinner, K. D., Wales, C. T., Gristock, R. A., Vo, H. T., So, N., & Jacobs, A. T., (2016). Flavokawains A and B from kava (Piper methysticum) activate heat shock and antioxidant responses and protect against hydrogen peroxide-induced cell death in HepG2 hepatocytes. *Pharm. Biol.*, *54*(9), 1503–1512.
52. Hseu, Y. C., Lee, M. S., Wu, C. R., Cho, H. J., Lin, K. Y., Lai, G. H., Wang, S. Y., et al., (2012). The chalcone flavokawain B induces G2/M cell-cycle arrest and apoptosis in human oral carcinoma HSC-3 cells through the intracellular ROS generation and downregulation of the Akt/p38 MAPK signaling pathway. *J. Agri. Food Chem.*, *60*(9), 2385–2397.
53. Seo, Y. H., Kim, S. S., & Shin, K. S., (2015). *In vitro* antifungal activity and mode of action of 2′, 4′-dihydroxychalcone against *Aspergillus fumigatus*. *Mycobiol.*, *43*(2), 150–156.

CHAPTER 6

Perspectives of Cashew Nut Shell Liquid (CNSL) in a Pharmacotherapeutic Context

TOMY MURINGAYIL JOSEPH,[1*] DEBARSHI KAR MAHAPATRA,[2] P. MEREENA LUKE,[1] JÓZEF T. HAPONIUK[1] and SABU THOMAS[3]

[1]*Chemical Faculty, Polymers Technology Department, Gdansk University of Technology, Gdansk – 80233, Poland, E-mail: say2tomy@gmail.com (T. M. Joseph)*

[2]*Department of Pharmaceutical Chemistry, Dadasaheb Balpande College of Pharmacy, Nagpur, Maharashtra – 440037, India, E-mail: dkmbsp@gmail.com*

[3]*School of Chemical Sciences and International and Inter-University Center for Nanoscience and Nanotechnology, Mahatma Gandhi University, Kottayam – 686560, Kerala, India, E-mail: sabuthomas@mgu.ac.in*

ABSTRACT

The cashew tree consists of the cashew nut fruit (which is a curved edible seed, housed in a honeycomb-like shell), the apple, leaf, and bark. The fruit consists of an outer shell, inner shell, and the kernel. By weight, cashew nuts comprise 50% shell, 25% kernel (including testa), and 25% shell liquid. The shell of the cashew nut (*Anacardium occidentale* L.) contains an alkylphenolic oil named cashew nut shell liquid (CNSL) amounting to nearly 25% of the total weight of the nut. This oil is composed of anacardic acid (3-*n*-pentadecylsalicylic acid), and lesser amounts of cardanol (3-*n*-pentadecylphenol), cardol (5-*n*-pentadecylresorcinol), methylcardol (2-methyl-5-*n*-pentadecylresorcinol) and a small amount of polymeric material. Other components include

1-hydroxy-2-carboxy-3-pentadecyl benzene, 1-hydroxy-2-carboxy-3-(8'-pentadecenyl) benzene, 1-hydroxy-2-carboxy-3-(8,'11'-pentadecadienyl) benzene, 1-hydroxy-2-carboxy-3-(8,'11,'14'-pentadecatrienyl)benzene, 2-methyl-5-pentadecyl resorcinol, 2-methyl-5-(8'-pentadecenyl)resorcinol, 2-methyl-5-(8,'11'-pentadecadienyl) resorcinol, 2-methyl-5-(8,'11,'14'-pentadecatrienyl)resorcinol, 3-pentadeca-anisole, 3-(8'-pentadecenyl) anisole, 1-methoxy-3-(8,'11'-pentadecadienyl)benzene, 1-methoxy-3-(8,' 11,'14'-pentadecatrienyl)benzene, 5-pentadecyl resorcinol, 5-(8'-pentadecyl) resorcinol, 5-(8,'11'-pentadecadienyl)resorcinol, and 5-(8,' 11,'14'-pentadecatrienyl)resorcinol. The present chapter deals with the source of production, processes, and geographical information, export/ import, and chemical composition, industrial utility, and environmental effects of CNSL in modern perspectives.

6.1 INTRODUCTION

The cashew tree consists of the cashew nut (which is a curved edible seed, housed in a honeycomb-like shell), the apple, leaf, and bark. The nut consists of an outer shell, inner shell, and the kernel. The nut was too hard to digest and was later expelled with the droppings. The thickness of the cashew nutshell (CNS) is about 1/8 inch (0.32 cm). The Cashew tree is of great economic significance to Nigeria and other tropical countries due to its valuable products [1]. These products are utilized in food, medicine, chemical, and allied industries. In addition to providing shade, the cashew tree is an embellished ornamental plant that suitably controls soil erosion. Usually, it takes about 3–4 years from planting time before fruiting. Thereafter, the tree can live up to 40 years and more [2]. The cashew fruit is unusual in comparison with other tree nuts since the nut is outside the fruit. The cashew apple which is about 10 cm long is an edible false fruit, attached to the externally born nut by a stem. Its color ranges from yellow to red, fibrous in nature, very juicy, sweet, pungent, and high in vitamins A and C [3]. In its raw state, CNS, which is leathery in nature, contains the vesicant oily liquid. The shell is separated from the kernel by the testa which is a thin skin surrounding the kernel. The soft honeycomb matrix, in between the outer and inner shell, contains a dark brown liquid, which is known as cashew nut shell liquid (CNSL). CNSL is dark viscous oil with a characteristic smell, quite unlike other vegetable oils. It is opaque and when applied as a thin film, it is reddish-brown in color. It is immiscible with water but is miscible with most organic solvents. CNSL has germicidal and fungicidal properties. It is used traditionally as a cure for fungal attack of the feet in India [4]. The physicochemical characteristics of CNSL are depicted in Table 6.1.

TABLE 6.1 Physicochemical Characteristics of CNSL

Characteristics	Requirement
Matter insoluble in toluene (% by wt.)	1
Viscosity (cps)	550
Loss in weight on heating (% by wt.)	1
Moisture (% by wt.)	1
Ash (% by wt.)	1
Specific gravity	0.95–0.97
Iodine value	250
Polymerization time (min.)	4

6.2 SOURCE OF PRODUCTION AND PROCESSES

In the industrial processing tasks, CNSL are often produced as a waste material. CNS contains CNSL up to about 25–35 wt.% of the nutshell weight which related to the method of extraction. Thermal treatment of cashew nuts and CNSL induces the partial decarboxylation of anacardic acid, which is completed by subsequent purifying distillation [5]. The result is 90% industrial-grade cardanol oil, which is amber-yellow, with a smaller percentage of cardol and methylcardol. Crude CNSL is corrosive but becomes less so by decarboxylation and removal of H_2S during the refining process. CNSL extracted by a cold-solvent process is called natural CNSL and hot-oil and/or roasting processed CNSL is called technical CNSL. Although natural CNSL has more anacardic acid than cardanol, it is decarboxylated to produce anacardol which when hydrogenated yields cardanol, the chief constituent of technical CNSL (Figure 6.1) [6].

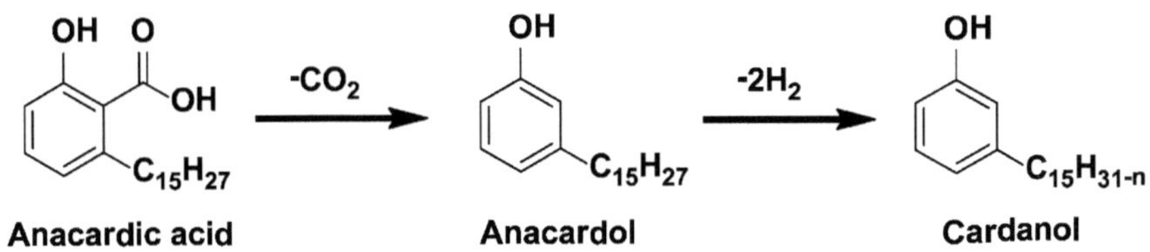

FIGURE 6.1 Decarboxylation and hydrogenation of natural CNSL.

Traditionally, a number of methods have been employed to extract CNSL from the nuts. Two processes are used mainly in the extraction of CNSL: hot oil process and roasting process in which CNSL oozes out from the shell. Extraction of the oil using solvents such as benzene, toluene, petroleum hydrocarbon solvents, or alcohols, or supercritical extraction of the oil using a mixture of CO_2 and isopropyl alcohol are the other reported techniques for extraction of CNSL [7].

By weight, cashew nuts comprise 50% shell, 25% kernel (including testa), and 25% shell liquid. The astringent shell liquid is removed prior to shelling to allow handling of the nut and prevent the liquid from damaging the nut kernel. Kernels damaged by the dark shell liquid will be discolored and may have a serious 'off flavor.' The CNSL is removed either by drum roasting in high temperature oil/CNSL or by steaming, and the volumes recovered differ. With drum roasting, production is 10–12% of a possible 25%. With steaming, production is 20–25%, as it is possible to use an expeller. The use of a centrifuge following drum roasting boosts recovery levels. CNSL can also be produced from shells by using the 'cold press' method in an extraction unit. Gloves are sometimes worn for protection in the shelling process, but more often than not, the nuts and hands are coated in ash or some other absorbent material for protection. CNSL has an economic value and can provide many phenolic compounds with far greater versatility than petrochemical phenols [8].

6.3 GEOGRAPHICAL INFORMATION

The cashew tree, native to Brazil, was introduced to Mozambique and then India in the sixteenth century by the Portuguese, as a means of controlling coastal erosion. It was spread within these countries with the aid of elephants that ate the bright cashew fruit along with the attached nut. The world production of cashew nuts is nearly 500,000 tons per year, with Brazil being the largest producer. It was not until the nineteenth century that plantations were developed and the tree then spread to a number of other countries in Africa, Asia, and Latin America. Other countries, which produce CNSL, are India and Tanzania with the potential of producing 8,000 tonnes per annum [9].

6.4 COMPOSITION OF CNSL

The earliest work published concerning the composition of cashew nut oil was by Staedeler. Since then, many researchers have investigated the constitution of the oil. Chemically, it is a mixture of several closely related organic compounds, each consisting of salicylic acid substituted with an alkyl chain. For anacardic acid, the combination of phenolic, carboxylic, and a 15-carbon alkyl side chain functional group makes it attractive in biological applications or as a synthon for the synthesis of a multitude of bioactive compounds [10].

The shell of the cashew nut (*Anacardium occidentale* L.) contains an alkylphenolic oil named CNSL amounting to nearly 25% of the total weight of the nut. This oil is composed of anacardic acid (3-*n*-pentadecylsalicylic acid), and lesser amounts of cardanol (3-*n*-pentadecylphenol), cardol (5-*n*-pentadecylresorcinol), methylcardol (2-methyl-5-*n*-pentadecylres-orcinol) and a small amount of polymeric material (Figure 6.2). The long aliphatic side-chains of CNSL are saturated, mono-olefinic (8), di-olefinic (8, 11), and tri-olefinic (8, 11, 14) with an average value of two double bonds per molecule. A large number of chemicals and products have been developed starting from CNSL by taking advantage of the three reactive sites, namely, phenolic hydroxyl, aromatic ring, and unsaturation (s) in the alkenyl side chain [11].

A group of researchers identified the components of anacardic acid as 1-hydroxy-2-carboxy-3-pentadecyl benzene, 1-hydroxy-2-carboxy-3-(8'-pentadecenyl) benzene, 1-hydroxy-2-carboxy-3-(8,'11'-pentadecadienyl)

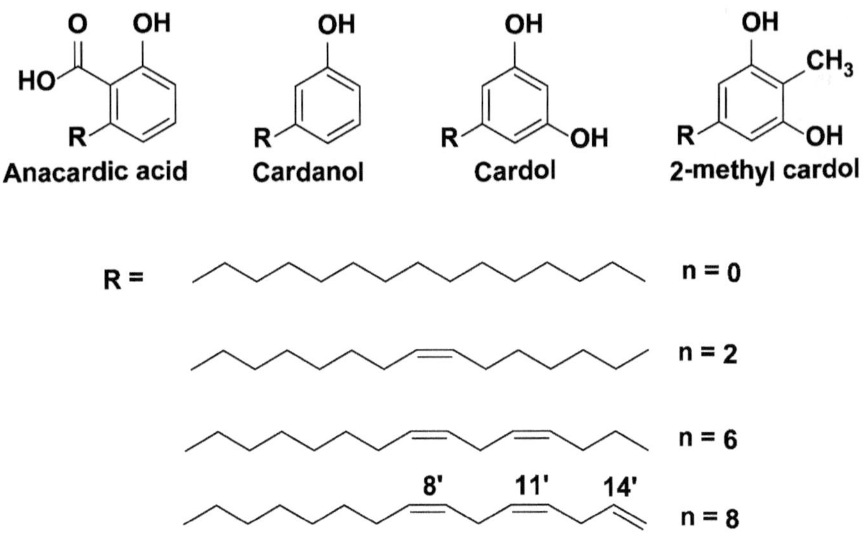

FIGURE 6.2 Structure of active constituents of CNSL.

benzene, and 1-hydroxy-2-carboxy-3-(8,'11,'14'-pentadecatrienyl)benzene [12].

In a report, the components of 2-methyl cardol as 2-methyl-5-pentadecyl resorcinol, 2-methyl-5-(8'-pentadecenyl)resorcinol, 2-methyl-5-(8,'11'-pentadecadienyl) resorcinol, and 2-methyl-5-(8,'11,'14'-pentadecatrienyl) resorcinol have been found [13].

Researchers identified the components of cardanol as 3-pentadeca-anisole, 3-(8'-pentadecenyl)anisole, 1-methoxy-3-(8,'11'-pentadecadienyl) benzene, and 1-methoxy-3-(8,'11,'14'-pentadecatrienyl)benzene. The same group of workers also identified the components of cardol as 5-pentadecyl resorcinol, 5-(8'-pentadecyl)resorcinol, 5-(8,'11'-pentadecadienyl)resorcinol, and 5-(8,'11,'14'-pentadecatrienyl)resorcinol [14].

6.5 INDUSTRIAL UTILITY

CNSL is a very important industrial raw material which has got multifarious uses. Considerable research work has been done on the utilization of CNSL in India and abroad. The main outlets for the utilization of CNSL and its derivatives are the brake lining industry, paint, and varnish industry, laminated products, foundry core oil, and rubber compounding. CNSL is a versatile raw material which has many industrial applications with 200 patents [15]. There is considerable scope for its utilization in the development of drugs, antioxidants, fungicides, bactericides, insecticides, surface-active agents, etc. The main factors that are affecting the utilization of CNSL are its dark color, variable quality, and above all its high price. It is a dark brown and partially polymerized by-product, especially when derived from the most diffused roasted mechanical processing of the cashew nuts. It may represent both a dangerous source of pollutant, and a low-cost, widely available, and renewable raw material for obtaining pure cardanol useful in fine chemical processes [16]. Among the renewable resource materials, CNSL is considered as an important starting material due to its unique structural features, abundant availability, and low cost. In the search for functional materials and chemicals from renewable resources, CNS are one of the agro wastes from cashew nut processing factories, have proven to be among the most versatile bio-based renewable materials. The main uses include waterproofing, boiler fuel, laminating resin, biofuel, paint, adhesive, electrical conductor, flame retarders, foundry chemical, automotive brake lining, insulating varnish, marine antifouling coating, epoxy resin, polyurethane (PU) polymer, topical medicine, etc. [17].

6.6 EXPORT/IMPORT DETAILS

India is the largest producer of CNSL (estimated at 45,000 mt). Brazil is the largest exporter, with average exports over the years of around 17,000 mt, although within the last few years, this has risen to 35,000 mt per year. Other origins such as Vietnam and Tanzania export volumes that are far behind the two main suppliers, India, and Brazil. In India, most factories use drum roasting but some are now moving to the steam process as it produces a better quality kernel [18]. The Mangalore factories that use steam produce 21,000 mt of India's 60,000 mt of CNSL, although their proportion of total kernel production is much lower. Karnataka State factories, which process large volumes of kennels, produce about 20,000 mt, and the balance of CNSL production is spread over India's other cashew producing areas. The export of CNSL from India received a significant boost last year following lobbying by the trade. CNSL was included in an export incentive scheme which allows offsetting import duties against export scrip [19].

Given that India processes approximately 1.1 million tones of in-shell cashews per year, potential CNSL production is about 160,000 mt, assuming that half the factories use steaming. However, as many of the processing units are small, seasonal, and based on low levels of capital investment, it is highly unlikely that this figure, or anything close to it, will be realized in the medium term. For India's supply of commercial CNSL to grow, major consolidation of the industry is likely needed, and this would require heavy and widespread capital investment. In addition, rising energy costs reduce the incentive to commercialize CNSL, rather than use it as a fuel. India has a significant secondary industry refining CNSL into cardanol and other usable products. Indian exports are approximately 75% CNSL and 25% cardanol [20].

In Brazil, CNSL is produced by all ten processors. Recovery is about 12% of in-shell weight. This gives overall production for 2009 of 38,400 mt, and for 2010, about 32,000 mt are projected. Production is clearly linked to the size of the cashew crop which has varied from 250,000 mt to 320,000 mt in recent years. Exports of CNSL have once or twice reached as high as 35,000 mt (2005), but in recent years have been just less than 20,000 mt per year. Producers earn more from exporting CNSL, but it also can be sold to the domestic biofuel sector, which is well developed in Brazil [21]. There are no signs of difficulty in commercializing the volumes of CNSL produced each year. Commercialization in other countries is low. Vietnam is capable of producing large quantities of CNSL, given its position as a major sheller. But while interest in CNSL exports, and exports themselves are growing, in 2007 only 21 mt were exported, and in 2008, anecdotal evidence suggests exports were only 600 mt. In total about 65,000 mt of CNSL are produced annually, with commercial sales closely linked to petroleum and petrochemical prices. Tanzania also produces CNSL, but export quantities are small. Trade sources mention Nigeria as an exporter, but there is no hard data to support this. It is traded under tariff heading 13021920 [22].

6.7 ENVIRONMENTAL IMPACT

CNSL may be a weak promoter of carcinogenesis but no mutagenic or carcinogenic activity has been reported. Epidemiological studies suggested that CNSL may contribute to oral sub-mucous fibrosis. In addition, its phenolic components exerted several biological activities, including antioxidative properties, inhibition of acetylcholinesterase, and membrane perturbation. There is no direct evidence regarding the toxicity of CNSL or its major phenolic components [23]. However, the effluent generated during the processing of the cashew nut could be considered potentially harmful to the environment due to its high phenol content. The hazardous effects of phenolic compounds have been extensively studied. A concentration above 1 mg/l can affect aquatic life, while recognizes deleterious effects at concentrations as low as 1 μg/ml. As previously mentioned, data on the chemical composition of the cashew nut industry effluent are scarce, but preliminary analysis showed high phenol content due to the cardol, cardanol, and anacardic acid of CNSL. The high toxicity observed for the isolated nols (cardol and cardanol) potentially contribute to the toxicity of the cashew nut industry effluent [24].

6.8 BIOMEDICAL APPLICATIONS

Anacardic acid present in CNSL is the most prominent component responsible for pharmacological activities. It is commonly found in plants of the Anacardiaceae family and is a dietary component found in cashew apple (*Anacardium occidentale*) and ginkgo (Ginkgo biloba) leaves and fruits. The traditional Ayurveda depicts nutshell oil as a medicinal remedy for alexeritic, amebicidal, gingivitis, malaria, and syphilitic ulcers. Anacardic acid is found in a number of medicinal plants that have potential activity against cancer cell lines. However, the enduring research and emerging evidence suggests that anacardic acid could be a potent target molecule with bactericide, fungicide, insecticide, anti-termite, and molluscicide properties and as a therapeutic agent in the treatment of the most serious pathophysiological disorders like cancer, oxidative damage, inflammation, and obesity. Moreover, anacardic acid was found to be a common inhibitor of several clinically targeted enzymes such as NFκB kinase, histone acetyltransferase (HATs), lipoxygenase (LOX-1), xanthine oxidase, tyrosinase, and ureases. An interesting observation in this context is its ability to modulate NF-κB by acting on its upstream pathways. Because NF-κB is known to be a key player in the progression of human cancers and chronic inflammation, its suppression by anacardic acid indicates a putative potential molecular target of this compound. However, this requires a comprehensive inspection for establishing the scientific rationale for the use of anacardic acid as an anti-cancer and anti-inflammatory agent prior to its use as a novel therapeutic agent for the treatment of human malignancies. Potential of anacardic acids and their semi-synthetic derivatives for antibacterial, antitumor, and antioxidant activities have been reported. The use of anacardic acid as a starting material for the synthesis of diverse biologically active compounds and complexes as well as the natural anacardic acid from CNSL and their semi-synthetic derivatives as lead compounds were reported [25].

6.9 CONCLUSION

The report has highlighted various chemicals obtainable from CNSL both directly and indirectly. This inexhaustible listing presents CNSL as a very important, reliable source of raw material for petrochemical industry. It is a good, promising supplement and/or alternate to petroleum, which is currently facing depletion globally. Varied CNSL compositions with varying modes of extraction are a vantage opportunity with potential for multiple applications. Though CNSL contains phenolics, its products are highly environmental friendly. However, efficient treatment strategy may be inevitable to reduce environmental impact associated with the production industry.

KEYWORDS

- **biomedical applications**
- **cashew nut shell**
- **cashew nut shell liquid**

- **histone acetyltransferase**
- **lipoxygenase**
- **semi-synthetic derivatives**

REFERENCES

1. Salehdeen, M. U., Abdulrazaq, Y., & Osinlu, C. A., (2019). Microbiological and chemical indicator of multicomponent nature of cashew nut shell hot water extract. *J Appl. Sci. Environ. Manag.*, *23*(5), 883–887.
2. Lubi, M. C., & Thachil, E. T., (2000). Cashew nut shell liquid (CNSL)-a versatile monomer for polymer synthesis. *Des. Monom. Polym.*, *3*(2), 123–153.
3. Menon, A. R. R., Pillai, C. K. S., Sudha, J. D., & Mathew, A. G., (1985). Cashew nut shell liquid-its polymeric and other industrial products. *J. Sci. Indust. Res.*, *11*(3), 189–197.
4. Gedam, P. H., & Sampathkumaran, P. S., (1986). Cashew nut shell liquid: Extraction, chemistry and applications. *Prog. Org. Coat.*, *14*(2), 115–157.
5. Tyman, J. H. P., (1979). Non-isoprenoid long chain phenols. *Chem. Soc. Rev.*, *8*(4), 499–537.
6. Tyman, J. H. P., Wilczynski, D., & Kashani, M. A., (1978). Compositional studies on technical cashew nutshell liquid (CNSL) by chromatography and mass spectroscopy. *Am. Oil Chem. Soc.*, *55*(9), 663–668.
7. Mahanwar, P. A., & Kale, D. D., (1996). Effect of processing parameters on refining of CNSL. *Indian J. Chem. Technol.*, *3*, 191–193.
8. VPaul, V. J., & Yeddanapalli, L. M., (1954). Olefinic nature of anacardic acid from Indian cashew-nut shell liquid. *Nature*, *174*(4430), 604–611.
9. Symes, W. F., & Dawson, C. R., (1953). Separation and structural determination of the olefinic components of poison ivy urushiol, cardanol and cardol. *Nature*, *171*(4358), 841.
10. Cornelius, J. A., (1966). Cashew nut shell liquid and related materials. *Tropical Sci.*, *8*, 79–84.
11. Joseph, T. M., Nair, S. M., Ittara, S. K., Haponiuk, J. T., & Thomas, S., (2020). Copolymerization of styrene and pentadecylphenylmethacrylate (PDPMA): Synthesis, characterization, thermomechanical and adhesion properties. *Polymers (Basel). 12*, doi:10.3390/polym12010097.
12. Cassady, J., (1980). Recent advances in the isolation and structural elucidation of antineoplastic agents from higher plants. In: Mechler, E., & Reinhard, E., (eds.), *Int. Research Cong. Natural Prod. Med. Agents.*
13. Sowmyalakshmi, S., Nur-E-Alam, M., Akbarsha, M. A., Thirugnanam, S., Rohr, J., & Chendil, D., (2005). Investigation on *Semecarpus lehyam*—a Siddha medicine for breast cancer. *Planta*, *220*(6), 910–918.
14. Rea, A. I., Schmidt, J. M., Setzer, W. N., Sibanda, S., Taylor, C., & Gwebu, E. T., (2003). Cytotoxic activity of *Ozoroa insignis* from Zimbabwe. *Fitoterapia*, *74*(7/8), 732–735.
15. Hamad, F., & Mubofu, E., (2015). Potential biological applications of bio-based anacardic acids and their derivatives. *Int. J. Mol. Sci.*, *16*(4), 8569–8590.

16. George, J., & Kuttan, R., (1997). Mutagenic, carcinogenic, and cocarcinogenic activity of cashew nut shell liquid. *Cancer Lett.*, *112*(1), 11–16.
17. Varghese, I., Rajendran, R., Sugathan, C. K., & Vijayakumar, T., (1986). Prevalence of oral sub-mucous fibrosis among the cashew workers of Kerala-south India. *Indian J. Cancer*, *23*(2), 101–104.
18. Trevisan, M. T. S., Pfundstein, B., Haubner, R., Würtele, G., Spiegelhalder, B., Bartsch, H., & Owen, R. W., (2006). Characterization of alkyl phenols in cashew (Anacardiumoccidentale) products and assay of their antioxidant capacity. *Food Chem. Toxicol.*, *44*(2), 188–197.
19. Facanha, M. A. R., Mazzetto, S. E., Carioca, J. O. B., & De Barros, G. G., (2007). Evaluation of antioxidant properties of a phosphorated cardanol compound on mineral oils (NH10 and NH20). *Fuel*, *86*(15), 2416–2421.
20. Stasiuk, M., & Kozubek, A., (2008). Membrane perturbing properties of natural phenolic and resorcinolic lipids. *FEBS Lett.*, *582*(25/26), 3607–3613.
21. Stasiuk, M., Bartosiewicz, D., & Kozubek, A., (2008). Inhibitory effect of some natural and semisynthetic phenolic lipids upon acetylcholinesterase activity. *Food Chem.*, *108*(3), 996–1001.
22. Veeresh, G. S., Kumar, P., & Mehrotra, I., (2005). Treatment of phenol and cresols in upflow anaerobic sludge blanket (UASB) process: A review. *Water Res.*, *39*(1), 154–170.
23. Newman, M. C., & Unger, M. A., (2003). *Fundamentals of Ecotoxicology* (p. 458). Lewis Publishers. Boca Raton, Florida.
24. Martins, R., Beatriz, A., Santaella, S. T., & Lotufo, L. V. C., (2009). Ecotoxicological analysis of cashew nut industry effluents, specifically two of its major phenolic components, cardol and cardanol. *Pan-American J. Aquatic Sci.*, *4*(3), 363–368.
25. Hemshekhar, M., Sebastin, S. M., Kemparaju, K., & Girish, K. S., (2012). Emerging roles of anacardic acid and its derivatives: A pharmacological overview. *Basic Clin. Pharmacol. Toxicol.*, *110*(2), 122–132.

CHAPTER 7

Role of Tandem Mass Spectrometry in Diagnosis and Management of Inborn Errors of Metabolism

KANNAN VAIDYANATHAN[1] and SANDHYA GOPALAKRISHNAN[2]

[1]*Professor, Department of Biochemistry and Head Molecular Biology, Amrita Institute of Medical Science and Research Center, Kochi, Kerala, India, E-mail: drkannanvaidyanathan@gmail.com*

[2]*Associate Professor, Department of Prosthodontics, Government Dental College, Kottayam, Kerala, India, E-mail: sandhya_gopal@rediffmail.com*

ABSTRACT

Various proteomics techniques have found wide applications in the field of inborn errors of metabolism (IEM). Detection of IEM by newborn screening (NBS) using tandem mass spectrometry (TMS) is extensively used across the globe. With the development of tandem MS, the sensitivity and specificity of the detection of IEM have increased tremendously. Recent years have also seen the advent of second-tier testing. This has further expanded the scope of IEM diagnosis and treatment. Second-tier testing has also benefitted from advances in proteomics. Other proteomic techniques have also led to various advances in the field of IEM. The most important development has been the identification of novel molecules which can be used as markers for detection of IEM. These markers have led to early diagnosis and in some instances better treatment options in IEM. Developments in these fields shall pave the way for much more rapid diagnosis of IEM and shall play a significant role in improving and enhancing the quality of life in such patients.

7.1 INTRODUCTION

Most of the metabolic disorders can lead to mental retardation and sometimes even death if untreated. The effects of toxic substances and their by-products increase, with time if the offending diet is not restricted or other suitable mechanisms for removing the offending and accumulating toxins are not taken. Hence, it is not unexpected that some disorders may be mild at onset, but with time deteriorate. Many metabolic disorders however have a very acute onset of clinical symptoms.

There are some disorders which are not very dangerous to the child, (Essential pentosuria, alkaptonuria, etc.). Some of these, like alkaptonuria, scare the parents because of the symptoms, blackish discoloration of diapers, but other than skeletal abnormalities which appear in the 4th–5th decade of life, they do not produce harm even in adult life. There is accumulation of alkaptone bodies in the patients' bones which like to blackish discoloration of vertebrae and other bones in this condition. However, the vast majority of metabolic disorders are dangerous and to publicize them as benign and easily treatable is a dangerous tendency which should be resisted. Most of the metabolic disorders can be treated, with special diets or avoidance of particular food stuff and other specific measures. However, treatment has to be started very early in life, and has to be continued for a very long time, and in most cases for a lifetime.

Even though there is no real consensus as to how to metabolic disorders have to be classified, one easy way of classifying them, which has been adapted from Scriver et al. is given here:

1. Amino acid disorders;
2. Organic acidemias;
3. Urea cycle disorders;
4. Carbohydrate disorders;
5. Mitochondrial fatty acid disorders;
6. Mitochondrial disorders;
7. Peroxisomal disorders;
8. Lysosomal storage disorders;
9. Purine and Pyrimidine disorders;
10. Porphyrias;
11. Metal metabolism disorders.

This classification is by no means a complete one, but a very convenient one.

In our set-up, the lack of public knowledge has been increased by a lack of government involvement and propaganda. In the Indian scenario, it has been seen that programs like polio and TB vaccination, AIDS, and many others including Hepatitis B vaccine more recently have reached the public mainly through aggressive central government involvement. IEM also need a similar kind of approach [1].

Deficiency of an enzyme in a pathway can produce the following effects:

1. There is accumulation of the substrate of that enzyme;
2. The pathway maybe directed to an alternate pathway, leading to accumulation of abnormal metabolites, or uncommon metabolites;
3. Normal metabolic products of the main pathway also are reduced.

For example, phenylketonuria (PKU) is caused by deficiency of phenyl alanine hydroxylase (PAH). This enzyme blocks the metabolism of the amino acid, phenyl alanine. As a result of this blockage, the amino acid tyrosine is not generated from phenyl alanine. Tyrosine is normally needed for the synthesis of the skin pigment, melanin, and it is not formed in this condition. Several toxic products, collectively known as phenyl ketones, are formed, by alternate pathways. These produce neurotoxicity. The clinical effects of PKU are due to the combined effect of tyrosine deficiency and phenylketone accumulation.

These different effects are partially responsible for the diverse clinical manifestations of the IEM. Most of the IEMs can lead to mental retardation and sometimes even death if untreated. The toxic intermediates accumulate, if they are not removed, and the clinical features exacerbate with time. For treatment to be instituted at an early stage, a proper and early diagnosis is mandatory. In India, many people now understand the importance of early diagnosis. New laboratories are now being established and newer programs are being started for screening babies for IEM. Funding agencies also understand the need for mass screening programs and more and more clinicians are coming forward to participate in such programs. What we are lacking at the moment is a nation-wide statistic on the prevalence of IEM, which need to be effectively carried out. A proper screening program will be able to recognize a child in the initial stages itself.

Establishing proper quality laboratory work-up for IEM is a big challenge in developing countries. Often the child is born with a poor socio-economic status; and may not be able to afford some of the sophisticated tests needed for diagnosis. Failure to understand the seriousness of the problem coupled with a retired attitude towards the disease can further reduce the chances of long-term beneficial effects. Knowledge of genetics and patterns of inheritance may be available for an educated family or even in a middle-class family, however not so much for parents from poor socio-economic status [1].

In recent times, tandem mass spectrometry (TMS, MS/MS) and molecular diagnosis by mutation detection have become popular in the West for the diagnosis of metabolic disorders, besides their routine metabolic screening programs. In India also, some centers do offer similar facilities (Examples are NIMHANS, Bangalore, AIIMS, New Delhi, PGIMER, Chandigarh, as well as private laboratories like NeoGen Laboratory, Bangalore).

Proteomics is the study of the complete protein profile expressed in an organism. The word proteomics was coined in 1995 by Marc Wilkins [2, 3]. Currently, an increased emphasis is placed on the application of proteomics techniques like MS in the diagnosis of various diseases [4]. Inborn errors of metabolism constitute an important group of genetic pediatric disorders. Even though they are individually rare, collectively they constitute a group of diseases, with a projected incidence of 1 in 1500 [5].

TMS is the most important tool used for IEM screening [6]. MS/MS is used for newborn screening (NBS) in all developed countries and many developing countries around the world. NBS helps to identify IEM at an early stage, so as to plan effective treatment thereby reducing morbidity and mortality due to IEM. Therefore, it is very important that every country has guidelines for NBS. With the proper application of MS, we can expect much improvement in the detection rates of IEM. Currently, more than 70 disorders can be detected in the newborn period itself by NBS by means of TMS with a simple heel-prick blood sample [7].

7.2 TANDEM MASS SPECTROMETRY (TMS) IN NEWBORN SCREENING (NBS)

MS has become one of the most potent tools for the investigation and monitoring the treatment of inborn errors of metabolism. The past century has witnessed tremendous developments in MS. TMS (Figure 7.1) is employed for neonatal screening. The technique of TMS is well known and has been extensively reviewed earlier; hence, this shall not be discussed in much detail here. However, a few salient points need to be mentioned.

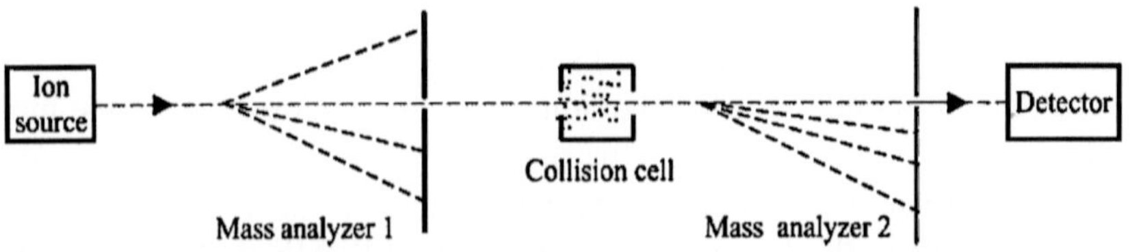

FIGURE 7.1 Schematic diagram for tandem mass spectrometer. Two mass spectrometers are placed in tandem with a collision cell in between. First mass spectrometer used to select ions of a particular m/z ratio (parent ion, precursor ion). In the collision cell, ions collide with background gas molecules and are broken into smaller ions (daughter ions, product ions). Second mass spectrometer acquires mass spectrum of daughter ions.

The sensitivity and specificity of TMS is estimated to be 99% and 99.995%, respectively for aminoacidurias, organic acidurias, and fatty acid oxidation disorders [8]. In an interesting study from Belgium, C5 carnitine was elevated and this was initially thought to be false positive for the diagnosis of isovalericaciduria. However, a clinical investigation reveals that an external preparation was found to contribute to C5 carnitine load and removal of the same resulted in normalizing of C5 carnitine level [9].

MS has led to the identification of many new diseases. New methods for lipid analysis have opened up the fields ofinbornerrorsof cholesterol synthesis, bile acid synthesis and leukotriene synthesis. Developments in TMS allowed it to be used for determination of the amino acid sequence and post-translational modifications of proteins [10]. Short/branched chain acyl-CoA dehydrogenase deficiency (SBCADD) has been differentiated from isovalericaciduria [11]. Janzen et al. used UPLC with C18 column and gradient elution with MS/MS in ESI+ mode for the identification of C5 acyl carnitine species. This technique helps to reduce false positives and also helps differentiation between isovalericaciduria and 2 methyl butyryl CoA dehydrogenase deficiency [12]. Scott et al. have used MS/MS for the simultaneous detection of lysosomal storage disorders and mucopolysaccharidosis (Fabry, Pompe, and MPS-I) [13]. UPLC along with ESI-MS/MS can be used for the diagnosis of different types of mucopolysaccharidoses [14]. MS is important for structural glycomics and study of diseases like-congenital disorders of glycosylation, lysosomal storage diseases, autoimmune diseases, and cancer [15]. LaMarca et al. developed an LC-MS/MS method for the simultaneous determination of succinylacetone, tyrosine, phenylalanine, methionine, and NTBC for the diagnosis and treatment of tyrosinemia Type I [16]. 4-hydroxy-butyrate elevation and succinic semialdehyde dehydrogenase deficiency was diagnosed in three infants from China using organic acid analysis by GC/MS. 3-hydroxyisovaleric acid was used to detect biotin deficiency by UPLC-MS/MS [17].

The detection rates of IEM have improved tremendously since the introduction of MS/MS for NBS. In Austria, in the year 1966, NBS was introduced. More than 2600 children were diagnosed with IEM among 4 million babies screened in the five decades ever since. MS/MS was introduced in 2002 which substantially increased detection rates [18]. Similarly, MS/MS improved detection rates of IEM by 50% in Mexican patients [19].

The NBS using MS/MS has not only led to early diagnosis, but also better outcome for children with IEM, with respect tomorbidity and mortality. In a cohort study, Wilcken et al. analyzed the outcome at 6 years for patients detected by screening or by clinical diagnosis among >2 million infants born from 1994 to 1998 (1,017,800, all unscreened) and 1998 to 2002 (461,500 screened, 533,400 unscreened) recording intellectual and physical condition, school placement, other medical problems, growth, treatment, diet, and hospital admissions. They report fewer deaths and fewer clinically significant disabilities compared to unscreened population [20].

Even though expanded NBS programs are available in developed countries, no proper screening programs are available in many developing countries [21]. NBS by MS/MS is available for some diseases; but not for all. Several reasons are there, including lack of testing for rare diseases as well an undue caution by regulators for new tests [22].

Ross has suggested that consent should be included while screening children for IEM [23]. Specific criteria need to be met before the introduction of any new test to the NBS panel [24]. Wilcken cautions against the increasing use of "expanded" NBS and the possibility of false-positive results and related issues [25].

However, in 2009 a top-level committee analyzed the performance of NBS by MS/MS in the US and recommended the continued use of the technique. They gave guidelines for the diagnosis of each IEM by MS/MS [26]. MS/MS screening for rare diseases was specified to be a cost-effective method in Australia [27]. Carroll and Downs also report that NBS is cost-effective [28].

Padilla and Therrell report on the NBS program in the Asia-Pacific region. The countries were NBS is practiced are Australia, New Zealand, Japan, and Singapore (from the 1960s), Taiwan, Hong Kong, China (Shanghai), India, and Malaysia (since 1980s for selected diseases like congenital hypothyroidism), Korea, Thailand, and the Philippines (since the 1990s) and Indonesia, Mongolia, Sri Lanka, Myanmar, and Pakistan (in the 2000s). No information is available from Nepal, Cambodia, Laos, and the other Pacific Island nations [29].

7.3 SECOND-TIER TESTING

Second-tier testing can be otherwise referred to as confirmatory testing, when initial NBS has produced positive results. Since the specificity and sensitivity of MS/MS is very high, the chances of false-positive and false-negative results are very low. Still, in some cases, the results needed to be verified by re-testing. Second-tier tests have reduced the incidence of false positives in addition to increasing the type of disorders which could be detected by NBS. Broader range of metabolites like RNA, DNA, proteins, and other metabolites can also be used for NBS. However, second-tier tests are comparatively more complex [30]. Shigematsu et al. studies the advantages of second-tier screening for isovalericaciduria and methyl malonicaciduria [31]. Chalcraft and Britz-McKibbin report on the use of second tier testing by capillary electrophoresis and ESI-MS/MS [32]. Turgeon et al. report the combined analysis of succinyl acetone (SUAC), amino acids, and acylcarnitines [33].

UPLC-MS/MS method was described for simultaneous analysis of 48 acylcarnitines for the identification of different IEM. This has been developed as a second-tier test for extended NBS [34]. Another method for rapid acylcarnitine profiling has been developed using paper spray MS. The method is described to have high sensitivity and specificity [35]. Multiple enzyme assays (second-tier multiplex enzyme assay) for galactosemia can be carried out on dried blood spots using UPLC-MS/MS [36].

Finally, MS/MS, though a very useful screening test for most IEM may not be adequate for all disorders, for example, citrin deficiency (for urea cycle disorders) and carnitine uptake defect. Second-tier molecular testing may be needed in these cases [37].

7.4 OTHER PROTEOMIC TOOLS AND INBORN ERRORS OF METABOLISM

TMS is now accepted worldwide as a screening method for NBS and also for the diagnosis and management of inborn errors of metabolism. However, there are also other proteomic techniques which have found application in IEM.

Ostermann et al. used a combination of MALDI and high-resolution accurate mass (HR/AM) MS using a linear ion trap-Orbitrap for the quantification of acyl carnitines and organic acids in dried blood spots. They report that this method can be employed as a complementary tool to electrospray ionization MS for the detection of organic acids and acyl carnitines. The advantages include simple sample preparation and simultaneous identification of hundreds of organic acids in very small sample size [38].

Imaging mass spectrometry (IMS) has been used as a tool for assessing lipid abnormalities. This technique may be used for analyzing the distribution of various biomolecules and identifying biomolecules in cells or tissues without labeling [39]. Sleat et al. used affinity chromatography followed by TMS to identify protein profile in lysosomal storage disorders, and especially were able to validate two proteins, CLN5, and sulfamidase, in the adult form of neuronal ceroidlipofuscinosis (NCL) [40].

Matysik et al. have described a GC/MS method for the simultaneous determination of a variety of sterols (24-, 25-, 27-hydroxycholesterol; 7-ketocholesterol, lanosterol, lathosterol, 7-dehydrocholesterol, desmosterol, stigmasterol, sitosterol, and campesterol) in human plasma [41]. GC/MS was used to detect methylmalonicaciduria in a Chinese population. Over a 5 year period, 398 cases were detected, out of which 160 patients presented during neonatal period [42]. Cangemi et al. developed a GC/MS method for the determination of galactose-1-phosphate in RBC and suggest that this might be suitable in the treatment of classical galactosemia [43].

Isotope dilution LC-MS/MS was used for the identification of SUAC which is the diagnostic test for hepatorenal tyrosinemia (HT) [44]. Lin et al. report that LC-MS/MS is useful for the detection of maple syrup urine disease (MSUD) as it identifies branched chain amino acids (BCAA) and allo-isoleucine (Allo-Ile) [45]. An LC-MS/MS method to detect and quantify lyso-Gb(3), a marker for Fabry disease, has been described [46]. Giordano et al. developed an HPLC-ESI-MS/MS method for determination of 40 underivatized amino acids and related compounds quickly and with high sensitivity and specificity [47]. Moore and Cowan developed a technique for the determination of glutarylcarnitine by MS/MS which could be used for the diagnosis of classical and low excretor types of glutaricacidemia type I [48].

Herrmann and Herrmann have reviewed the importance of mitochondrial proteomics in the detection of various genetic disorders [49]. "Shotgun" and "top-down" global approaches along with LC/MS and GC/MS are employed as tools for "lipidomics" [50].

7.5 ROLE OF PROTEOMICS FOR DEVELOPMENT OF NOVEL MARKERS IN IEM

In many cases, the use of proteomics has led to the successful development of potential markers. This is an important development because novel markers have the potential to develop therapeutic strategies in addition to offering promising diagnostic work-up. Many of these markers described have been the result of individual research studies, and hence are not completely validated. Hence, care should be given before introducing them to clinical practice. However, these are promising markers, and studies such as those mentioned below would help in an early diagnosis of many diseases.

7.5.1 LYSOSOMAL AND OTHER STORAGE DISORDERS

Ong et al. used MS to study Gaucher's disease. Specific mutations are known to produce defective lysosomal enzyme folding in the endoplasmic reticulum and lead to protein degradation and loss of function. They show that nine proteins were down-regulated and two were up-regulated. Increased levels of a protein, FKBP10 accelerated degradation of mutant glucocerebroside and decreased level led to more of the enzyme entering calnexin pathway, with a positive effect on folding. They suggest that this protein could hence be used as a therapeutic target [51].

Proteomic techniques help in the identification of a variant of juvenile neuronal ceroidlipofuscinoses due to deficiency of CLN5 protein. Earlier CLN9 deficiency was known to be causative of this disease. CLN5 deficiency produces adhesion defects, increased growth, apoptosis, and decreased level of ceramide, sphingomyelin, and glycosphingolipids. CLN8p closely correlates with CLN5p, and they are activators of dihydroceramide synthase. CLN8(-/-) cells also demonstrated reduction in C16/C18:0/C24:0/C24:1 ceramide species as measured by MS. CLN5 and CLN8 are proven to be closely related protein species [52].

Leigh syndrome is caused due to mutations in the protein coding gene, NDUFS4; (NADH dehydrogenase ubiquinone Fe-S protein 4). This protein is part of the mitochondrial Complex I, comprising 45 sub-units encoded by mitochondrial and nuclear genes. A recessive mouse phenotype was developed by inclusion of a transposable element into Ndufs4, and the resultant metabolite analysis of the model revealed increased hydroxyacylcarnitine species leading to imbalanced NADH/NAD(+) ratio which inhibited mitochondrial beta oxidation [53]. Proteomic analysis showed that the genes encoding acetyl-coA carboxylase beta, M-cadherin, calpain III, creatine kinase, glycogen synthase (GS), and sarcoplasmic reticulum calcium ATPase 1 (SERCA1) were down-regulated in patients with McArdle disease [54]. Statistically significant decreases were observed for five proteins following enzyme replacement therapy in patients with Fabry disease, namely, alpha(2)-HS glycoprotein, vitamin D-binding protein, transferrin, Ig-alpha-2 C chain, and alpha-2-antiplasmin [55].

Fogli et al. used MALDI-TOF/MS to analyze CSF N-glycan profile and they suggest that this can be used to study leukodystrophies arising due to eIF2B mutations. They suggest that the method could also be extended to study other neurological disorders involving developmental gliogenesis/synaptogenesis abnormalities [56].

7.5.2 WILSON'S DISEASE

The levels of expression of mitochondrial matrix proteins including isovaleryl coenzyme A dehydrogenase, agmatinase, and cytochrome b5 were downregulated in early stages of Wilson's disease. As mitochondrial injuries progressed, expression levels of malate dehydrogenase 1, annexin A5, S-adenosylhomocysteine hydrolase, transferrin, and sulfite oxidase 1 were differentially regulated. S-adenosylhomocysteine hydrolase was under-expressed and is hypothesized to play a role in neurological pathology of Wilson's disease. The study was done on a mouse model of Wilson's disease (LEC rats) [57]. Shotgun proteomic analysis of Atp7b(-/-) mouse model of Wilson's disease revealed increased expression of DNA repair machinery and nucleus-localized glutathione peroxidase (SelH), and reduced expression ofligand-activated nuclear receptors FXR/NR1H4 and GR/NR3C1 and nuclear receptor-interacting partners [58]. Remodeling of RNA processing machinery may be involved in the pathogenesis of Wilson's disease [59].

7.5.3 DISORDERS OF CHOLESTEROL AND STEROID HORMONE METABOLISM

Lathosterolosis and Smith-Lemli-Opitz syndrome (SLOS) are congenital disorders of cholesterol synthesis which occur due to mutations in lathosterol 5-desaturase (SC5D) and7-dehydrocholesterol reductase (DHCR7) respectively. Proteomic analysis showed alterations in multiple pathways like mevalonatemetabolism, oxidative stress, apoptosis, protein biosynthesis, glycolysis, and intracellular trafficking [60].

Proteomic study has been undertaken in hyper-androgenism syndromes like polycystic ovarian syndrome (PCOS) and congenital adrenal hyperplasia (CAH). Proteins involved in many physiological processes as the functional state of immune system, the regulation of the cytoskeleton structure, oxidative stress, coagulation process, and insulin resistance were found to be involved [61].14 proteins were identified in CAH and PCOS, 15 in PCOS and 35 exclusively in CAH [62].

NPC1, a trans-membrane protein involved in lysosomal cholesterol transport and NPC2, an intra-lysosomal cholesterol transport protein are recognized to be involved in Niemann-Pick C (NPC) disease. Proteomic study of lysosomal proteins led to the understanding of a number of proteins involved in lipid metabolism like prosaposin and beta-hexosaminidase subunits, as well as proteases and glycosidases. Further studies may disclose the role of these proteins in pathogenesis of NPC [63].

7.5.4 OTHER DISEASES

Kelch-like 3 (KLHL3) and cullin 3 (CUL3), components of E3 ubiquitin ligase complex were found to be linked in the pathogenesis of pseudohypoaldosteronism type II (PHAII). Shibata et al. used MS and co-immunopreceipitation techniques to show that these genes have role in electrolyte homeostasis and hence have a role in actual pathogenesis of PHAII [64]. Sialin is a multi-functional transporter that mediates aspartate and glutamate neuro-transmission and may be involved in a variety of disorders like Salla disease [65]. Significant variations in gluconeogenesis, glycolysis, and fatty acid pathways are seen in liver and kidneys of patients with primary hyperoxaluria type I [66]. Burillo et al. identified two proteins in carotid atherosclerosis, gelsolin like capping protein (CapG) and glutathione-S-transferase omega 1 (GSTO1) [67].

Proteomic study on patients with congenital disorder of glycosylation type 1a (CDG1a) revealed 14 proteins which are expressed in patients and not seen in controls. The most notable proteins are those involved in immune response, coagulation mechanism, and tissue protection against oxidative stress and include alpha(2)-macroglobulin, afamin, fibrin, and fibrinogen [68]. An LC-MS/MS method was developed for the simultaneous measurement of delta aminolevulinic acid and porphobilinogen in patients with acute porphyria [69]. Thirty differentially expressed proteins were identified in Hutchinson-Gilford progeria syndrome (HGPS); these genes were involved in methylation, ionic calcium-binding, cytoskeleton, duplication, and regulation of apoptosis. These genes may therefore be involved in the aging process [70].

Proteomic studies (2DE-MS/MS) on multiple acyl-CoA dehydrogenase deficiency (MADD) identified differentially expressed proteins associated to binding/folding functions, mitochondrial antioxidant enzymes and proteins associated to apoptotic events. 35 mitochondrial proteins showed significant changes compared to controls reflecting mitochondrial protein plasticity in the disease [71]. 24 salivary auto-antibodies were detected in Sjogren's syndrome, out of which 4 were validated, anti-transglutaminase, anti-histone, anti-SSA, and anti-SSB. The new biomarkers may be used for early identification of Sjogren's syndrome [72].

Patients with ethylmalonic encephalopathy have under-representation of two proteins in the detoxification/oxidative stress pathway, mitochondrial superoxide dismutase (SOD2) and aldehyde dehydrogenase X (ALDH1B). Other proteins which were under-represented are sulfide: quinoneoxidoreductase (SQRDL), apoptosis inducing factor (AIFM1), lactate dehydrogenase (LDHB), chloride intracellular channel (CLIC4) and dimethylarginine dimethylaminohydrolase 1 (DDAH1). The authors postulate that the involvement of the variety of proteins might explain the wide-ranging symptoms seen in this condition [73].

Fifteen carbonylated membrane proteins were solely identified in infected G6PD A-red blood cells which reveal selective oxidation of host proteins following malarial infection. Three pathways in host RBC are oxidatively damaged. These include traffic/assembly of exported parasite proteins in RBC cytoskeleton and surface, the oxidative stress defense proteins, and finally the stress response proteins. Hemichromes associated with membrane proteins were also identified which supports a role for specific oxidative alterations in protection against malaria which occurs by G6PD polymorphisms [74].

Forty protein peaks, differentially expressed in NCL were identified from two-dimensional protein fragmentation (PF2D) maps and twenty-four proteins were identified by MALDI-TOF-MS or LC-ESI-MS/MS [75]. Salivary transthyretin (TTR) was used as a biomarker for patients with familial amyloidotic polyneuropathy (FAP) using MALDI-FTICR [76].

7.6 URINOMICS

Urinomics is the study of proteins and metabolites in urine samples. Abnormal urine proteome profile has been detected in many diseases including IEM.

Obtaining a urine sample from a child or neonate is much easier and it is also non-invasive. Hence, diagnosis and management of IEM is going to be much easier with further developments in urinomics. The coming years shall witness more research and developments in this exciting area. The knowledge of proteomic techniques has led to the identification of a variety of urinary markers for the detection of renal Fanconi syndromes [77]. Urinary proteomics can be employed for the early diagnosis of Fabry disease [78]. Urinary proteomic analysis by MALDI-TOF/MS of Anderson-Fabry disease showed significant increase in alpha-1-antitrypsin, alpha-1-microglobulin, prostaglandin H2 d-isomerase, complement-c1q tumor necrosis factor-related protein, and Ig kappa chain V-III [79]. Caubet et al. have reviewed the potential clinically useful urinary biomarkers in ureteropelvic junction (UPJ) obstruction and renal Fanconi syndrome in pediatric population [80].

Unprocessed oligosaccharides and glycoconjugates were detected in urine in patients with glycoproteinosis, Pompe's disease and sialic acid storage diseases by MALDI-TOF-MS [81]. Urinary analysis of free oligosaccharides by MALDI/TOF/TOF aids the diagnosis of lysosomal storage disorders [82]. Using label-free quantitative proteomics, two potential markers, prosaposin, and GM2 activator protein (GM2AP), were identified in urine of pediatric patients for pre-symptomatic kidney disease in Type I diabetes and Fabry disease [83]. An LC-MS/MS method for the simultaneous determination of homovanillic acid, VMA, orotic acid and homogentisic acid in urine has been developed [84]. Acylglycines are important metabolites for the diagnosis of a number of IEM. A UPLC/MS method was developed for the assay of acylglycines in urine samples [85]. Urinary steroids were analyzed in patients with 21 hydroxylase deficiency by GC-MS/MS. Many novel steroids some of which may be used as biomarkers were identified [86]. Large number of putative IEM markers was identified in urine by ESI-MS/MS technique by Rebollido-Fernandez et al. [87]. An analytical method for the detection of pterins in urine for the identification of both typical and atypical PKU has been described [88].

Metabolomics is more important in the diagnosis of IEM compared to other branches of clinical medicine, since a large number of small metabolites are excreted in IEM [89]. Urinary metabolomics is a useful tool for various disorders neonatal and infancy including IEM [90]. Selicharová et al. usedproteomicsand metabolomics analyses of human hepatocytes in primary cell culture. This technique helped to search the spectrum of proteins and associated metabolites which are affected by the interruption of methyl groupmetabolism. They studied the effect of hyperhomocysteinemia at two concentrations, 0.1 mM and 2.0 mM, and used the inhibitor, BHMT, Betaine Homocysteine N Methyltransferase. The higher concentration produced up-regulation of phosphatidylethanolamine carboxykinase and ornithine aminotransferase, cellular proliferation was affected, secretome composition was altered and signs of apoptosis were seen. In addition, fibrinogen gamma dimers were detected and defective maturation of apolipoprotein A1 was seen [91].

7.7 INDIAN STATUS

We have reported on the prevalence pattern of inborn errors of metabolism earlier [92–99]. Our studies have found that organic acidurias may be the frequent and common IEM in India. Further studies are needed to find out the actual incidence of various IEM in India. Also, proteomic studies are needed for development of novel markers. TMS technique for NBS as practiced in Western countries in not a routine procedure in India.

7.8 CHALLENGES IN TMS IMPLEMENTATION

There are several challenges limiting the use of TMS in clinical situations for NBS. These include cost of the instrument, technical expertise, space requirement, etc. There is need for large-scale data interpretation software and technical know-how for the use of the same. Also, the laboratories conducting NBS should have policies and procedures for post-analytical systems assessment. These should be addressing, among other things, procedures for immediate reporting of results that are considered out of range or are indicative of a clinical emergency; procedures for obtaining a second, freshly collected specimen for confirmatory analysis for each abnormal screening result, etc.

7.9 CONCLUSIONS

Historically, analysis for inborn errors of metabolism has been provided predominantly by research laboratories, each offering analyses only for disorders in line with their scientific interest. With the increasing attentiveness of genetics in medicine and nearly one thousand IEMs identified so far, Clinical Biochemical Genetics is now recognized as a laboratory discipline concerned with the estimation and diagnosis of patients and their families with inherited metabolic disease. Further, this discipline helps in monitoring of treatment and also differentiating heterozygous carriers from non-carriers by series of metabolite and enzymatic analysis. Proteomics is definitely the future of medical diagnosis. Developments in this field are occurring rapidly. Specifically in the field of IEM diagnosis and treatment, proteomics is going to play a dynamic role in future. Currently, TMS is widely used in all developed countries and many developing countries for NBS for IEM.

In this commentary, we have highlighted the role of proteomics in detecting novel markers for IEM. Future attempts should be focused on validating these markers so as develop them into tools for early diagnosis of IEM. When used together with TMS, these markers shall aid in the early recognition of IEM. The ultimate aim of many of these research works is to improve both morbidity and mortality for patients with IEM. It is hoped that many of these markers shall satisfy these requirements. Right now, we have a technology which can change the way we look at disease states. It needs to be revealed, whether the technology can deliver the promises it offers, specifically for IEM. Theoretically, it may be possible to identify new markers for IEM by MS; however, they need to undergo the demanding process of validation before being available as useful markers for the detection/diagnosis of IEM. Such markers also need to be clinically useful, cheap, and relevant if it has to be included into the NBS panel. Much work needs to be done in this direction.

KEYWORDS

- **inborn error of metabolism**
- **mass spectrometry**
- **metabolomics**
- **newborn screening**
- **novel markers**
- **proteomics**
- **second tier testing**
- **tandem mass spectrometry**
- **urinomics**

REFERENCES

1. Vaidyanathan, K., & Vasudevan, D. M., (2007). *The Challenge of Metabolic Disorders in India: Amala Research Bulletin, 27*, 278–284.
2. Wilkins, M. R., Pasquali, C., Appel, R. D., Ou, K., Golaz, O., Sanchez, J. C., et al., (1996). From proteins to proteomes: Large scale protein identification by two-dimensional electrophoresis and amino acid analysis. *Nature Biotechnology, 14*, 61–65.
3. Graves, P. R., & Haystead, T. A. J., (2002). Molecular biologist's guide to proteomics. *Microbiol. Mol. Biol. Rev., 66*, 39–63.
4. Hanash, S., (2003). Disease proteomics. *Nature, 422*, 226–232.
5. Raghuveer, T. S., Garg, U., & Graf, W. D., (2006). Inborn errors of metabolismin infancy and early childhood: An update. *Am. Fam. Physician, 73*, 1981–1990.
6. Banta-Wright, S. A., & Steiner, R. D., (2004). Tandem mass spectrometry in newborn screening: A primer for neonatal and perinatal nurses. *J. Perinat. Neonat. Nurs., 18*, 41–58.
7. Somasundaram, K., Nijaguna, M. B., & Kumar, D. M., (2009). Serum proteomics of glioma: Methods and applications. *Expert Rev. Mol. Diagn., 9*, 695–707.
8. Mak, C. M., Lee, H. C., Chan, A. Y., & Lam, C. W., (2013). Inborn errors of metabolism and expanded newborn screening: Review and update. *Crit. Rev. Clin. Lab Sci., 50*, 142–162.
9. Boemer, F., Schoos, R., De Halleux, V., Kalenga, M., & Debray, F. G., (2013). Surprising causes of C5-carnitine false positive results in newborn screening. *Mol. Genet. Metab.*, pii. S1096-7192(13)00372-7. doi: 10.1016/j.ymgme.2013.11.005. [Epub ahead of print].
10. Clayton, P. T., (2001). Applications of mass spectrometry in the study ofinborn errors of metabolism. *J. Inherit. Metab. Dis., 24*, 139–150.
11. Van, C. S. C., Baker, M. W., Williams, P., Jones, S. A., Xiong, B., Thao, M. C., et al., (2013). Prevalence and mutation analysis of short/branched chain acyl-CoA dehydrogenase deficiency (SBCADD) detected on newborn screening in Wisconsin. *Mol. Genet. Metab., 110*, 111–115.
12. Janzen, N., Steuerwald, U., Sander, S., Terhardt, M., Peter, M., & Sander, J., (2013). UPLC-MS/MS analysis of C5-acylcarnitines in dried blood spots. *Clin. Chim. Acta, 421*, 41–45.
13. Scott, C. R., Elliott, S., Buroker, N., Thomas, L. I., Keutzer, J., Glass, M., et al., (2013). Identification of infants at risk for developing Fabry, Pompe, or mucopolysaccharidosis-I from newborn blood spots by tandem mass spectrometry. *J. Pediatr., 163*, 498–503.
14. Zhang, H., Young, S. P., & Millington, D. S., (2013). Quantification of glycosaminoglycans in urine by isotope-dilution liquid chromatography-electro sprays ionization tandem mass spectrometry. *Curr. Protoc. Hum. Genet.*, Chapter 17: Unit 17.12.
15. Wuhrer, M., (2013). Glycomics using mass spectrometry. *Glycoconj. J., 30*, 11–22.
16. La Marca, G., Malvagia, S., Materazzi, S., Della, B. M. L., Boenzi, S., Martinelli, D., et al., (2012). LC-MS/MS method for simultaneous determination on a dried blood spot of multiple analytes relevant for treatment monitoring in patients with tyrosinemia type I. *Anal. Chem., 84*, 1184–1188.
17. Horvath, T. D., Matthews, N. I., Stratton, S. L., Mock, D. M., & Boysen, G., (2011). Measurement of 3-hydroxyisovaleric acid in urine from marginally biotin-deficient humans by UPLC-MS/MS. *Anal. Bioanal. Chem., 401*, 2805–2810.

18. Pollak, A., & Kasper, D. C., (2013). Austrian newborn screening program: A perspective of five decades. *J. Perinat. Med.*, 1–8.
19. Ibarra-González, I., Fernández-Lainez, C., Belmont-Martínez, L., Guillén-López, S., Monroy-Santoyo, S., & Vela-Amieva, M., (2013). Characterization of inborn errors of intermediary metabolism in Mexican patients. *An Pediatr (Barc), pii*, S1695-4033(13)00380-9. doi: 10.1016/j.anpedi.2013.09.003. [Epub ahead of print].
20. Wilcken, B., Haas, M., Joy, P., Wiley, V., Bowling, F., Carpenter, K., et al., (2009). Expanded newborn screening: Outcome in screened and unscreened patients at age 6 years. *Pediatrics, 124*, e241–248.
21. Parini, R., & Corbetta, C., (2011). Metabolic screening for the newborn. *J. Matern. Fetal Neonatal Med., 24*(2), 6–8.
22. Wilcken, B., (2011). Newborn screening: How are we traveling, and where should we be going? *J. Inherit. Metab. Dis., 34*, 569–574.
23. Ross, L. F., (2010). Mandatory versus voluntary consent for newborn screening? *Kennedy Inst. Ethics J.*, 20, 299–328.
24. Dhondt, J. L., (2010). Expanded newborn screening: Social and ethical issues. *J. Inherit. Metab. Dis., 33*(2), S211–217.
25. Wilcken, B., (2010). Expanded newborn screening: Reducing harm, assessing benefit. *J. Inherit. Metab. Dis., 33*(2), S205–S210.
26. Dietzen, D. J., Rinaldo, P., Whitley, R. J., Rhead, W. J., Hannon, W. H., Garg, U. C., et al., (2009). National academy of clinical biochemistry laboratory medicine practice guidelines: Follow-up testing for metabolic disease identified by expanded newborn screening using tandem mass spectrometry; executive summary. *Clin. Chem., 55*, 1615–1626.
27. Norman, R., Haas, M., Chaplin, M., Joy, P., & Wilcken, B., (2009). Economic evaluation of tandem mass spectrometry newborn screening in Australia. *Pediatrics, 123*, 451–457.
28. Carroll, A. E., & Downs, S. M., (2006). Comprehensive cost-utility analysis of newborn screening strategies. *Pediatrics, 117*, S287–S295.
29. Padilla, C. D., & Therrell, B. L., (2007). Newborn screening in the Asia Pacific region. *J. Inherit. Metab. Dis., 30*, 490–506.
30. Chace, D. H., & Hannon, W. H., (2010). Impact of second-tier testing on the effectiveness of newborn screening. *Clin. Chem., 56*, 1653–1655.
31. Shigematsu, Y., Hata, I., & Tajima, G., (2010). Useful second-tier tests in expanded newborn screening of isovaleric acidemia and methylmalonic aciduria. *J. Inherit. Metab. Dis., 33*(2), S283–288.
32. Chalcraft, K. R., & Britz-McKibbin, P., (2009). Newborn screening of in born errors of metabolism by capillary electrophoresis-electrospray ionization-mass spectrometry: A second-tier method with improved specificity and sensitivity. *Anal. Chem., 81*, 307–314.
33. Turgeon, C., Magera, M. J., Allard, P., Tortorelli, S., Gavrilov, D., Oglesbee, D., et al., (2008). Combined newborn screening for succinylacetone, amino acids, and acylcarnitines in dried blood spots. *Clin. Chem., 54*, 657–664.
34. Gucciardi, A., Pirillo, P., Di Gangi, I. M., Naturale, M., & Giordano, G., (2012). A rapid UPLC-MS/MS method for simultaneous separation of 48 acylcarnitines in dried blood spots and plasma useful as a second-tier test for expanded newborn screening. *Anal. Bioanal. Chem., 404*, 741–751.

35. Yang, Q., Manicke, N. E., Wang, H., Petucci, C., Cooks, R. G., & Ouyang, Z., (2012). Direct and quantitative analysis of underivatized acylcarnitines in serum and whole blood using paper spraymass spectrometry. *Anal. Bioanal. Chem., 404*, 1389–1397.
36. Ko, D. H., Jun, S. H., Park, K. U., Song, S. H., Kim, J. Q., & Song, J., (2011). Newborn screening for galactosemia by a second-tier multiplex enzyme assay using UPLC-MS/MS in dried blood spots. *J. Inherit. Metab. Dis., 34*, 409–414.
37. Wang, L. Y., Chen, N. I., Chen, P. W., Chiang, S. C., Hwu, W. L., Lee, N. C., et al., (2013). Newborn screening for citrin deficiency and carnitine uptake defect using second-tier molecular tests. *BMC Med. Genet., 14*, 24.
38. Ostermann, K. M., Dieplinger, R., Lutsch, N. M., Strupat, K., Metz, T. F., Mechtler, T. P., et al., (2013). Matrix-assisted laser desorption/ionization for simultaneous quantitation of (acyl) carnitines and organic acids in dried blood spots. *Rapid Commun. Mass Spectrom., 27*, 1497–1504.
39. Goto-Inoue, N., Hayasaka, T., Zaima, N., Nakajima, K., Holleran, W. M., Sano, S., et al., (2012). Imagingmass spectrometry visualizes ceramides and the pathogenesis of dorfman-chanarin syndrome due to ceramide metabolic abnormality in the skin. *PLoS One., 7*, e49519.
40. Sleat, D. E., Ding, L., Wang, S., Zhao, C., Wang, Y., Xin, W., et al., (2009). Mass spectrometry-based protein profiling to determine the cause of lysosomal storage diseases of unknown etiology. *Mol. Cell Proteomics, 8*, 1708–1718.
41. Matysik, S., Klünemann, H. H., & Schmitz, G., (2012). Gas chromatography-tandemmass spectrometry method for the simultaneous determination of oxysterols, plant sterols, and cholesterol precursors. *Clin. Chem., 58*(11), 1557–1564.
42. Liu, Y. P., Ma, Y. Y., Wu, T. F., Wang, Q., Li, X. Y., Ding, Y., et al., (2012). Abnormal findings during newborn period of 160 patients with early-onset methylmalonicaciduria. *Zhonghua Er Ke Za Zhi., 50*, 410–414.
43. Cangemi, G., Barco, S., Barbagallo, L., Di Rocco, M., Paci, S., Giovannini, M., et al., (2012). Erythrocyte Galactose-1-phosphate measurement by GC-MS in the monitoring of classical galactosemia. *Scand J. Clin. Lab Invest., 72*, 29–33.
44. Zytkovicz, T. H., Sahai, I., Rush, A., Odewale, A., Johnson, D., Fitzgerald, E., et al., (2013). Newborn screening for hepatorenaltyrosinemia-I by tandemmass spectrometry using pooled samples: A four-year summary by the New England newborn screening program. *Clin. Biochem., 46*, 681–684.
45. Lin, N., Ye, J., Qiu, W., Han, L., Zhang, H., & Gu, X., (2013). Application of liquid chromatography-tandemmass spectrometry in the diagnosis and follow-up of maple syrup urine disease in a Chinese population. *J. Pediatr. Endocrinol. Metab., 26*, 433–439.
46. Boutin, M., Gagnon, R., Lavoie, P., & Auray-Blais, C., (2012). LC-MS/MS analysis of plasma lyso-Gb3 in fabrydisease. *Clin. Chim. Acta, 414*, 273–280.
47. Giordano, G., Di Gangi, I. M., Gucciardi, A., & Naturale, M., (2012). Quantification of underivatised amino acids on dry blood spot, plasma, and urine by HPLC-ESI-MS/MS. *Methods Mol. Biol., 828*, 219–242.
48. Moore, T., Le, A., & Cowan, T. M., (2012). An improved LC-MS/MS method for the detection of classic and low excretorglutaricacidemia type 1. *J. Inherit. Metab. Dis., 35*, 431–425.

49. Herrmann, P. C., & Herrmann, E. C., (2012). Mitochondrial proteome: Toward the detection and profiling of disease associated alterations. *Methods Mol. Biol., 823*, 265–277.
50. Griffiths, W. J., Ogundare, M., Williams, C. M., & Wang, Y., (2011). On the future of "omics": Lipidomics. *J. Inherit. Metab. Dis., 34*, 583–592.
51. Ong, D. S., Wang, Y. J., Tan, Y. L., Yates, J. R. III, Mu, T. W., & Kelly, J. W., (2013). FKBP10 depletion enhances glucocerebrosidase proteostasis in Gaucher disease fibroblasts. *Chem. Biol., 20*, 403–415.
52. Haddad, S. E., Khoury, M., Daoud, M., Kantar, R., Harati, H., Mousallem, T., et al., (2012). CLN5 and CLN8 protein association with ceramide synthase: Biochemical and proteomic approaches. *Electrophoresis, 33*, 3798–3809.
53. Leong, D. W., Komen, J. C., Hewitt, C. A., Arnaud, E., McKenzie, M., Phipson, B., et al., (2012). Proteomic and metabolomic analyses of mitochondrial complex I-deficient mouse model generated by spontaneous B2 short interspersed nuclear element (SINE) insertion into NADH dehydrogenase (ubiquinone) Fe-S protein 4 (Ndufs4) gene. *J. Biol. Chem., 287*, 20652–20663.
54. Nogales-Gadea, G., Consuegra-García, I., Rubio, J. C., Arenas, J., Cuadros, M., Camara, Y., et al., (2012). A transcriptomic approach to search for novel phenotypic regulators in McArdle disease. *PLoS One, 7*, 31718.
55. Moore, D. F., Krokhin, O. V., Beavis, R. C., Ries, M., Robinson, C., Goldin, E., et al., (2007). Proteomicsof specific treatment-related alterations in Fabry disease: A strategy to identify biological abnormalities. *Proc. Natl. Acad. Sci. U.S.A, 104*, 2873–2878.
56. Fogli, A., Merle, C., Roussel, V., Schiffmann, R., Ughetto, S., Theisen, M., et al., (2012). CSF N-glycan profiles to investigate biomarkers in brain developmental disorders: Application to leukodystrophies related to eIF2B mutations. *PLoS One, 7*, e42688.
57. Lee, B. H., Kim, J. M., Heo, S. H., Mun, J. H., Kim, J., Kim, J. H., et al., (2011). Proteomic analysis of the hepatic tissue of Long-Evans cinnamon (LEC) rats according to the natural course of Wilson disease. *Proteomics, 11*, 3698–3705.
58. Wilmarth, P. A., Short, K. K., Fiehn, O., Lutsenko, S., David, L. L., & Burkhead, J. L., (2012). A systems approach implicates nuclear receptor targeting in the Atp7b (-/-) mouse model of Wilson's disease. *Metallomics, 4*, 660–668.
59. Burkhead, J. L., Ralle, M., Wilmarth, P., David, L., & Lutsenko, S., (2011). Elevated copper remodels hepatic RNA processing machinery in the mouse model of Wilson's disease. *J. Mol. Biol., 406*, 44–58.
60. Jiang, X. S., Backlund, P. S., Wassif, C. A., Yergey, A. L., & Porter, F. D., (2010). Quantitative proteomics analysis of inborn errors of cholesterol synthesis: Identification of altered metabolic pathways in DHCR7 and SC5D deficiency. *Mo. Cell Proteomics, 9*, 1461–1475.
61. Misiti, S., Stigliano, A., Borro, M., Gentile, G., Michienzi, S., Cerquetti, L., et al., (2010). Proteomic profiles in hyperandrogenic syndromes. *J. Endocrinol. Invest., 33*, 156–164.
62. Borro, M., Gentile, G., Stigliano, A., Misiti, S., Toscano, V., & Simmaco, M., (2007). Proteomic analysis of peripheral T lymphocytes, suitable circulating biosensors of strictly related diseases. *Clin. Exp. Immunol., 150*, 494–501.
63. Sleat, D. E., Wiseman, J. A., Sohar, I., El-Banna, M., Zheng, H., Moore, D. F., et al., (2012). Proteomic analysis of mouse models of Niemann-Pick C disease reveals

alterations in the steady-state levels of lysosomal proteins within the brain. *Proteomics, 12,* 3499–3509.

64. Shibata, S., Zhang, J., Puthumana, J., Stone, K. L., & Lifton, R. P., (2013). Kelch-like 3 and Cullin 3 regulate electrolyte homeostasis via ubiquitination and degradation of WNK4. *Proc. Natl. Acad. Sci. U.S.A, 110*, 7838–7843.
65. Miyaji, T., Omote, H., & Moriyama, Y., (2010). A vesicular transporter that mediates aspartate and glutamate neurotransmission. *Biol. Pharm. Bull., 33,* 1783–1785.
66. Hernández-Fernaud, J. R., & Salido, E., (2010). Differential expression of liver and kidney proteins in a mouse model for primary hyperoxaluria type I. *FEBS J., 277,* 4766–4774.
67. Burillo, E., Recalde, D., Jarauta, E., Fiddyment, S., Garcia-Otin, A. L., Mateo-Gallego, R., et al., (2009). Proteomic study of macrophages exposed to oxLDL identifies a CAPG polymorphism associated with carotid atherosclerosis. *Atherosclerosis, 207*, 32–37.
68. Richard, E., Vega, A. I., Pérez, B., Roche, C., Velázquez, R., Ugarte, M., et al., (2009). Congenital disorder of glycosylation Ia: New differentially expressed proteins identified by 2-DE. *Biochem. Biophys. Res. Commun., 379*, 267–271.
69. Zhang, J., Yasuda, M., Desnick, R. J., Balwani, M., Bishop, D., & Yu, C., (2011). A LC-MS/MS method for the specific, sensitive, and simultaneous quantification of 5-aminolevulinic acid and porphobilinogen. *J. Chromatogr. B Analyt. Technol. Biomed. Life Sci., 879*, 2389–2396.
70. Wang, L., Yang, W., Ju, W., Wang, P., Zhao, X., Jenkins, E. C., et al., (2012). A proteomic study of Hutchinson-Gilford progeria syndrome: Application of 2D-chromotography in a premature aging disease. *Biochem. Biophys. Res. Commun., 417*, 1119–1126.
71. Rocha, H., Ferreira, R., Carvalho, J., Vitorino, R., Santa, C., Lopes, L., et al., (2011). Characterization of mitochondrial proteome in a severe case of ETF-QO deficiency. *J. Proteomics, 75*(1), 221–228.
72. Hu, S., Vissink, A., Arellano, M., Roozendaal, C., Zhou, H., Kallenberg, C. G., et al., (2011). Identification of autoantibody biomarkers for primary Sjögren's syndrome using protein microarrays. *Proteomics, 11*, 1499–1507.
73. Palmfeldt, J., Vang, S., Stenbroen, V., Pavlou, E., Baycheva, M., Buchal, G., et al., (2011). Proteomicsreveals that redox regulation is disrupted in patients with ethylmalonicence phalopathy. *J. Proteome Res., 10*, 2389–2396.
74. Méndez, D., Linares, M., Diez, A., Puyet, A., & Bautista, J. M., (2011). Stress response and cytoskeletal proteins involved in erythrocyte membrane remodeling upon Plasmodium falciparum invasion are differentially carbonylated in G6PD A-deficiency. *Free Radic. Biol. Med., 50*(10), 1305–1313.
75. Wang, P., Ju, W., Wu, D., Wang, L., Yan, M., Zou, J., et al., (2011). A two-dimensional protein fragmentation-proteomic study of neuronal ceroid-lipofuscinoses: Identification and characterization of differentially expressed proteins. *J. Chromatogr. B Analyt. Technol. Biomed. Life Sci., 879*, 304–316.
76. Da Costa, G., Guerreiro, A., Correia, C. F., Gomes, R. J., Freire, A., onteiro, E., et al., (2010). A non-invasive method based on saliva to characterize transthyretin in familial amyloidotic polyneuropathy patients using FT-ICR high-resolution MS. *Proteomics Clin. Appl., 4*, 674–678.
77. Ludwig, M., & Sethi, S. K., (2011). Novel techniques and newer markers for the evaluation of "proximal tubular dysfunction." *Int. Urol. Nephrol., 43*(4), 1107–1115.

78. Cuccurullo, M., Beneduci, A., Anand, S., Mignani, R., Cianciaruso, B., Bachi, A., et al., (2010). Fabry disease: Perspectives of urinaryproteomics. *J. Nephrol., 23*(16), S199–212.
79. Vojtová, L., Zima, T., Tesař, V., Michalová, J., Přikryl, P., Dostálová, G., et al., (2010). Study of urinary proteomes in Anderson-fabry disease. *Ren. Fail, 32*, 1202–1209.
80. Caubet, C., Lacroix, C., Decramer, S., Drube, J., Ehrich, J. H., Mischak, H., et al., (2010). Advances in urinary proteome analysis and biomarker discovery in pediatric renal disease. *Pediatr. Nephrol., 25*, 27–35.
81. Faid, V., Michalski, J. C., & Morelle, W., (2008). A mass spectrometric strategy for profiling glycoproteinoses, Pompe disease, and sialic acid storage diseases. *Proteomics Clin. Appl., 2*, 528–542.
82. Xia, B., Asif, G., Arthur, L., Pervaiz, M. A., Li, X., Liu, R., et al., (2013). Oligosaccharide analysis in urine by maldi-tofmass spectrometryfor the diagnosis of lysosomal storage diseases. *Clin. Chem., 59*, 1357–1368.
83. Manwaring, V., Heywood, W. E., Clayton, R., Lachmann, R. H., Keutzer, J., Hindmarsh, P., et al., (2013). The identification of new biomarkers for identifying and monitoring kidney disease and their translation into a rapidmass spectrometry-based test: Evidence of presymptomatic kidney disease in pediatric Fabry and type-I diabetic patients. *J. Proteome. Res., 12*, 2013–2021.
84. Hsu, W. Y., Chen, C. M., Tsai, F. J., & Lai, C. C., (2013). Simultaneous detection of diagnostic biomarkers of alkaptonuria, ornithine carbamoyl transferase deficiency, and neuroblastoma disease by high-performance liquid chromatography/tandem mass spectrometry. *Clin. Chim. Acta, 420*, 140–145.
85. Stanislaus, A., Guo, K., & Li, L., (2012). Development of an isotope labeling ultra-high performance liquid chromatography mass spectrometric method for quantification of acylglycines in human urine. *Anal. Chim. Acta, 750*, 161–172.
86. Christakoudi, S., Cowan, D. A., & Taylor, N. F., (2012). Steroids excreted in urine by neonates with 21-hydroxylase deficiency. 3. Characterization, using GC-MS and GC-MS/MS, of androstanes and androstenes. *Steroids, 77*, 1487–1501.
87. Rebollido-Fernandez, M. M., Castiñeiras, D. E., Bóveda, M. D., Couce, M. L., Cocho, J. A., & Fraga, J. M., (2012). Development of electro sprays ionization tandem mass spectrometry methods for the study of a high number of urine markers of inborn errors of metabolism. *Rapid Commun. Mass Spectrom., 26*, 2131–2144.
88. Allegri, G., Costa, N. H. J., Ferreira, G. L. N., Costa, D. O. M. L., Scalco, F. B., & De Aquino, N. F. R., (2012). Determination of six pterins in urine by LC-MS/MS. *Bioanalysis, 4*, 1739–1746.
89. Ramautar, R., Berger, R., Van, D. G. J., & Hankemeier, T., (2013). Human metabolomics: Strategies to understand biology. *Curr. Opin. Chem. Biol., 17*, 841–846.
90. Fanos, V., Antonucci, R., Barberini, L., & Atzori, L., (2012). Urinary metabolomics in newborns and infants. *Adv. Clin. Chem., 58*, 193–223.
91. Selicharová, I., Kořínek, M., Demianová, Z., Chrudinová, M., Mládková, J., & Jiráček, J., (2013). Effects of hyperhomocysteinemia and betaine-homocysteine S-methyltransferase inhibition on hepatocyte metabolites and the proteome. *Biochim. Biophys. Acta., 1834*, 1596–1606.

Part II
Research Papers

CHAPTER 8

Elevated Cytosolic Phospholipase $A_2\alpha$ as a Target for Treatment and Prevention the Progression of Neurodegenerative Diseases

RACHEL LEVY, YULIA SOLOMONOV, KESENIA KASIANOV, YAFA MALADA-EDELSTEIN, and NURIT HADAD

Department of Clinical Biochemistry and Pharmacology, Faculty of the Health Sciences Ben-Gurion University of the Negev and Soroka Medical University Center, Beer Sheva, Israel, E-mail: ral@bgu.ac.il (R. Levy)

ABSTRACT

Intravenous injections of an antisense against the main pro-inflammatory enzyme; cytosolic phospholipase $A_2\alpha$ ($cPLA_2\alpha$) reduced $cPLA_2\alpha$ upregulation specifically at the site of inflammation. To study the role of $cPLA_2\alpha$ in neurodegenerative diseases a specific antisense against $cPLA_2\alpha$ (AS) was brain infused to inhibit $cPLA_2\alpha$ upregulation in the brain. Brain infusion of the antisense drug in a mouse model of amyloid brain infusion, representing a mouse model of Alzheimer's disease (AD), was found to be efficacious in preventing $cPLA_2\alpha$ upregulation in the brain and in the prevention of the disease. Reduction of the elevated expression of $cPLA_2\alpha$ in the spinal cord of human SOD1G93A transgenic (hmSOD1) mice (a mouse model for amyotrophic lateral sclerosis (ALS)) by brain infusion of AS at week 15 (shortly before the appearance of the disease symptoms) for a duration of 6 weeks, delayed the loss of motor neuron function in comparison with hmSOD1 mice and with sense brain infused hmSOD1 mice. Since specific reduction of $cPLA_2\alpha$ in the brain and spinal cord significantly attenuated the development of the diseases in mouse models of AD and ALS, cPLA2α may offer an efficient target for treatment neurodegenerative diseases.

8.1 INTRODUCTION

The role of inflammation in the pathogenesis of a vast array of diseases including neurodegenerative diseases has been well documented. Elevated cytosolic phospholipase $A_2\alpha$ ($cPLA_2\alpha$) expression and activity were detected in the inflammatory sites in a vast array of inflammatory diseases, including neurodegenerative diseases [1, 2]. $cPLA_2\alpha$ is a requisite component in the cascade of events leading to the production of eicosanoids during acute and chronic inflammation. Intravenous injections of an antisense against $cPLA_2\alpha$ reduced $cPLA_2\alpha$ upregulation specifically at the site of inflammation. Intravenous injections of the AS in mouse models of peritonitis, arthritis, colitis, and insulin resistance, proved efficacious for the treatment or prevention of these diseases [4–6]. Increased $cPLA_2\alpha$ immunoreactivity and transcript were observed in Alzheimer's disease (AD) in the brain [1] often associated with amyloid deposits, suggesting its role in the pathogenesis of the disease. Increased expression and activity of $cPLA_2\alpha$ has been detected in neurons, astrocytes, and in microglia in the spinal cord, brainstem, and cortex of sporadic ALS patients [7] and in the spinal cord of G93A human mutant transgenic (hmSOD1) mice, suggesting that $cPLA_2\alpha$ may have an important role in the pathogenesis of the disease in all ALS patients.

8.2 MATERIALS AND METHODS

1. For the AD mouse model Ab 1–42 brain infusion (alone or together with the antisense or the sense) to C57Bl/6 mice was done using a microcosmic pump attached to cannula stereotaxically implanted into the right lateral cerebral ventricle. Mice behavioral patterns were detected using Y-maze.
2. For the ALS mouse model, B6.Cg-Tg(SOD1G93A) 1 Gur/J hemizygous transgenic male mice were obtained from Jackson Laboratory (Bar Harbor, ME, U.S.A). A Rotarod (Rotamex-5, Columbus Instruments, Columbus, OH, USA) test was used to evaluate the motor performance of the mice using an accelerating paradigm of 0.12 rpm/s.
3. Antisense oligonucleotide against $cPLA_2\alpha$ were engineered using the computer-based approach RNADraw V1.1 (Mazura Multimedia, Stockholm, Sweden). An oligo-deoxynucleotide antisense (tcaaaggtctcattccaca) and its corresponding sense with phosphorothioate modifications on the last three bases at both 5' and 3' ends

were used as described in our previous article [4]. Mice (around 25 g weight) received 10 μg/day phospho oligo-deoxynucleotides diluted in saline.

4. Microglia were isolated from brains of mice C57BL 1 day old pups and grown in DMEM-F12 medium (10% FCS, 1% non-essential amino-acids, 11.4 μM β-mercaptoethanol, 10 mM HEPES, 1 mM sodium pyruvate 2 mM L-glutamine, 100 U/ml penicillin, 100 μg/ml streptomycin and 12.5 U/ml nystatin into poly-L-lysine coated flasks and kept at 37°C in a humidified atmosphere of 5% CO_2 (Figure 8.1).

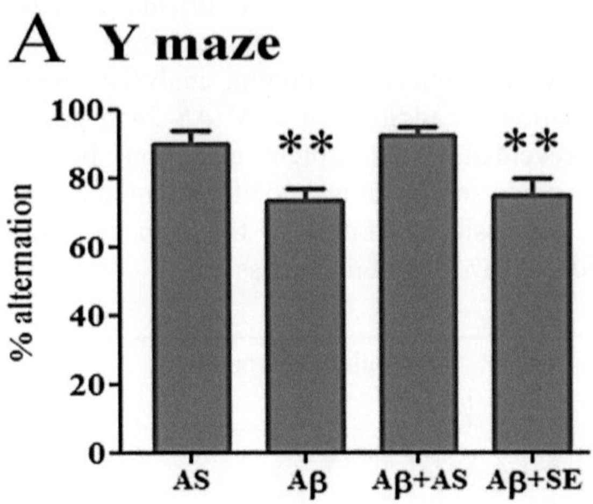

FIGURE 8.1 Onset of pathology in Ab brain infusion mouse model dependent on elevated cPLA$_2$α. Each group contained 10 mice. **p<0.05 significant decrease in mice infused with Ab or Ab + sense compared with control mice (with buffer) or mice infused with AS + Aβ.

8.3 RESULTS

1. Behavioral deficit was detected by reduction in spontaneous behavioral alterations using a Y-maze analysis at 8 weeks of Amyloid beta$_{1-42}$ (Aβ) brain infusion. Prevention of elevated cPLA$_2$α protein expression by brain infusion of AS with Aβ, prevented the behavioral deficit in comparison with none treated mice or mice treated with sense. Brain infusion of AS alone did not have a similar effect on mice behavior analyzed by Y-maze.
2. To determine the role of cPLA$_2$α in the progression of the disease, 10 μg AS or sense were injected to the spinal cord one a week from

week 15 (shortly before the onset of motor neuronal dysfunction) for 6 weeks. As shown in Figure 8.2, AS treatment prolonged survival in mice by 14 days (<0.05) and significantly delayed the onset of motor neuron dysfunction by 3 weeks. There was a significant ($p<0.001$) difference between the two groups.

3. In order to determine whether the reduction in glia activation determined by CD40 expression in AS treated hmSOD1 mice is due to the reduction of inflammation in the spinal cord or due to regulation by cPLA$_2$α, primary brain rat microglia were studied. As shown in Figure 8.3, addition of 10 ng/ml IFNγ to primary microglia for 24 h caused a significant ($p<0.0001$) elevation of cPLA$_2$α and of CD40 protein expression, as shown in the double immuno-fluoresce staining analysis. Preventing cPLA2α up-regulation by addition of 2μM AS 24 h prior to addition of IFNγ prevented CD40 protein induction. Incubation with the corresponding sense that had no effect on the elevation of cPLA$_2$α protein expression by either of the inducers, did not affect the elevation of CD40 protein expression.

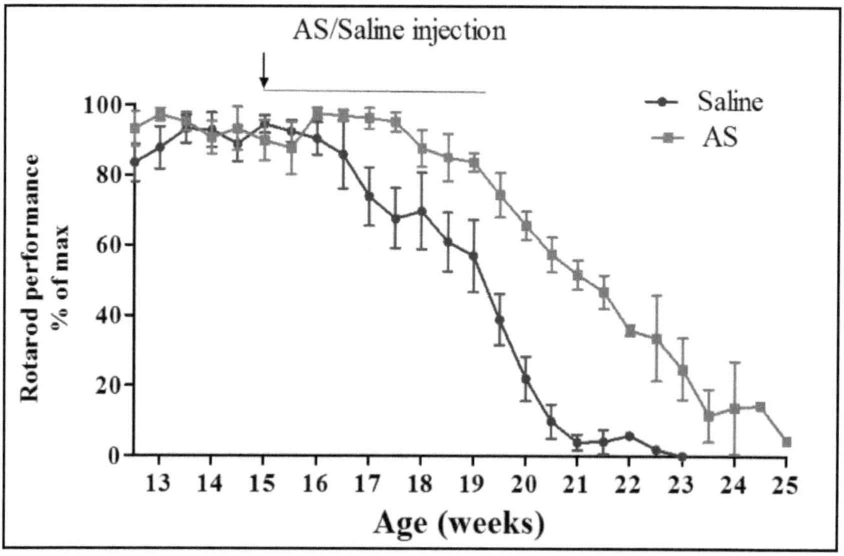

FIGURE 8.2 AS Brain infusion to hmSOD1 mice shortly before the onset of motor neuron, dysfunction delayed the development of the disease. To the spinal cord of 15 weeks old hmSOD1 mice 10 ug AS or the corresponding sense were injected once a week during 6 weeks (n = 7 in each group). AS brain infusion prolonged survival and delayed loss of motor function analyzed by Rotarod *$p<0.001$-significance relative to sense treated hmSOD1 mice.

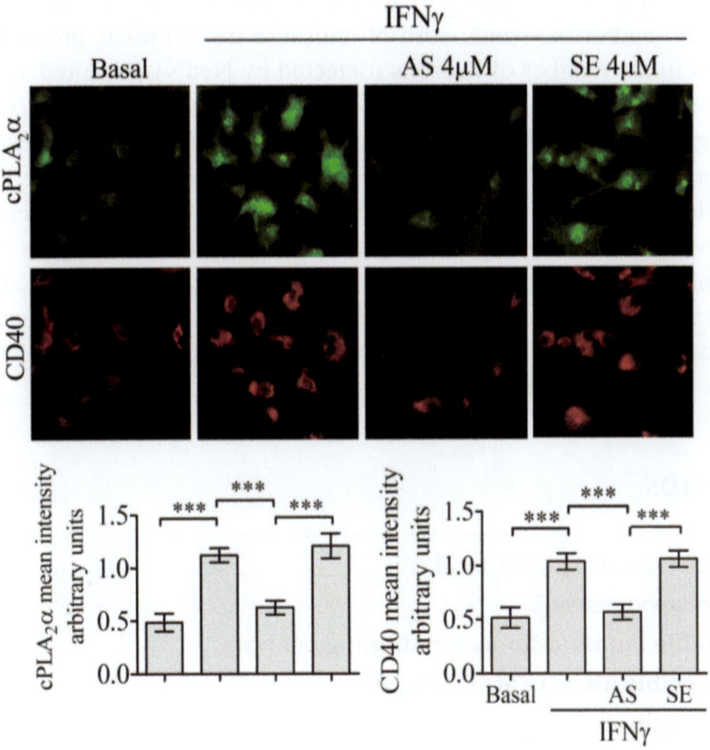

FIGURE 8.3 Elevated CD40 expression by IFNγ in primary microglia cells is regulated by cPLA$_2$α. A representative double-immunofluorescence staining of cPLA$_2$α (green) and CD40 (red) in unstimulated or stimulated microglia by 25 ng/ml IFNγ in the absence or presence of 2 μM AS or 2 μM sense (SE). Scale bars = 50μm. The intensity of CD40 or cPLA$_2$α were quantitated for cell and expressed in the bar graph as arbitrary units. ***$p<0.0001$-significance differences.

8.4 DISCUSSION

Inhibition of cPLA$_2$α upregulation in the cortex of Ab brain infusion (a mouse model of AD) as shown in our previous study [8] prevented the behavioral deficit. Similarly, inhibition of cPLA$_2$α upregulation in the spinal cord of hmSOD1 mice (a mouse model of ALS) at the onset of the disease symptoms [9] significantly delayed the development of the disease. Increased expression and activity of cPLA$_2$α have been detected in all cell types in the spinal cord, brainstem, and cortex of both sporadic and familial ALS, suggesting that cPLA$_2$α may have an important role in the pathogenesis of the disease in all ALS patients. The antisense treatment that reduced $cPLA_2\alpha$ upregulation in the brain and/or the spinal cord of antisense treated mice, prevented the reduction in the number of neurons (detected by NeuN), inhibited astrocyte activation (detected by GFAP) and microglial activation (detected by Iba-1 and/or by CD40). In addition, antisense treatment blunted the upregulation of the pro-inflammatory enzymes: inducible nitric oxide synthase (iNOS) and cyclooxygenase-2 (COX-2). The activation of microglia detected by CD40 overexpression is regulated by $cPLA_2\alpha$. In conclusion, antisense drug treatment is an exciting and emerging specialty area, not as yet in common use. Various antisense drugs for a variety of diseases and disorders are now in clinical phase testing, evincing the potential and promise of antisense drugs as a treatment strategy.

KEYWORDS

- **antisense drug treatment**
- **cyclooxygenase-2**
- **double immuno-fluoresce staining analysis**
- **inducible nitric oxide synthase**
- **microglia**
- **pro-inflammatory enzymes**

REFERENCES

1. Stephenson, D. T., Lemere, C. A., Selkoe, D. J., & Clemens, J. A., (1996). Cytosolic phospholipase A2 (cPLA2) immune reactivity is elevated in Alzheimer's disease brain. *Neurobiol. Dis., 3*, 51–63.
2. Clemens, J. A., et al., (1996). Reactive glia express cytosolic phospholipase A2 after transient global forebrain ischemia in the rat. *Stroke, 27*, 527–535.
3. Stephenson, D., Rash, K., Smalstig, B., Roberts, E., Johnstone, E., Sharp, J., Panetta, J., et al., (1999). Cytosolic phospholipase A2 is induced in reactive glia following different forms of neurodegeneration. *Glia, 27*, 110–128.
4. Raichel, L., Berger, Y., Kachko, L., Hadad, N., Karter, M., Solodkin, I., Williams, I. R., et al., (2008). Reduction of $cPLA_2\alpha$ over expression-an efficient anti-inflammatory therapy for collagen induced arthritis. *Eur. J. Immunol., 38*, 1–12.
5. Hadad, N., Burgazliev, O., Elgazar-Carmon, V., Solomonov, Y., Wueest, S., Item, F., Konrad, D., et al., (2013). Induction of cytosolic phospholipase $A_2\alpha$ is required for

adipose neutrophil infiltration and hepatic insulin resistance early in the course of high fat feeding. *Diabetes, 62*, 3053–3063.
6. Rosengarten, M., Hadad, N., Solomonov, Y., Lamprecht, S., & Levy, R., (2016). Cytosolic phospholipase $A_2\alpha$ has a crucial role in the pathogenesis of DSS-induced colitis in mice. *Eur. J. Immunol., 46*, 400–408.
7. Shibata, N., Kakita, A., Takahashi, H., Ihara, Y., Nobukuni, K., Fujimura, H., Sakoda, S., & Kobayashi, M., (2010). Increased expression and activation of cytosolic phospholipase A_2 in the spinal cord of patients with sporadic amyotrophic lateral sclerosis. *Acta Neuropathol., 119*, 345–354.
8. Sagy-Bross, C., Kasianov, K., Solomonov, Y., Braiman, A., Friedman, A., Hadad, N., & Levy, R., (2015). The role of cytosolic phospholipase $A_2\alpha$ in Amyloid Precursor Protein upregulation induced by amyloid $beta_{1-42}$-implication for neurodegeneration. *J. Neurochem., 132*, 559–571.
9. Solomonov, Y., Hadad, N., & Levy, R., (2016). Reduction of cytosolic phospholipase $A_2\alpha$ up regulation delays the onset of symptoms in SOD1G93A mouse model of amyotrophic lateral sclerosis. *J. Neuroinflammation, 13*, 134–146.

CHAPTER 9

Application of Biospeckle Laser Method for Drug Testing on Parasites

MOHAMMAD ZAHEER ANSARI,[1] HUMBERTO CABRERA,[2] HILDA C. GRASSI,[3] ANA VELÁSQUEZ,[3] EFRÉN D. J. ANDRADES,[3] and A. MUJEEB[1]

[1]*International School of Photonics, Cochin University of Science and Technology, Kochi – 682 022, Kerala, India,*
E-mails: mohamedzaheer1@gmail.com; mdzaheer@cusat.ac.in (M. Z. Ansari)

[2]*Optics Laboratory, The Abdus Salam International Center for Theoretical Physics (ICTP), Strada Costiera 11, Trieste – 34151, Italy,*
E-mail: hcabrera@ictp.it

[3]*Faculty of Pharmacy and Bioanalysis, University of the Andes, Merida – 5101, Venezuela*

ABSTRACT

We report on the application of the biospeckle laser method (BLM) to test drug effects on *Trypanosoma cruzi* (*T. cruzi*) parasites. The method, enabled to assess the activity of parasites under different drug concentrations, thus demonstrating the effectiveness of the proposed methodology. The spatial-temporal correlation and speckle grain size were measured in order to assess the immediate action of the drug on parasites. The achieved results allowed the validation of the methodology as a fast and non-invasive for testing the effectiveness of Epirubicin on *T. cruzi* parasites. The proposed methodology was also validated by other well-known digital processing approaches.

9.1 INTRODUCTION

American Trypanosomiasis (Chagas disease) in humans is caused by the protozoan parasite *T. cruzi* and very frequently affects the population of the region. The available drugs for the treatment of the infection by *T. cruzi* are Nifurtimox and Benznidazol [1]. However, long follow-up is necessary, accompanied by side effects and low effectiveness. In this regard, the BLM have been used for testing the activity of different biological media including microorganisms [2–6]. The BLM is based on laser dynamic speckle interferometry [2]. It is well known that when a laser is reflected from a rough surface, a speckle pattern will be created at a given detection plane. If the surface is moving and when is being illuminated with coherent light, the speckle pattern of the scattered light also varies in time. Therefore, characterization of the speckle dynamics can provide information about the surface activity.

To analyze the evolution of dynamic speckle patterns, the autocorrelation of the irradiance as a function of time has been employed [4]. Other approaches include the generation of the time history speckle pattern (THSP) and its characterization using different descriptors [2], the generalized difference method [5], a modified version of the time correlation [6], the characterization of the time evolution of the texture of the speckle pattern [7], the speckle pattern contrast imaging [8], empirical mode decomposition [9], the Fujii difference method [10], and the temporal difference method [11, 12].

Additionally, other optical methods have been used for testing biological motility including a laser-diffraction capillary assay developed by Schmidt et al. [13]. Recently, Tan et al. have demonstrated the use of optical coherence tomography to evaluate the dynamic cell behavior [14]. However, those techniques require expensive equipment and also well-trained personnel. The BLM is a fast testing method and easy to implement and the assays based on its principles require minimal resources.

Drug susceptibility testing is a necessary step prior to the treatment of clinical infections. However, in case of severe parasitic infections fast and reliable results are necessary in a short period of time to determine the appropriate drug that must be used. Anthracyclines have been used in the design of anti-protozoal drugs, especially against trypanosomes [16, 17] and have been tested using traditional methods requiring long test periods. New approaches are needed in order to reduce the required time and the sample volume consumption.

The aim of this work is to use the BLM for drug susceptibility test of Epirubicin [17] on *T. cruzi* parasites. For this purpose, the biospeckle experiment was designed considering two well-known targets of the drug:

an interaction on the cell surface which is an almost instantaneous effect and an interaction with DNA which depends on internalization and transport of the drug to reach the target in the cell nucleus [18, 19].

We mainly proposed the dynamic speckle activity segmentation by spatial-temporal speckle correlation technique as an alternative and fast algorithm for detection of different degrees of motility of *T. cruzi* parasites during the incubation period. This approach enabled to obtain presumptive results of trypanocide action of the pharmaceutical product in a very short period of time. Additionally, it is very well-known that the average speckle size increases linearly with the distance from the scattering surface to the observation plane and decreases as the illuminated area increases [19]. Therefore, the computation of the second-order statistics of the speckle dynamics, namely the speckle grain size measurement provided information about the surface roughness and diffusion area [20]. Finally, speckle grain size evolution was used to characterize the physical processes occurring during immediate action of the drug on the parasites. Additionally, the methodology was validated by other digital speckle methods and good agreement was achieved comparing all the obtained results.

9.2 MATERIALS AND METHODS

9.2.1 SPATIAL-TEMPORAL SPECKLE CORRELATION TECHNIQUE

Spatial-temporal correlation evaluation of speckle signals gives information related with the speckle dynamics [21, 22]. The algorithm uses a correlation analysis of the temporal sequence of speckle patterns. For the computation, each speckle pattern in the sequence is divided into equal fragment matrices, and thereby calculating the correlation of each fragment pair of two patterns taken. Thus, the activity of the given fragment is characterized in terms of square matrix of correlation coefficients [21].

Each m, n'th correlation coefficient is given by:

$$Cm,n\,(t) = Cm,n\,(t_0 + k\tau)$$

$$= \left| \frac{\left\langle \left(S_{i,j}^{t_0} - \left\langle S_{i,j}^{t_0} \right\rangle \right) \left(S_{i,j}^{t_0+k\tau} - \left\langle S_{i,j}^{t_0+k\tau} \right\rangle \right) \right\rangle_{m,n}}{\left[\sqrt{\left\langle \left(S_{i,j}^{t_0} - \left\langle S_{i,j}^{t_0} \right\rangle \right)^2 \right\rangle} \cdot \sqrt{\left\langle \left(S_{i,j}^{t_0+k\tau} - \left\langle S_{i,j}^{t_0+k\tau} \right\rangle \right)^2 \right\rangle} \right]_{m,n}} \right| \quad (9.1)$$

where, i, j are the pixel number of the m, n'th fragments of the speckle pattern; $i = 1..., I; j = 1..., J; m = 1..., M; n = 1..., N; S_{i,j}$ is the i, j'th pixel intensity; k is the frame sequence ($k = 0, 1.$); t_0 is the time of the initial frame; τ is the temporal distance between two adjacent frames. The measured coefficients represent the normalized intensities of correlation peaks located in the center of each fragment, and temporal evolution of each m, n'th peak intensity corresponds to a biospeckle activity of the fragment.

Speckle homogeneity test was carried out to ensure that the biospeckle properties of each sub-image are homogeneous. In this case, the intensity is calculated as a mean of all intensities of all peaks and the correlation coefficient can be expressed as [22, 23]:

$$C(t) = \left| \frac{\left\langle \left(S_{im,jn}^{t_0} - \left\langle S_{im,jn}^{t_0} \right\rangle \right) \left(S_{im,jn}^{t_0+k\tau} - \left\langle S_{im,jn}^{t_0+k\tau} \right\rangle \right) \right\rangle}{\left[\sqrt{\left\langle \left(S_{im,jn}^{t_0} - \left\langle S_{im,jn}^{t_0} \right\rangle \right)^2 \right\rangle} \cdot \sqrt{\left\langle \left(S_{im,jn}^{t_0+k\tau} - \left\langle S_{im,jn}^{t_0+k\tau} \right\rangle \right)^2 \right\rangle} \right]} \right| \tag{9.2}$$

where, $im = 1, \ldots, I, \ldots, 2I, \ldots, MI$ and $jn = 1, \ldots, J, \ldots, 2J, \ldots, NJ$.

9.2.2 *SPECKLE GRAIN SIZE EVOLUTION*

A normalized auto-covariance function of the speckle intensity pattern $I(x, y)$ obtained in the observation plane (x, y) of the camera was measured to evaluate the speckle grain size. This function corresponds to the normalized autocorrelation function of the intensity. Its width provides a reasonable measurement of the average width of the speckle grain [20]. The autocorrelation function C can be calculated by:

$$c(x,y) = \frac{FT^{-1}\left\{ \left| FT\left[I(x,y) \right] \right|^2 \right\} - \left\langle I(x,y) \right\rangle^2}{\left\langle I(x,y)^2 \right\rangle - \left\langle I(x,y) \right\rangle^2} \tag{9.3}$$

where, FT is the Fourier Transform and ⟨⟩ is a spatial average. The horizontal dimension of the speckle grain denoted by dx is the full width at half maximum (FWHM) of the horizontal profile of $c(x, y)$.

9.2.3 ABSOLUTE VALUES OF DIFFERENCE (AVD)

The speckle data was also evaluated using absolute values of difference (AVD). In each well containing the assay, the region of interest (ROI) was selected. Each ROI was analyzed by means of the AVD calculation, through the creation of a THSP and then a co-occurrence matrix (COM). The AVD processing can be expressed as [24, 25]:

$$AVD = \sum_{i,j} COM_{i,j} \left| i - j \right|, \tag{9.4}$$

where, COM is the co-occurrence matrix related to the THSP and i and j represent the dimension of the COM matrix, respectively [2].

The qualitative biospeckle index adopted was the GD that can be expressed as [24, 25]:

$$GD = \sum_{k} \sum_{l} \left| I_k - I_{k+l} \right| \tag{9.5}$$

The GD outcomes were used to compare between the motility index of the assay before and after the incubation of the drug. The level of activity can be represented in colors, with blue for low activity and red for high activity.

9.3 EXPERIMENTAL

9.3.1 PREPARATION OF THE ASSAY

In the experiment, *liver infusion tryptose (LIT)* was used as a medium for the growth of *Epimastigote* forms of *T. cruzi* at 28°C. The reaction mixture (100 μl) containing in a well was prepared by adding 5 μl of either a solution of Epirubicin (2 mg/ml in saline) or saline to 95 μl of LIT culture medium containing 2.4×10^5 parasites. A 100 $\mu g.ml^{-1}$ concentration of Epirubicin was used in the assay well. The assay was carried out at room temperature (20–25°C).

9.3.2 EXPERIMENT SETUP

The scheme of the experimental setup is shown in Figure 9.1. As a coherent light source, we used a 1-mW He-Ne laser operating at 632.8 nm emission which was expanded by a 10-cm focal lens to create a 10-mm spot on the well plate with parasites. A high-resolution CCD camera (Thorlabs USB.2, 30 fps, 6.45-μm square pixels with a resolution of 1280 × 1024 Pixels) recorded videos with duration of 1 minute. The parameters of the CCD camera are: optical system of focal lengths range 3.5–75 mm with maximum aperture of up to f/0.95, as well as 18–108 mm f/2.5 zoom lens was used with the CCD camera. Successive speckles were collected with a frame rate of 30 frames per second.

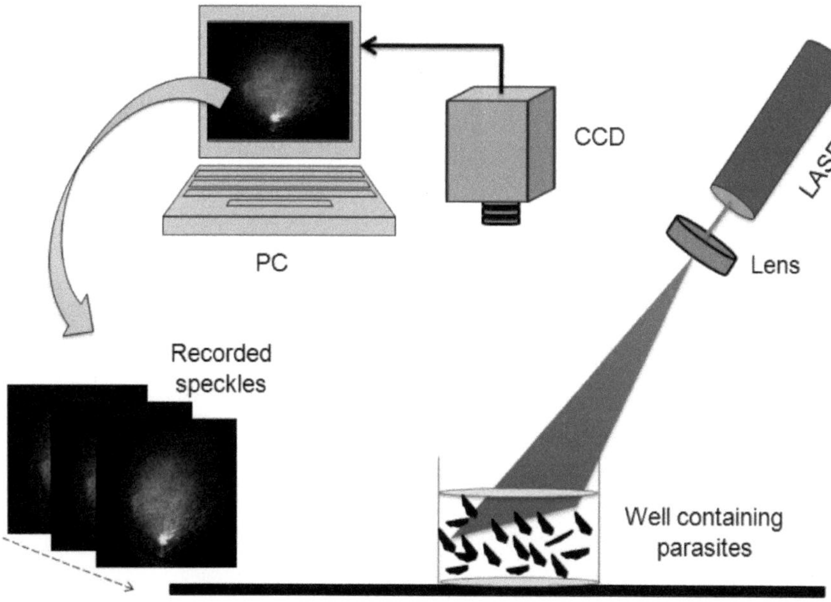

FIGURE 9.1 Schematic of the experimental setup used for the assessment of parasite motility.

9.4 RESULTS AND DISCUSSIONS

In this section, the trypanocidal effect of the drug on *T. cruzi* parasites was characterized using the second order statistics parameter. When the parasites are illuminated with a laser, a speckle pattern is obtained and a sequence of images is registered. Through the evaluation of the dynamic speckles using the methods discussed in Section 2.2, it becomes easier to identify different degrees of motility of the parasites during the Epirubicin effect.

9.4.1 SPECKLE CORRELATION ANALYSIS

After the action of the drug on the parasites, the time-varying correlation of speckle images was performed. The temporal analysis leads to evaluation of the effect of Epirubicin drug on *T. cruzi* parasites. The temporal correlation coefficient evolution *C*(*t*) for parasites undergoing different incubation periods of the drug is shown in Figure 9.2. Decorrelation is reduced with the increase of incubation period of the drug on the parasites. *C*(*t*) curves provide information about the motility index associated with the parasites before the treatment with Epirubicin as well as quantifies the motility in terms of the immediate drug action (t = 1 min) and after 15 minutes of incubation with the drug. Note that, before treatment parasites show a greater degree of motility but the correlation *C*(*t*) decreases fastly with time. The activity of the parasites just during the incubation in the first minute and after 15 minutes of incubation decreases which leads to a slower decrease of the

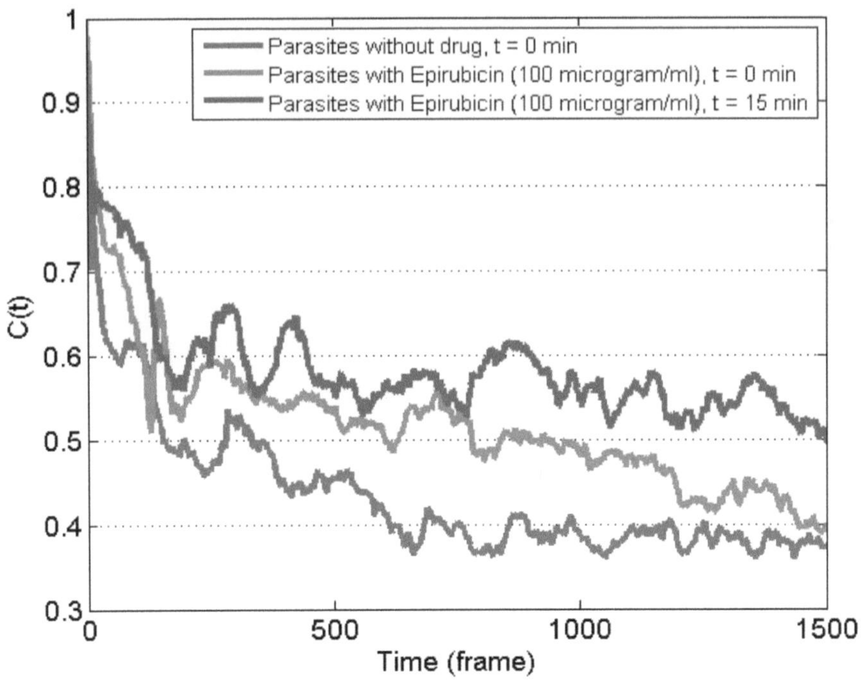

FIGURE 9.2 Evolution of the correlation coefficient versus correlation time (in frames) for parasites during different incubation times.

correlation coefficient and the reduced decorrelation. Thus due to the drug action, parasites are less active and reduce their movements which can be described as immediate effect of the drug on the parasites.

9.4.2 EVALUATION OF SPECKLE GRAIN SIZE

When considering second-order statistics of speckle images, the size of the speckle grain can provide information about surface roughness and diffusion surface as dx and dy are inversely proportional to the diameter of the diffusing area [20]. Thus, photons which are backscattered by the small surface as a result of the reduced activity of the parasites due to the action of the drug lead to larger speckle sizes. The evolution of the speckle grain size of dx is shown in Figure 9.3. Similar curves were obtained with a speckle grain size of dy (not shown here). Parasites show a relatively smaller speckle size before treatment with the drug. Before to be treated

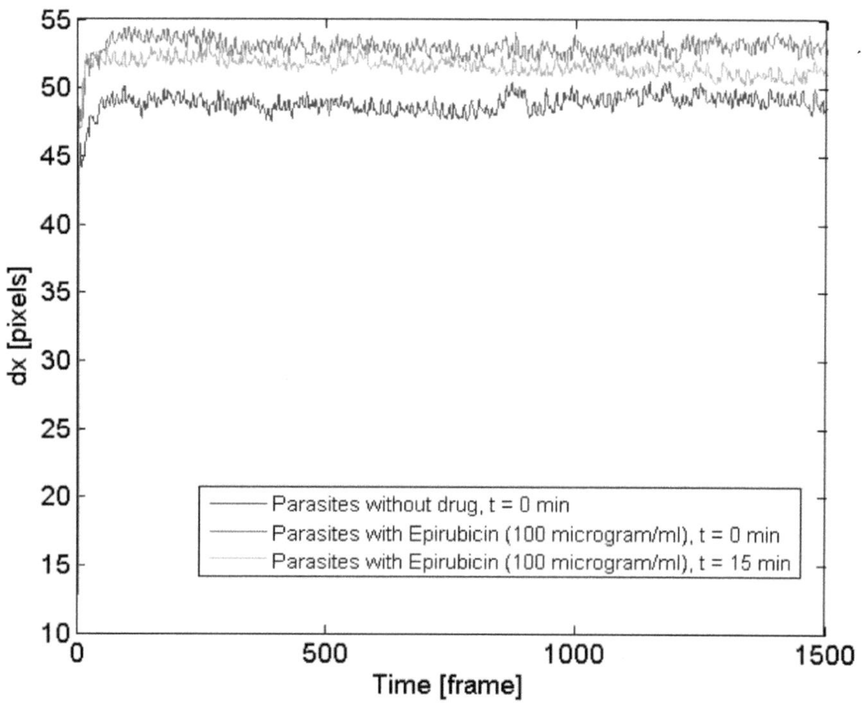

FIGURE 9.3 Variation of horizontal speckle grain sizes dx as a function of time (frames) during different incubation periods.

with the drug parasites show a higher degree of mobility, therefore the incident photons are backscattered by larger diffusion spots and thus produce a smaller speckle size [20]. This interpretation was previously observed by the spatial-temporal aspects of speckle patterns (Figure 9.2). Due to the drug incubation process, surface dimensions are continuously decreasing, leading to reduced surface scattering area. This, in turn generated a relatively larger speckle size (red line of Figure 9.3). The effect of the drug can be correlated with the increase of speckle grain size. As can be seen, Epirubicin has an instantaneous effect on the speckle dynamics as would be expected from an interaction of the drug on the cell surface, without entering the cell [18]. However, after 15 minutes of drug action the value of speckle grain size is shorter than in case of the results at the first minute, indicating that besides the effect of the Epirubicin, there was an additional effect may be related with the evaporation of the medium what introduces an erroneous interpretation. Therefore, care must be taken when designing the experiment in order to avoid the evaporation of the liquid. Anyway, still there is an appreciable difference when compared to the speckle grain size of the parasites without drug.

9.4.3 ABSOLUTE VALUES OF DIFFERENCE (AVD) EVOLUTION

Figure 9.4(a) shows the quantitative evaluation of parasites motility under the action of Epirubicin drug (100 μg/ml). The parasites activity before and after the drug incubation has been measured using the COM matrices. The COM matrix presents the number of occurrences of a particular intensity level followed by its effect in the characteristic dynamic image, called the THSP. The activity of the sample can be quantified as a measure of the spread values around the main diagonal. For the sample showing high activity, the corresponding COM matric resembles a cloud-like shape and the values lie far away from the principal diagonal. The values that represent the activity lie rather on or near the diagonal if the sample presents low activity.

The parasites not treated with the drug show a higher degree of motility which can easily be followed using the COM matrix as shown in Figure 9.4(a). After the drug action, the activity of parasites decreases as a result of the instantaneous interaction of the drug on the cell surface [26, 27]. Similarly and confirming our hypothesis, after fifteen minutes, one can observe a slight decrease in the activity of the assay, attributed to the evaporation of the saline solution as in case of Figure 9.3 and previous reports [28, 29].

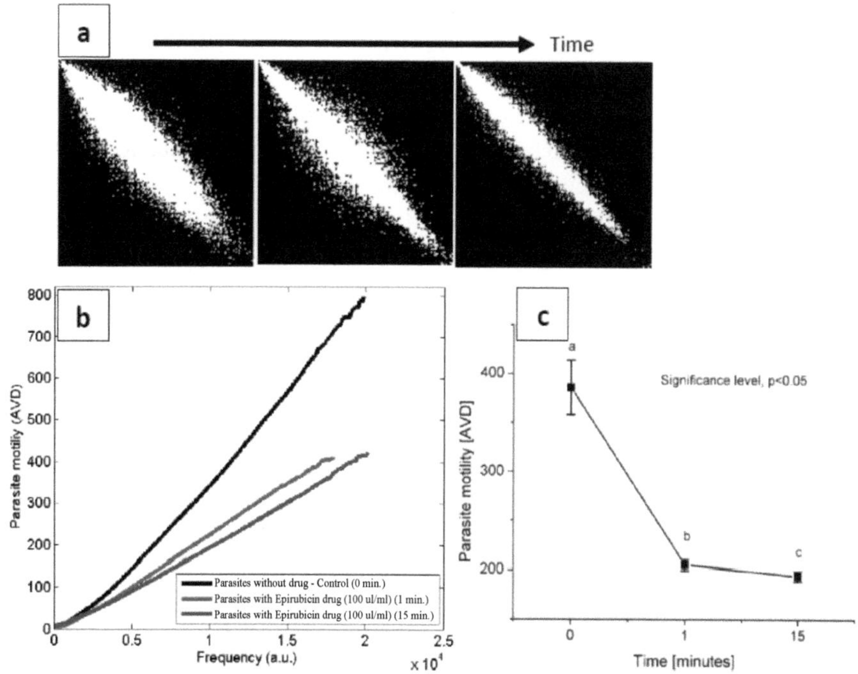

FIGURE 9.4 Quantitative evaluation of parasites motility under the action of Epirubicin drug (100 μg/ml). (a) COM matrices displaying different degrees of activity characterized by the number of occurrences of intensity values on the THSP images (respectively at 1 min (without drug control), 1 min of Epirubicin action, and 15 min of Epirubicin action); (b) corresponding AVD values, showing its characteristics variation with frequency; (c) Normalized AVD values quantifying the motility as a function of incubation time (minutes). Error bar corresponds to the standard error (SE). The letters denote significant differences ($p<0.05$) between the mean values of AVD using a statistical Tukey's test.

Figure 9.4(b) shows the corresponding AVD values, showing its characteristics variation with frequency. As shown in Refs. [30, 31], higher frequency contributes more to the AVD values and thereby quantifying the activity/motility of the assay.

Figure 9.4(c) presents the normalized AVD values as the motility measure of the parasites as a function of incubation time (minutes). As can be seen, there is an instantaneous and drastic change (decrease) in the parasites activity under the action of Epirubicin during the first minute of incubation. The assay shows a further decrease in its activity during 15 minutes of incubation. All the values are statistically significant ($p<0.05$) using a statistical Tukey's test.

Next, for a qualitative evaluation of the assay, an algorithm, the generalized differences (GD) as the spatial activity index was used and the results are presented in Figure 9.5.

Figure 9.5 shows the spatial activity distribution of the assay before and after the incubation with Epirubicin (100 μg/ml). One can easily differentiate between the motility of the parasites; the parasites without drug show a greater degree of motility in comparison to that treated with the drug (Figure 9.5a). The GD values for the parasites incubated with the drug have been decreased (Figure 9.5b and c) and that is due to the instantaneous action of the Epirubicin.

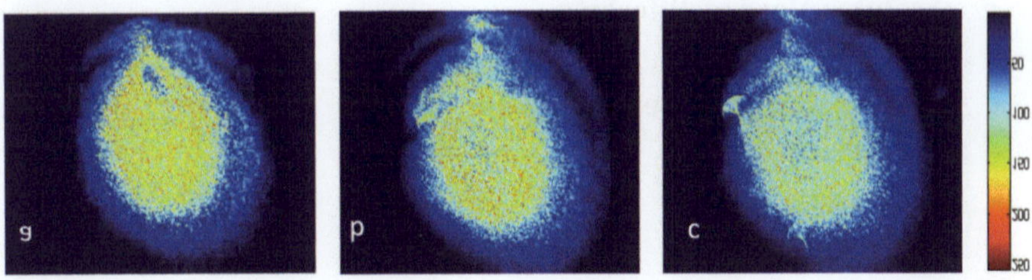

FIGURE 9.5 Motility index, the GD of the parasites incubated with Epirubicin drug (100 μg/ml). (a) t = 1 min (without drug), (b) t = 1 min (with Epirubicin) and (c) t = 15 min (with Epirubicin). The activity of the parasites incubated with the drug has been discriminated with that of treated without Epirubicin. The color bar represents the activity level with blue (low activity) to red (high activity).

Collectively, the results suggest that all analyzed digital methods were able in minor or major degree to detect and distinguish the instantaneous biological action of the drug on parasites. In general, it was possible to obtain a linear tendency which shows that the measured parameter of the biospeckle pattern of the parasites decreases as the time increases. The four evaluation methods used: Spatial-temporal correlation technique, speckle grain size, AVD, and GD susceptibility tests, show similar results, being able to distinguish among different situations with parasites and drug or without drug.

9.5 CONCLUSION

A biospeckle methodology was designed for the evaluation of drug susceptibility on *T. cruzi* parasites. Digital image processing was performed by four different methods that achieved the same results. The effective action of Epirubicin on *T. cruzi* parasites was clearly demonstrated in all cases. The results evidenced a very fast action of the drug that may be related to its instantaneous effect on the membrane of the parasites. All the methods used here reproduced the results with quite good correlation, thus demonstrating its applicability. The validated methodology open new research directions and it starts a possible way to perform a fast method to evaluate drug susceptibility on a biological system including bacteria, viruses, and microorganisms.

KEYWORDS

- **absolute values of difference**
- **biospeckle**
- **liver infusion tryptose**
- **speckle correlation**
- **speckle grain size**
- **standard error**

REFERENCES

1. Rivera, G., García, B., Pichardo, O., Torres, B. N., & Monge, A., (2009). New therapeutic target for drug design against *Trypanosoma cruzi*; advances and perspectives. *Curr. Med. Chem.,16*(25), 3286.–3293.
2. Arizaga, R., Trivi, M., & Rabal, H., (1999). Speckle time evolution characterization by the co-occurrence matrix analysis. *Optics and Laser Technology, 31*(2), 163–169.
3. Waterman-Storer, C. M., Desai, A., Bulinski, J. C., & Salmon, E. D., (1998). Fluorescent speckle microscopy, a method to visualize the dynamics of protein assemblies in living cells. *Current Biology, 8*(22), 1227–1230, S1.
4. Aizu, Y. A., & Asakura, T., (1991). Bio-speckle phenomena and their application to the evaluation of blood flow. *Optics and Laser Technology,23*(4), 205–219.
5. Arizaga, R. A., Cap, N. L., Rabal, H. J., & Trivi, M., (2002). Display of local activity using dynamical speckle patterns. *Optical Engineering, 41*(2), 287–295.
6. Pomarico, J. A., & DiRocco, H. O., (2004). Compact device for assessment of microorganism motility. *Review of Scientific Instruments, 75*(11), 4727–4731.
7. Limia, M. F., Nunez, A. M., Rabal, H., & Trivi, M., (2002). Wavelet transform analysis of dynamic speckle patterns texture. *Applied optics, 41*(32), 6745–6750.
8. Briers, J. D., & Webster, S., (1996). Quasi-real time digital version of flow monitoring using laser speckle contrast analysis (LASCA). *J. Biomed Opt., 2*, 174–179.
9. Federico, A., & Kaufmann, G. H., (2006). Evaluation of dynamic speckle activity using the empirical mode decomposition method. *Optics Communications, 267*(2), 287–294.

10. Ansari, M. Z., & Nirala, A. K., (2015). Biospeckle assessment of torn plant leaf tissue and automated computation of leaf vein density (LVD). *The European Physical Journal Applied Physics, 70*(2), 21201.
11. Martí-López, L., Cabrera, H., Martínez-Celorio, R. A., & González-Peña, R., (2010). Temporal difference method for processing dynamic speckle patterns. *Optics Communications, 283*(24), 4972–4977.
12. Minz, P. D., & Nirala, A. K., (2014). Intensity based algorithms for biospeckle analysis. *Optik-International Journal for Light and Electron Optics, 125*(14), 3633–3636.
13. Schmidt, S., Widman, M. T., & Worden, R. M., (1997). A laser-diffraction capillary assay to measure random motility in bacteria. *Biotechnology Techniques, 11*(6), 423–426.
14. Tan, W., Oldenburg, A. L., Norman, J. J., Desai, T. A., & Boppart, S. A., (2006). Optical coherence tomography of cell dynamics in three-dimensional tissue models. *Optics Express, 14*(16), 7159–7171.
15. Zuma, A. A., Cavalcanti, D. P., Maia, M. C., De Souza, W., & Motta, M. C. M., (2011). Effect of topoisomerase inhibitors and DNA-binding drugs on the cell proliferation and ultra structure of *Trypanosoma cruzi. International Journal of Antimicrobial Agents, 37*(5), 449–456.
16. Jobe, M., Anwuzia-Iwegbu, C., Banful, A., Bosier, E., Iqbal, M., Jones, K., & Ward, E., (2012). Differential *in vitro* activity of the DNA topoisomerase inhibitor idarubicin against *Trypanosoma rangeli* and *Trypanosoma cruzi. Memórias do Instituto Oswaldo Cruz, 107*(7), 946–950.
17. Triton, T. R., & Yee, G., (1982). The anticancer agent adriamycin can be actively cytotoxic without entering cells. *Science, 217*(4556), 248–250.
18. Chiquero, M. J., Pérez-Victoria, J. M., O'Valle, F., González-Ros, J. M., Del Moral, R. G., Ferragut, J. A., & Gamarro, F., (1998). Altered drug membrane permeability in a multidrug-resistant *Leishmania tropica* line. *Biochemical Pharmacology, 55*(2), 131–139.
19. Nassif, R., Pellen, F., Magné, C., Le Jeune, B., Le Brun, G., & Abboud, M., (2012). Scattering through fruits during ripening: Laser speckle technique correlated to biochemical and fluorescence measurements. *Optics Express, 20*(21), 23887–23897.
20. Muravsky, L. I., Maksymenko, O. P., & Frankevych, L. F., (2006). Studying of botanical specimen ageing with spatial-temporal speckle correlation technique." In: *3rd Int. Scientific Conf. Influence of Electromagnetic Field on Agricultural Environment "AGROLASER 2006"* (pp. 83–89). Lublin, Poland, Papers and short communications.
21. Ansari, M. Z., & Nirala, A. K., (2013). Biospeckle activity measurement of Indian fruits using the methods of cross-correlation and inertia moments. *Optik-International Journal for Light and Electron Optics, 124*(15), 2180–2186.
22. Zdunek, A., Muravsky, L. I., Frankevych, L., & Konstankiewicz, K., (2007). New nondestructive method based on spatial-temporal speckle correlation technique for evaluation of apples quality during shelf-life. *International Agrophysics, 21*(3), 305.
23. Ansari, M. Z., Ramírez-Miquet, E. E., Otero, I., Rodríguez, D., & Darias, J. G., (2016). Real time and online dynamic speckle assessment of growing bacteria using the method of motion history image. *Journal of Biomedical Optics, 21*(6), 066006.
24. Braga, R. A., González-Peña, R. J., Viana, D. C., & Rivera, F. P., (2017). Dynamic laser speckle analyzed considering in homogeneities in the biological sample. *Journal of Biomedical Optics, 22*(4), 045010.

25. Chiquero, M. J., Pérez-Victoria, J. M., O'Valle, F., González-Ros, J. M., Del Moral, R. G., Ferragut, J. A., & Gamarro, F., (1998). Altered drug membrane permeability in a multidrug-resistant *Leishmania tropica* line. *Biochemical Pharmacology, 55*(2), 131–139.
26. Triton, T. R., & Yee, G., (1982). The anticancer agent adriamycin can be actively cytotoxic without entering cells. *Science, 217*(4556), 248–250.
27. Ansari, M. Z., Grassi, H. C., Cabrera, H., & Andrades, E. D. J., (2016). Real time monitoring of drug action on *T. cruzi* parasites using a biospeckle laser method. *Laser Physics, 26*(6), 065603.
28. Ramírez-Miquet, E. E., Cabrera, H., Grassi, H. C., Andrades, E. D. J., Otero, I., Rodríguez, D., & Darias, J. G., (2017). Digital imaging information technology for biospeckle activity assessment relative to bacteria and parasites. *Lasers in Medical Science, 32*(6), 1375–1386.
29. Ansari, M. Z., & Nirala, A. K., (2013). Assessment of bio-activity using the methods of inertia moment and absolute value of the differences. *Optik, 124*(6), 512–516.
30. Ansari, M. Z., & Nirala, A. K., (2015). Assessment of Fevicol (adhesive) drying process through dynamic speckle techniques. *AIMS Bioeng., 2*(2), 49–59.
31. Ansari, M. Z., Grassi, H. C., Cabrera, H., Velásquez, A., & Andrades, E. D., (2016). Online fast biospeckle monitoring of drug action in *Trypanosoma cruzi* parasites by motion history image. *Lasers in Medical Science, 31*(7), 1447–1454.

CHAPTER 10

Hirak Bhasma: A Potential Ayurvedic Antibacterial Drug Assessed by *In Vitro* Pre-Clinical Studies

SUTAPA SOM CHAUDHURY,[1] BHUBAN RUIDAS,[1]
PRASANTA KUMAR SARKAR,[2] and
CHITRANGADA DAS MUKHOPADHYAY[1]

[1]*Center for Healthcare Science and Technology,
Indian Institute of Engineering Science and Technology, Shibpur,
Howrah – 711103, West Bengal, India,
E-mails: somchaudhurysutapa@gmail.com (S. S. Chaudhury),
chitrangadam@chest.iiests.ac.in (C. D. Mukhopadhyay)*

[2]*Department of Rasashastra, J. B. Roy State Ayurvedic Medical College
and Hospital, Kolkata – 700004, West Bengal, India*

ABSTRACT

The growing evidences of multi drug resistance are of genuine concern to combat nosocomial diseases in current time. Herein this study Hirak Bhasma (HB), the widely used herbometallic drug in Ayurveda, was prepared as a nano-drug component and evaluated thoroughly *in vitro* for the pre-clinical evaluation of an effective anti-bacterial drug candidate. HB showed rich mineral constituents along with some below detection level toxic elements such as Si, Ni and Al which enhance its potential as an antibacterial drug. The nano-dust preparation of the HB came with the presence of a higher organic compositions and mineral constituents having a particle size in nanometer range as was evidenced by the physic-chemical characterization via inductively coupled plasma optical expression spectrometry (ICP-OES), Fourier transform infrared spectra (FTIR), and field emission scanning electron microscopy (FE-SEM) techniques respectively. Antimicrobial potential of

HB was tested in clinically isolated pathogenic strains of *Escherichia coli, Staphylococcus aureus* and *Candida intermedia* through disc diffusion assay, broth turbidity measurements and cell imaging via FE-SEM. The non-cytotoxic dose of HB was found to be 50 μg/ml by the reduction of tetrazolium dye MTT (3-(4,5-dimethylthiazol-2-yl)-2,5-diphenyltetrazolium bromide) (MTT assay) on the human breast cancer cell line MCF7. HB was effective in increasing the intracellular reactive oxygen species (ROS) in the bacterial cells while being helpful to keep the intracellular redox balance in human cells. This proves that HB has the potential to be the effective anti-bacterial drug for the treatment of diseases caused by bacteremia with a minimal or no side effect. Moreover, in the context of the evaluation of the ROS level, HB appeared as a promising drug candidate for the anti-inflammatory and immune-modulatory effects although it has not shown any antifungal effect.

Since long the herbometallic ayurvedic compound HB claims its high potential as adaptogenic antibacterial agent, analgesic, antimicrobial, alternative antioxidant, anti-inflammatory, immunomodulator and so on; but there is lack of scientific evidences. This work reports the lagging scientific documentation for HB as antibacterial and anti-inflammatory agent.

10.1 INTRODUCTION

In the current scenario, the multi-drug resistance (MDR) of pathogenic bacterial strains has already rung an alarm to the treatment of nosocomial infections [1]. The overuses of broad-spectrum antibiotics and their prolonged applications are making the situation worsen day by day [2, 3]. Since last, few decades' combinatorial therapies are being proposed to combat MDR bacterial strains [4]. But the use of antibiotics in combinations is also in vain because the question has already been raised whether we are currently in the post-antibiotic era [5]. The increasing resistance to the antibiotics available until date necessitates the use of biocompatible natural compounds as an alternative medicine to combat bacterial pathogenesis without any side effects. Nowadays natural compound based Ayurvedic preparations are widely accepted as an effective antimicrobial drug [6, 7]. In Ayurveda, the traditional Indian medical system, Hirak Bhasma (HB) is such a preparation. This Ayurvedic medicine is mainly composed of Hirak, i.e., diamond dust. HB is prescribed for treating immunity disorders, crippling rheumatoid arthritis, bone marrow depression, cancer, and so on [8]. Here, we report a novel preparation of HB as a nano-drug component and its thorough *in vitro* evaluation for the pre-clinical assessment as an effective anti-bacterial drug candidate. Although, it has been claimed that HB is highly potential as an adaptogenic, antibacterial, anti-inflammatory, and immunomodulatory drug, no scientific evidence is available in support of this claim [9]. Metallic sources including sulfide-bearing minerals, metal oxides, and alumina silicates like biotite mica, chalcopyrite, and others have been used in traditional preparation as metallic antimicrobial sources. Interestingly, the antimicrobial function of metallic preparations is best achieved when the metal ions are effectively blended with the organic molecules [10, 11]. None the organic compounds alone or the metals singly can construct a highly effective antimicrobial drug. Ag, Au nanoparticles have also been in the limelight since few decades but anyone metal formula as a sole composition of an antimicrobial drug candidate may appear with the drawbacks of intense side effects [12, 13]. Destruction of pathogenic microbes *via* an elevated level of reactive oxygen species (ROS) is one of the prime mechanisms of these metal-based drug candidates, but this comes with the uncontrolled damage of the neighboring healthy cells in patients [14]. Here lies the importance of a perfect blend of organic compounds along with the trace metal components as in the HB. Being a balanced source of mineral components such as Zn, Cr which is effective against pathogenic microbes, the nanoformulation of HB may pave the path towards an alternative medicine in this premise [15, 16]. Also, HB is prescribed in Ayurveda to rejuvenate the body and mind [17]. This seems to be an advantage to regain the potentiation of the body in the disease condition such as bacteremia. But these claims need a scientific support. This work reports the lagging scientific documentation for HB as an antibacterial agent.

10.2 MATERIALS AND METHODS

10.20.1 PREPARATION OF HIRAK BHASMA (HB)

The HB (incinerated diamond powder) was procured from J. B. Roy State Ayurvedic Medical College and Hospital, Kolkata, West Bengal, India. The raw diamond powder was collected from Diamond Market, Surat, India. Briefly, the purification treatment of raw diamond powder was done by heating the powder to red hot and then quenching immediately in horse gram (Dolichosbiflorus) decoction. The procedure was repeated for 21 times. The obtained powder was taken in a mortar; levigation was done by horse gram (Dolichosbiflorus) decoction for six hours; pellets were prepared and dried. Those were taken in between two earthen saucers and the junction was sealed. The arrangement is called Sharava Samputa in parlance of Ayurveda. It was placed inside a muffle furnace and heating was done at 750°C for 1 h. After self-cooling, the pellets were collected and made into powder form. Levigation by horse gram decoction and heat treatment were done for 25 times. The powdered drug was further purified by sequential soxhlet extraction process with hexane, dichloromethane, and methanol and stored in a glass bottle (Scheme 10.1 and Figure 10.1). The final yield was about 10–15% of the initial raw material.

FIGURE 10.1 The image of procured HB powder.

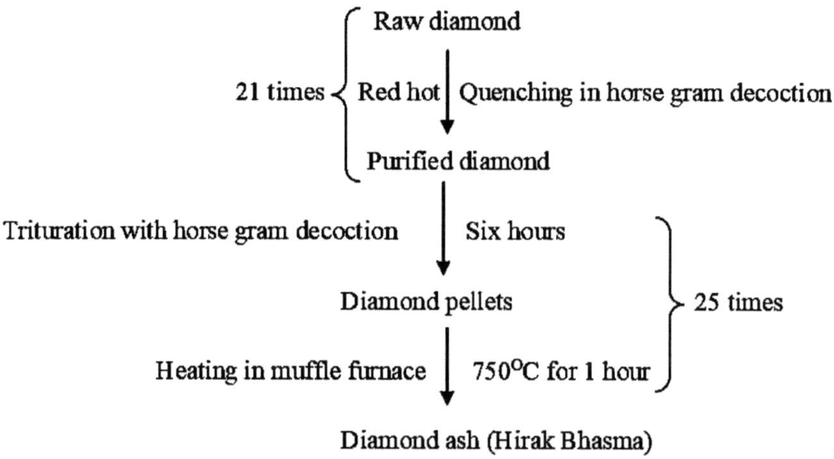

SCHEME 10.1 Flow chart representing preparation of Hirak Bhasma.

The solvents for the soxhlet extraction and all the other reagents were of analytical grade and commercially procured from Sigma-Aldrich (MO, USA) if not mentioned in the text.

10.2.2 PHYSICO-CHEMICAL CHARACTERIZATION OF HB

The total cation concentration of HB was screened with the aid of inductively coupled plasma-optical emission spectra (inductively coupled plasma optical emission spectra (ICP-OES); Thermo iCAP 7000 series). By analyzing the Fourier transform infrared spectrum (FT-IR) (Perkin-Elmer Spectrum 100 FT-IR spectrometer) the presence of organic fractions were detected in HB compound. The average particle size and morphology of HB was recorded under thefield emission scanning electron microscopy (FE-SEM) (Carl Zeiss Supra SEM instrument). The pH of the HB solution was recorded with the aid of a pH meter (Eutech pH 700).

10.2.3 DRUG RESISTANCE PROFILE OF THE CLINICALLY ISOLATED BACTERIA

In this study, two clinically isolated bacterial strains of *Escherichia coli* and *Staphylococcus aureus*, and one fungal strain of *Candida intermedia* were chosen to prove the antimicrobial action of HB. The MDR characteristic of these microbes was also confirmed following a standard protocol reported elsewhere [7]. Briefly, stock solutions of 100 μg/ml oxacillin, ampicillin, chloramphenicol, gentamycin, tetracycline, and levofloxacin each were prepared and diluted to a final concentration of 40 μg/ml. Approximately 20 ml LB agar (HiMedia Laboratories, India) containing these antibiotics each at their final concentration were plated. Then, a 100 μl of 10^8 CFU/ml log phase culture of the respective bacteria were spread on those agar plates and grown at 37°C for 24 h. Bacterial growth was measured in terms of the percentage of the colony-forming unit with respect to the initial inoculum. A plate containing 40 μg/ml of HB was compared with respect to those containing antibiotics.

10.2.4 ANTIMICROBIAL ACTIVITY ANALYSIS

The *in vitro* antibacterial ability of the HB was appraised using the disc diffusion method following the protocol by Zaidan et al. [20]. HB extract was tested (at 25 μg/ml–10 mg/ml) against clinically isolated *E. coli*, *S. aureus*, and *C. intermedia* by incubating them at 37°C for 24 h in LB agar and yeast extracts peptone dextrose (YPD) agar (HiMedia Laboratories, India), respectively keeping ampicillin as the positive control. Broth turbidimetric analyses were performed following a modified method by Berridge and colleagues to determine minimal inhibitory concentration (MIC) *via* the measurement of OD_{600} [21]. From the tubes where the respective cultures were first inhibited (determined by the absence of visible growth) by HB a 100 μl broth was taken and spread on a fresh LB agar plate each. Thus, the minimal bactericidal concentration (MBC) was determined. In addition, a bacterial cell viability assay was performed with an improved MTT assay, i.e., 3-(4,5-dimethylthiazol-2-yl)-2,5-diphenyltetrazolium bromide reduction assay (MTT assay kit, Invitrogen, Thermo Fisher Scientific Corporation) [22].

10.2.5 IMAGE DOCUMENTATION OF HB TREATED MICROBES

FE-SEM was performed to visualize morphological changes of drug treated microbes at different time interval compared to untreated microbes as control. Drug treated bacterial smear was heat-fixed and dipped in formaldehyde for 2 h at RT. After washing and proper dehydration, samples were dried in vacuum for 2–3 hours at RT and imaged by FE-SEM.

10.2.6 CYTOTOXICITY ASSAY

The non-cytotoxic dose of HB was checked by MTT assay on the human breast cancer cells MCF-7 reported elsewhere [23]. Briefly, 10,000 cells maintained in high glucose Dulbecco's modified eagle media (DMEM) (HiMedia Laboratories, India) along with 10% FBS and 1X penicillin-streptomycin were seeded per well of 96 well plates. The HB was added and tested over the concentration range of 10 μg/ml to 100 μg/ml. After 24 h treatment, the media was changed and 10 μl MTT solutions (1 mg/ml) were added to each well. It was incubated for 4 h and the resulting Formosan crystals were solubilized in dimethyl sulfoxide (DMSO) prior to the absorbance measurement at 570 nm wavelength.

10.2.7 INTRACELLULAR ROS MEASUREMENT

The increase in intra-cellular ROS with an increasing concentration of HB was measured by the fluorescence DCFDA method. For this, MCF-7 cells were treated with HB with an increasing concentration of 10–100 μg/ml for 12 h at 37°C and 5% CO_2. Following a washing step with 1X ice-cold PBS the cells were again incubated with 100 μM DCFDA (Invitrogen, Thermo Fisher Scientific Corporation) for 30 min in the dark at 37°C. The fluorescence intensity was measured spectroscopically with the excitation and emission wavelengths at 485 nm and 520 nm, respectively [24].

10.2.8 CHANGES IN CELLULAR ROS IN BACTERIAL CELLS

Change in the intracellular ROS after drug treatment in microbes was evaluated following the protocol by Su et al. *via* the DCFHDA method, to determine the mechanism of the drug action [25]. Briefly, 4 ml LB broth was inoculated with 2% log-phase bacterial culture and treated with 10–100 μg/ml HB at 37°C for 1 h. To this bacterial culture 10 μM DCFHDA (Invitrogen, Thermo Fisher Scientific Corporation) was added and incubated in dark for 30 min before recording the emission spectra at 529 nm with the excitation wavelength at 504 nm.

10.3 RESULT AND DISCUSSION

The presence of major inorganic elements like Fe, Al, and Ca was documented by the ICP-OES study (Table 10.1). The bacterial growth inhibiting influential elements were Ni, Si, Cr, and Zn, which were present as parts per million (ppm) levels in the final nano preparation of HB. These implied low toxicity of HB as a drug candidate as expected (Figure 10.7). Moreover, these heavy metals altogether may impact a synergistic effect to bring out the antibacterial action against pathogenic *E. coli* and *S. aureus* tested.

The Bhasma are claimed to be biologically produced nanoparticles [19]. The organic compounds present in HB were detected *via* the absorption peaks (between 650–1020 cm^{-1} due to C-H and C-N group, between 1410–1420 cm^{-1} due to C-O, C-CH_2 and N-H group and between 3200–3600 cm^{-1} due to OH group) obtained from FTIR study (Figure 10.2A). This proved the presence of organic components along with the cationic metals (i.e.,Fe, Ca, Cr, Zn, Al, etc., detected in ICP-OES) in HB preparation. The average diameter of the HB nanoparticles was about 30 nm as depicted by the FE-SEM analyses (Figure 10.2B).

TABLE 10.1 ICP-OES Analysis has Shown the Total Cation Concentration of the HB

Element	Concentration (ppm)
Fe	730.48
Al	270.592
Ca	70.355
Ni	16.05
Si	8.03
Cr	28.007
Zn	32.79

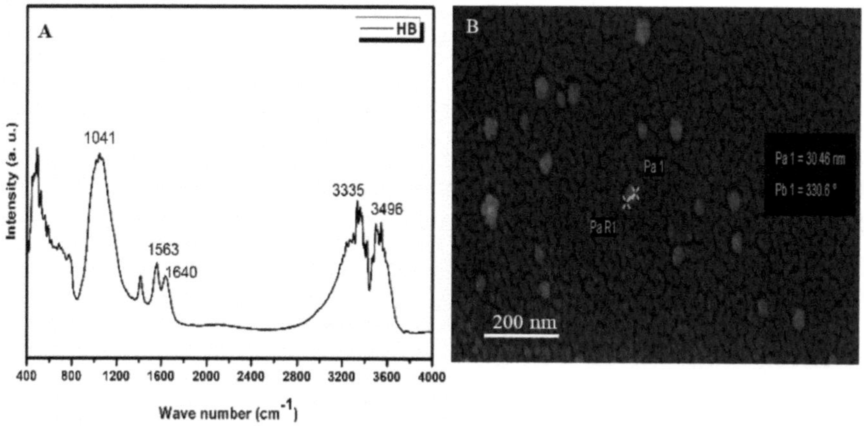

FIGURE 10.2 FTIR profile of HB (A) and FE-SEM micrograph showing the particle size of HB (B).

HB has shown an antibacterial activity at a lower concentration of 50 µg/ml in the disc diffusion method (Figure 10.3A and B). HB did not show any anti-fungal activity on testing with *C. intermedia* (data not shown).

By testing the sensitivity of the bacterial cultures for commercially available antibiotics, it was observed that the clinically isolated strains of *E. coli* and *S. aureus* were moderately resistant to the penicillin group of antibiotics and also to the protein synthesis inhibiting antibiotics like chloramphenicol, tetracycline, and gentamycin. Although in comparison with the above-mentioned antibiotics, HB has shown a greater efficiency as a bactericidal drug candidate (Table 10.2).

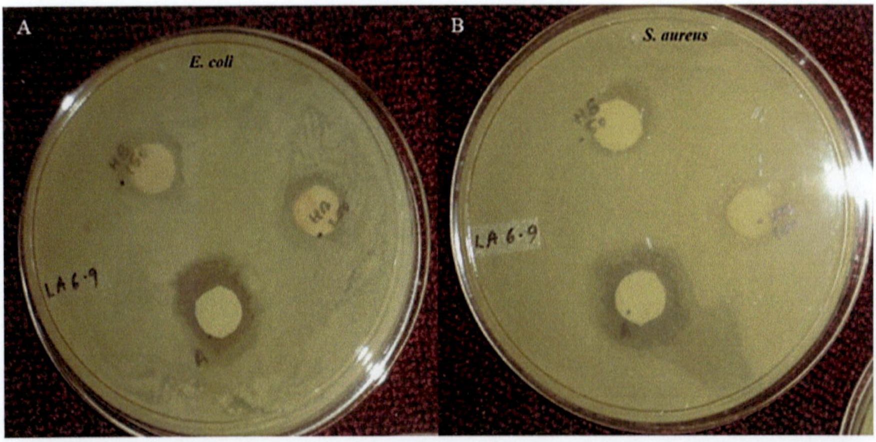

FIGURE 10.3 Zones of inhibition against *E. coli* (A) and *S. aureus* (B) for the concentrations of 50–200 μg/ml of HB with respect to the positive control ampicillin (200 μg/ml).

TABLE 10.2 The Antibiotic Resistance Profile of the Clinically Isolated Bacterial Strains Used in the Study

Antibiotics Tested	% Viability of Bacterial Strains (CFU/ml)	
	E. coli	*S. aureus*
Oxacillin	45.9	26.08
Ampicillin	26.5	20.8
Chloramphenicol	43	65.7
Gentamicin	39	45.2
Tetracycline	23.8	29.3
Levofloxacin	69.7	87
HB	12.6	17.5

The MIC of HB was 46 μg/ml against *E. coli* and 115 μg/ml against *S. aureus,* respectively when tested over a concentration range from 10–500 μg/ml in broth dilution method of MIC (Table 10.3). The MBC of HB for *E. coli* and *S. aureus* was determined as 50 μg/ml and 130 μg/ml, respectively (data not shown here). Interestingly, HB was equally effective against both the Gram-positive *S. aureus* and Gram-negative *E. coli.* This signified the good penetration ability of HB irrespective of the thick peptidoglycan layer of Gram-positive bacterial cells. The nano-sized particles of HB with an average 30 nm diameter were advantageous to penetrate the bacterial cells.

TABLE 10.3 The MIC of HB was Found as 46 μg/ml and 115 μg/ml Against *E. coli* and *S. aureus*, respectively as Measured by the OD_{600} Value in the Broth Dilution Method*

Concentration of HB (μg/ml)	Negative Control	OD_{600} *E. coli*	OD_{600} *S. aureus*
10	0.004	0.59	0.48
20	0.09	0.47	0.39
30	0.008	0.33	0.27
40	0.04	0.18	0.22
50	0.06	—	0.17
100	0.009	—	0.12
200	0.01	—	—
300	0.01	—	—
400	0.004	—	—
500	0.006	—	—

* LB broth was kept as negative control.

The improved MTT assays for the above-mentioned bacteria (Figure 10.4) and SEM images of the drug-treated microbes also supported this anti-bacterial activity of HB (Figure 10.5A–D).

The probable mechanism for the antibacterial activity of HB might be the increment in intracellular ROS level in the bacteria (Figure 10.6A). Also, HB was able to increase the intracellular ROS level in MCF-7 cancer cells (Figure 10.6B). This suggested the triggering of the downstream cell signaling cascades leading to the destruction of pathogenic gram-positive and gram-negative bacterial cell walls as well as the controlled demolition of cancer cells. Thus, we can say that HB may influence the potentiation of human body by regulating the immunomodulation and destroying the pathogens present in the body in disease condition.

HB was proved to be a safe drug candidate as its IC_{50} value was quite high (50 μg/ml) when tested on MCF-7 cell line (Figure 10.7).

10.4 CONCLUSION

In conclusion, the presence of various mineralogical constituents of HB has probably enhances the effective action against the bacterial strains synergistically. Most of the element individually plays an important role with effective interference in living system. Thus, the presence of

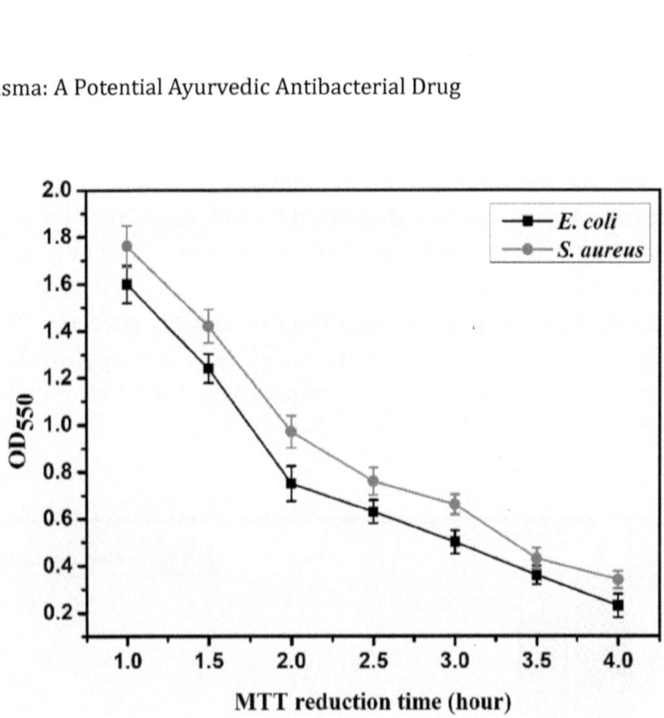

FIGURE 10.4 Time course profile of MTT reduction time for *E. coli* and *S. aureus*.

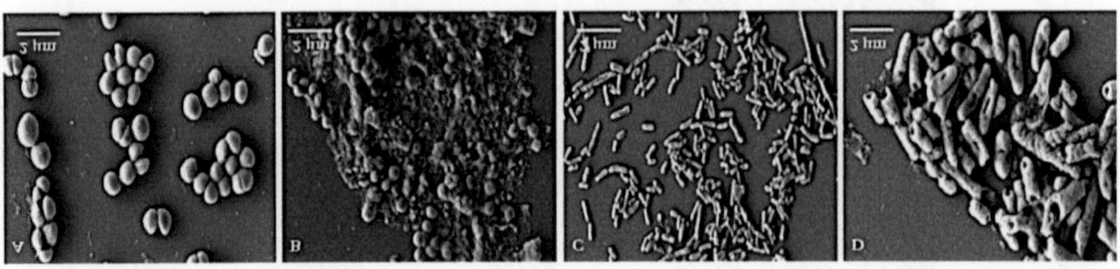

FIGURE 10.5 FE-SEM image of control *S. aureus* (A), HB-treated *S. aureus* at 24 hours (B), control *E. coli* (C) and HB-treated *E. coli* at 24 hours, (D) Scale bar is of 2 μm.

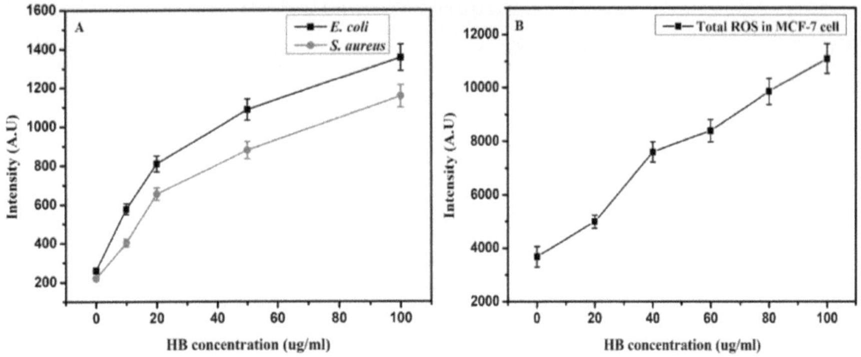

FIGURE 10.6 Increase in intracellular ROS level in *E. coli* and *S. aureus* (A) and in MCF-7 cells (B), with the increase in HB concentration.

different elements together might have played significant role to enhance ROS generation which in turns induces the oxidative stress leading to cell death. Moreover, Enhancement in ROS level may influence the internal signaling cascade directly which is actively associated with cell wall synthesis in both Gram-positive and Gram-negative bacteria. Therefore, a significant ROS enhancement in bacterial cells is a key mechanism of HB mediated cell wall destruction irrespective of the Gram character of the bacteria.

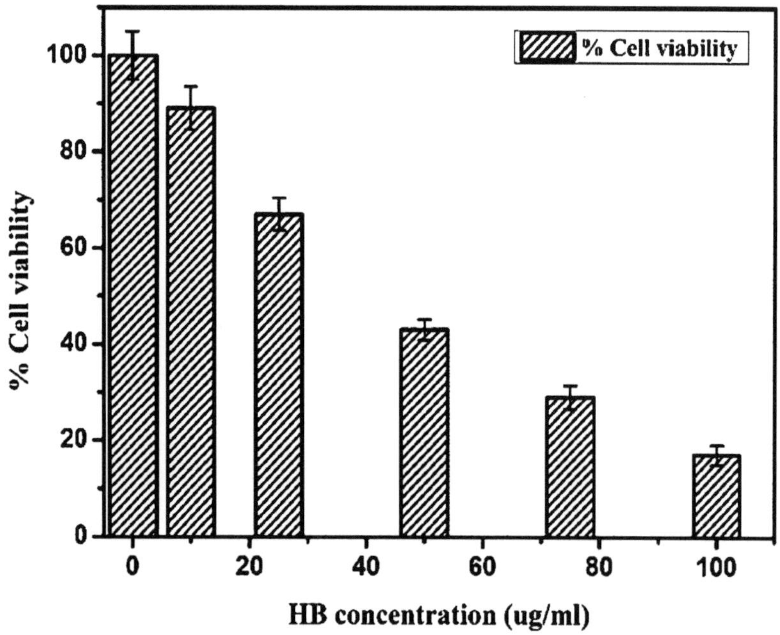

FIGURE 10.7 The cytotoxicity of HB. The IC_{50} value of 50 μg/ml has shown the non-toxic nature of HB.

10.5 FUTURE PROSPECT

The age-old claims of HB to be effective against many diseases including cancer, bacteremia, inflammatory diseases, and neurological disorders and to rejuvenate the body and mind needs more precise scientific documentations. Our report here may file the evidences for the antibacterial profiling of HB leaving a hint on the anticancer and immunomodulatory role of HB. We suggest a future endeavor of the immunomodulatory role of HB in neurodegenerative diseases like Alzheimer's and Parkinson's disease (PD) which still lack a therapeutic drug and are coping up with the symptomatic drugs marketed so far.

KEYWORDS

- **field emission scanning electron microscopy**
- **Fourier transform infrared**
- **Hirak Bhasma**
- **inductively coupled plasma optical emission spectra**
- **minimal bactericidal concentration**
- **multidrug resistance**

REFERENCES

1. Matzov, D., et al., (2017). A bright future for antibiotics? *Annu. Rev. Biochem., 86*, 567–583.
2. Nikaido, H., (2009). Multidrug resistance in bacteria. *Annu. Rev. Biochem., 78*, 119–146.
3. Livermore, D. M., (2004). The need for new antibiotics. *Clinical. Microbiol., 10*, 1–9.
4. Perez, F., et al., (2007). Global challenge of multidrug-resistant *Acinetobacter baumannii. Antimicrobial Agents and Chemotherapy, 51*, 3471–3484.
5. Alanis, A. J., (2005). Resistance to antibiotics: Are we in the post-antibiotic era? *Arch Med. Res., 36*, 697–705.
6. Wijenayake, A., et al., (2014). The role of herbometallic preparations in traditional medicine: A review on mica drug processing and pharmaceutical applications. *J. Ethnopharmacol., 155*, 1001–1010.
7. Ruidas, B., et al., (2019). A novel herbo metallic nano drug has the potential for antibacterial and anticancer activity through oxidative damage. *Nanomedicine, 14*, 1173–1189.
8. Sharon, M., (2019). *History of Nanotechnology: From Prehistoric to Modern Times*. John Wiley & Sons: New Jersey.
9. Singh, J., (2017). *Classical Ayurvedic Medicine Hirak Bhasma-Vajra Bhasma. Ayur Times*. https://www.ayurtimes.com/heerak-bhasma-hirak-bhasma/ (accessed on 24 June 2020).
10. Wu, J., et al., (2011). The medicinal use of realgar (As4S4) and its recent development as an anticancer agent. *J. Ethnopharmacol., 135*, 595–602.
11. Lemire, J. A., et al., (2013). Antimicrobial activity of metals: Mechanisms, molecular targets and applications. *Nat. Rev. Microbiol., 11*, 371–385.
12. Choi, O., & Hu, Z., (2008). Size dependent and reactive oxygen species related nano silver toxicity to nitrifying bacteria. *Environ. Sci. Technol., 42*, 4583–4588.

13. Yeh, Y. C., et al., (2011). Gold nanoparticles: Preparation, properties, and applications in bionanotechnology. *Nanoscale, 4*, 1871–1880.
14. Cabiscol, E., et al., (2000). Oxidative stress in bacteria and protein damage by reactive oxygen species. *Int. Microbiol., 3*, 3–8.
15. Pal, D., & Gurjar, V. K., (2017). Nanometals in bhasma: Ayurvedic medicine. In: *Metal Nanoparticles in Pharma* (pp. 389–415). Springer, Cham.
16. Kalantari, N., (2008). Evaluation of toxicity of iron, chromium, and cadmium on *Bacillus cereus* growth. *Iran. J. Basic. Med. Sci., 10*, 222–228.
17. Crasto, A. M., (2014). *Bhasma: The Ancient Indian Nanomedicine*. New drug approvals. https://newdrugapprovals.org/2014/02/26/bhasma-the-ancient-indian-nanomedicine/ (accessed on 24 June 2020).
18. Kulkarni, D. A., (1998). *Vagbhattachariya's Rasa Ratna Samucchaya.* Meharchand Lachhmandas Publications: New Delhi.
19. Sarkar, P. K., & Chaudhuri, A. K., (2010). Ayurvedic Bhasma: The most ancient application of nanomedicine. *J. Sci. Ind. Res., 69*, 901–905.
20. Zaidan, M. R., et al., (2005). *In vitro* screening of five local medicinal plants for antibacterial activity using disc diffusion method. *Trop. Biomed., 22*, 165–170.
21. Berridge, N. J., & Barrett, J. A., (1952). Rapid method for the turbidimetric assay of antibiotics. *J. Gen. Microbiol., 6*, 14–20.
22. Wang, H., et al., (2010). An improved 3-(4, 5-dimethylthiazol-2-yl)-2, 5-diphenyl tetrazolium bromide (MTT) reduction assay for evaluating the viability of *Escherichia coli* cells. *J. Microbiol. Methods., 82*, 330–333.
23. Som Chaudhury, S., et al., (2019). A novel PEGylated block copolymer in new age therapeutics for Alzheimer's disease. *Mol. Neurobiol., 56*, 6551–6565.
24. https://www.abcam.com/ps/products/113/ab113851/documents/ab113851%20DCFDA%20Booklet_20181001_MSC%20v2a%20(website).pdf (accessed on 24 June 2020).
25. Su, H. L., et al., (2009). The disruption of bacterial membrane integrity through ROS generation induced by nanohybrids of silver and clay. *Biomaterials, 30*, 5979–5987.

CHAPTER 11

Pathophysiology of Fluorosis and Calcium Dose Prediction for Its Reversal in Children: Mathematical Modeling, Analysis, and Simulation of Three Clinical Case Studies

SUJA GEORGE, A. B. GUPTA, MAYANK MEHTA, and AKSHARA GOYAL

Department of Chemical Engineering, Malaviya National Institute of Technology Jaipur, Jaipur, Rajasthan – 302017, India,
E-mail: sgeorge.chem@mnit.ac.in (S. George)

ABSTRACT

Long-term ingestion of fluoride through food and drinking water sources cause dental, skeletal, and non-skeletal manifestations. Severe grades of skeletal fluorosis are indicated to be an irreversible process in the human body. However, few medical studies especially in children are available indicating the feasibility of its treatment and reversal, even though no mathematical analysis or modeling has been attempted by these researchers. Three published Indian double-blind clinical studies on the reversal of skeletal fluorosis in children has been analyzed to investigate the effects of increased fluoride exposure, and develop a predictive mathematical model for predicting the doses of calcium which is crucial for reversing the effects of fluoride ingestion. Bivariate analysis was used for quantitatively representing the steps involved in the pathophysiology of fluoride intake in human body. The probability of skeletal fluorosis occurrence was classified on the basis of physical parameters namely: age, weight, and fluoride intake. Multilayer perceptron (MLP) technique which adopted the neural network approach was used to obtain the normalized importance of various biochemical parameters. The resulting significant parameter, serum GAG, along with calcium retention equation was used to develop a model to predict the doses of calcium which can reverse the effects of fluoride named as reduced retention model (RRM). The calcium doses predicted by RRM at different drinking fluoride levels were in coherence with the doses recommended by clinicians and therefore predicted doses could be used as guidelines for clinical recommendations. The results of MLP analysis also support the possibility of the use of serum GAG as a parameter in diagnosing fluorosis at an early stage in developing children, which may prove useful in improving the efficacy of treatment.

11.1 INTRODUCTION

Fluorosis is a well-known endemic disease in many countries across the globe including Mexico, Australia, India, Africa, Thailand, China, and Sri Lanka. According to the various field surveys, it has been found that 15 States in India are endemic for fluorosis (fluoride level >1.5 mg/l in drinking water from groundwater supply) [1]. The presence of fluoride in the groundwater resources is due to the leaching of the fluoride-rich rocks such as Fluorspar CaF_2 (Sedimentary rocks, limestones, sandstones), Cryolite Na_3AlFPV_6 (Igneous rocks, Granite), Fluorapatite $Ca_3(PO)_2$ $Ca(FCl)_2$,etc., and when water percolates through these rocks, the fluoride is leached out from them. In India in the state of Rajasthan, all the 32 districts are fluorosis prone and the inhabitants are consuming water with high fluoride concentrations (8 to 10 mg/l) and some even up to 44 mg/l [2]. World Health Organization (WHO) has set the prescribed upper limit of drinking water fluoride concentration at 1.5 mg/l [3], whereas the Bureau of Indian Standards has put down Indian standards as 1.0 mg/l with further remarks as "lesser the better" [4].

11.2 PATHOPHYSIOLOGY OF FLUOROSIS

Fluorosis is normally found to have three different forms in the human body: clinical, skeletal, and dental [5]. Until the recent times, fluorosis was described as irreversible, but Gupta et al. [2] reported reversal of fluorosis among children by controlled supplements of calcium, vitamin D and vitamin C. The pathophysiology of fluorosis proposed by Gupta et al. [2] is given in Figure 11.1, which describes how the biochemical substances within our body fluctuate after fluoride ingestion to maintain chemical homeostasis.

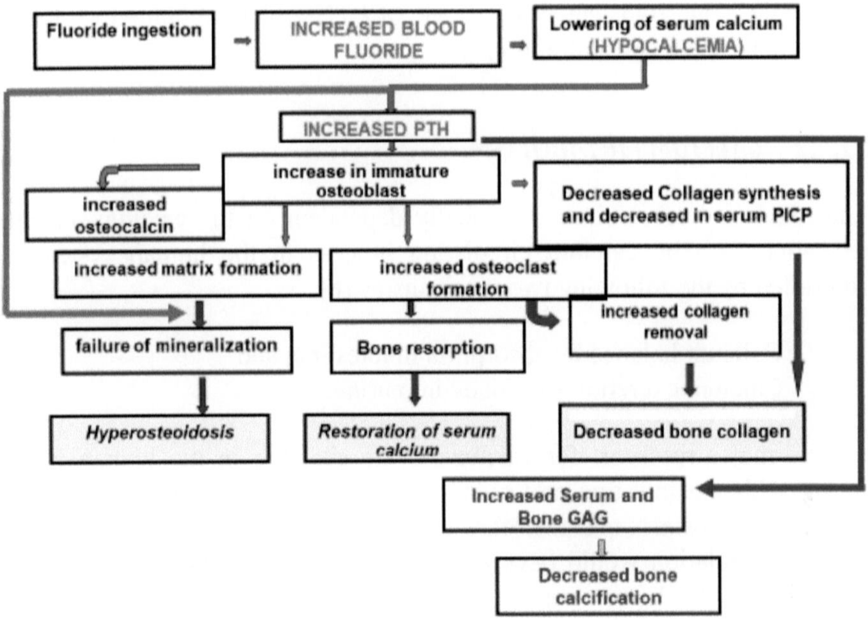

FIGURE 11.1 Pathophysiological process diagram for skeletal fluorosis.

11.2.1 EFFECT OF FLUORIDE INGESTION ON CIRCULATORY CALCIUM

Calcium is one of the key ions affected by fluoride intake; and the role of calcium supplementation is termed as an essential step in the treatment of fluorosis. Oral calcium supplementation not only helps in chelating the fluoride ion in the gut forming insoluble compound calcium fluoride (CaF_2), but also helps to maintain the calcium homeostasis in the body and as a net result reduces the conditions of calcium stress due to fluoride ingestion.

Fluoride Ingestion causes a decrease in the ionized calcium in the human blood (hypocalcemia). This hypocalcemia changes the internal milieu of the body in order to maintain the calcium levels and leads to secondary hyperparathyroidism. It is well known that ionic calcium is one of the important ions for the initiation and maintenance of the activity of the vital organs and musculoskeletal system.

Lowering of the ionized calcium is one of the important stimulants for the release of parathyroid hormone (PTH). When the blood ionized calcium is lowered by an amount as low as 0.02 mmol/l within 30 min, it elicits an immediate large, transient peak release of PTH which amounts to 6 to 16 times the baseline concentration [6].

11.2.2 CALCIUM METABOLISM

The regulation of calcium is basically dependent on the calcium-sensing receptors [7]. The calcium circulation process in the human system is controlled by the following two mechanisms [8]:

1. Calcium balance by absorption in intestine; and
2. Calcium excretion by kidney into urine.

The major reservoir for calcium is the skeleton in the human body. Along with the intestine and kidney, the skeleton has the major role in calcium homeostasis and therefore any regulatory changes in the calcium level in circulation would have a significant impact on bone metabolism. Any prolonged calcium stress in the human body which may sometimes be caused due to insufficient dietary calcium intake, may cause disturbance in the in-vivo availability of calcium as well as its absorption causing deleterious effects on the skeletal system or the bone mass. Hypocalcemia will also lead to increase in serum PTH [9].

11.2.3 EFFECT OF INCREASED PTH ON BONE

The presence of increased PTH in the blood stream is responsible for the following two major body mechanisms:

1. **Significant Loss of Bone Mass:** Bone loss is due to increase in the number of osteoblasts and the increase of their activity [9, 10] which causes the stimulation of bone resorption and depletion of bone mass as well as its formation. Osteoclast is a type of bone cell that resorbs bone tissue. Under the normal conditions, where there is an increase in bone resorption, it gets coupled with an effective compensatory increase in an equal magnitude of bone formation and therefore no net bone mass is depleted in the skeletal system. However, during the adjustor mechanism if there is a demand to mobilize calcium from skeletal system to counteract the effects of hypocalcemia, then the bone coupling process gets compromised.

2. **Depletion of Bone Formation:** It has been reported that despite the effectiveness in a significant increase in bone resorption, bone formation decreased and was significantly inhibited due to PTH [9]. Therefore, the combination of various combined actions such as calcium depletion, bone resorption and decrease in its formation led to a significant loss of bone mass [9].

To summarize the above mechanisms, prolonged fluoride ingestion will cause hypocalcemia, secondary hyperparathyroidism causing increased bone resorption, defective collagen formation, and defective bone formation which explains all the clinical, dental, and skeletal presentations of fluorosis. It is therefore hypothesized that all these effects are mediated through hyperparathyroidism secondary to increased fluoride ingestion.

11.3 MATHEMATICAL MODELING

A mathematical model has been developed for the pathophysiology proposed by supplementing with clinical data studies, that predicts for the dose of calcium supplementation required for preventing fluorosis based on drinking water fluoride levels as well as clinical and biochemical parameters such as serum calcium, blood fluoride, serum GAG, etc., of the patients suffering from fluorosis. The analysis also predicted that serum GAG to be the notable index of skeletal fluorosis. The model predicts to evaluate for the occurrence of skeletal fluorosis on the basis of physical parameters only, i.e., a child's age, its weight, and daily fluoride intake through drinking water and determines whether these parameters are sufficient for providing its dependency on occurrence of fluorosis.

11.3.1 FLUOROSIS DATA SETS

Only few clinical studies have been reported on the treatment and reversal of skeletal fluorosis, even though the problem has been recognized in India since 1937 [11]. In the present study, data derived from two clinical case studies from the villages of Jaipur, Rajasthan, India carried out by Gupta et al. [2, 12] and one from Delhi, India reported by Susheela and Bhatnagar [13] abridged in Table 11.1, were used for development of mathematical model to estimate calcium requirement for excess fluoride ingestion.

TABLE 11.1 Summary of Reported Data with Major Results of Case Studies: (1) Gupta et al. [2], (2) Gupta et al. [12] and (3) Susheela and Bhatnagar [13]

Case Studies and References	Study Location	No. of Cases		Age Range (Years)	Average Drinking Fluoride Level (mg/l)	Results			
		Exposure Group	Control Group			No. of Complete Reversal*	No. of Partial Reversal**	No. of Lost Cases	Biochemical Parameters
Case Study 1: Gupta et al. [2]	Rampura, Jaipur, India	15	15	3–15	4.6	13	0	2	Increase in the level of SIP, SAA, LAA, PTH, and decrease in the level of serum Ca, SAP, serum GAG, serum F, blood F, urinary F. No reversal was observed in the control group.
	Ramsagarki Dhani, Jaipur India	15	15	5–14	2.4	12	1	2	
	Shivdaspura, Jaipur, India	15	15	7–12	5.6	12	1	2	
	Raipuria, Jaipur, India	15	15	6–15	13.6	11	4	0	
Case Study 2: Gupta et al. [12]	Shivdaspura, Jaipur, India	20	—	3–12	4.5	9	5	6	Increase in level of serum Ca, SAA, and LAA and decrease in the level of SAP, serum F, blood F, urinary F
	Vanasthali, Jaipur, India	20	—	3–12	8.5	13	2	5	
Case Study 3: Susheela and Bhatnagar [13]	Delhi, India	10	—	8–60	7	10	0	0	Decrease in the level of serum F and urinary F

*Complete reversal refers to cases when fluorosis got reversed to grade 0.

** Partial reversal refers to cases when fluorosis got reversed but not to grade 0.

11.3.1.1 *CASE STUDY 1: VILLAGES OF JAIPUR, RAJASTHAN*

The data were obtained from the PhD thesis titled Environmental Health Perspective of Fluorosis in Children by Gupta et al. [2]. It pertained to four fluorosis endemic villages of Rajasthan namely Rampura, Ram Sagar Ki Dhani, Shivdaspura, and Raipuria having average drinking water fluoride levels of 4.6 mg/l, 2.4 mg/l, 5.6 mg/l and 13.6 mg/l, respectively, the details of which are shown in Table 11.1. The data are related to 50 fluorosis affected children aged up to 12 years, comprising different socioeconomic groups. In the present study, the following physical and biochemical variables were selected based on their relevance, from above thesis [2] to determine the severity of fluorosis.

1. The physical parameters are:
    - age;
    - weight;
    - drinking fluoride level;
    - grade of skeletal fluorosis.
2. The biochemical parameters are:
    - calcium in serum;
    - ascorbic acid in serum (SAA);
    - leucocyte ascorbic acid (LAA);
    - glucosamineglycans (GAG) in serum;
    - alkaline phosphate in serum (SAP);
    - fluoride in serum;
    - fluoride in urine.

11.3.1.2 *CASE STUDY 2: SHIVDASPURA AND VANASTHALI, JAIPUR, RAJASTHAN*

The data were obtained from the study carried out by Gupta et al. [2] in Shivdaspura and Vanasthali having an average drinking fluoride level of 4.5 mg/l and 8.5 mg/l respectively, and are presented in Table 11.1. In both the regions 20 fluorosis affected children belonging to the age group of 3–12 years with body weight ranging from 12 to 25 kg were studied. Children were given 500 mg ascorbic acid in two equally divided doses; 250 mg calcium; and

800 IU Vitamin D3 per day in four equally divided doses in a double-blind study. The pre and post-treatment data (time of treatment was 3 months) were reported. In the present study, the following variables were used:

- Fluoride in serum;
- Calcium in serum;
- Alkaline phosphate in serum;
- Drinking fluoride level.

11.3.1.3 CASE STUDY 3: DELHI

The data were obtained from the study carried out by Susheela and Bhatnagar [13] in the hospitals of Delhi. It included 6 male and 4 female patients. Upon diagnosis of the disease, the patients were introduced to two interventions: (1) consumption of de-fluoridated water, (2) consumption of diet rich in calcium and other antioxidants for a treatment time of one year. In the present study, we derived data of drinking fluoride level, fluoride in serum, and fluoride in urine before and during their intervention.

11.3.2 STATISTICAL ANALYSIS

Data reported in the case studies have been used for model development, which was analyzed using statistical tools such as bivariate analysis, multilayer perceptron (MLP), and logistic regression analysis in SPSS.

11.3.2.1 BIVARIATE ANALYSIS

Bivariate analysis examines the relationship between two variables, which may be continuous or categorical in nature. Bivariate analysis was used to verify for a couple of associations suggested by the pathophysiology of fluorosis [2], first between blood fluoride and serum calcium and second between serum calcium and serum alkaline phosphatase (SAP).

Pearson product-moment coefficient (PPMC (ρ)) is a measure of the correlation which quantifies the strength and trend of a linear association between two variables. Value of $\rho > 0$ indicates a positive relation (or direct proportionality), while $\rho < 0$ implies a negative relation (or inverse proportionality) and $\rho = 0$ indicates that no linear relationship exists between the two variables. Further, if the numerical value of ρ is close to 1 or –1 it indicates a strong linear relationship. It may be noted that ρ is not a measure of causality, i.e., by observing the value of ρ, which of the two variables influences the other cannot be deduced.

11.3.2.2 MULTILAYER PERCEPTRON (MLP): A NEURAL NETWORK APPROACH

The MLP procedure produces a predictive model for a dependent/target variable, based on the values of the predictor variables. MLP was applied to interpret the change in the values of biochemical parameters; serum calcium, serum fluoride, SAP, SAA, LAA, serum GAG and fluoride in urine with respect to varying drinking fluoride levels, which is the target variable. The MLP procedure was applied using SPSS in the automatic architecture mode. This selection builds a network with one hidden layer and optimizes the number of units in the hidden layer. The frequency weights are ignored by this procedure. Automatic architecture selects hyperbolic tangent and Softmax activation function for all units in the Hidden layer and Output layer respectively. The error function is cross-entropy error.

Hyperbolic tangent function takes real-valued arguments and transforms them to the range (–1, 1) and has the following form.

$$\gamma(c) = \tanh(c) = \frac{exp(c) - exp(-c)}{exp(c) + exp(-c)} \tag{11.1}$$

Softmax function takes a vector of real-valued arguments and transforms it to a vector whose elements fall in the range (0, 1) and sum to 1 and has the following form.

$$\gamma(c_k) = \frac{\exp(c_k)}{\Sigma_j \exp(c_j)} \tag{11.2}$$

Batch Training was used because the data set was small and the process directly minimizes the total error. Moreover, it uses information from all records in the training dataset and updates the weights many times until the stopping rule is met.

11.3.2.3 LOGISTIC REGRESSION ANALYSIS

Logistic Regression is a computational algorithm used for classifying data, i.e., to decide class membership y of an unknown data item x based on a data set:

$$D = (x_i, y_i) \tag{11.3}$$

Data item 'x_1' is known to have class membership 'y_i.' Here y_i is a binary variable, i.e., a variable which can take only two values, either 0 (generally represents absence of an event) or 1 (represents presence of the event).

Logistic regression model relates the expected value of y for a given value of x by:

$$P\left(y=1|_x, H(x)\right) = \frac{1}{1+e^{-(H(x))}} \tag{11.4}$$

and,

$$P(y=0|_x, H(x)) = 1 - P(Y=1|_x, H(x)) \tag{11.5}$$

where, H(x) is a logit function defined by:

$$H(x) = log\ (odds) = \alpha_0 + \alpha_1 x \tag{11.6}$$

where, α_0 and α_1 are the regression coefficients and odds is the ratio of number of occurrences of y = 1 to its non-occurrences. There are mainly three methods to determine these coefficients namely enter method, forward stepwise method and backward stepwise method. Logistic regression analysis was employed in this work to develop the model to predict occurrence of skeletal fluorosis in children on the basis of age, weight, and fluoride intake.

11.4 RESULTS AND DISCUSSIONS

This section first describes the results of bivariate, MLP analysis, and reduced retention model (RRM) estimated by the data set of case studies 1, 2, and 3 described in Section 11.2. Afterwards, it illustrates the results of LR model developed using the data set of case study 1.

11.4.1 BIVARIATE, MLP ANALYSIS AND REDUCED RETENTION MODEL (RRM)

The pathophysiological process involved in fluorosis has been described by Gupta et al. [2] which states that increased blood fluoride is followed by lowering of serum calcium which is further followed by increase in immature osteoblasts. This leads to increased serum and bone GAG which further cause decrease in calcification of bone tissues. Considering the logical sequence of manifestation of fluorosis proposed here along with the fact that SAP can be considered as an indicator of the osteoblastic bone activity [2], a bivariate hypothesis was postulated "Subjects with higher level of blood fluoride have lower level of serum calcium which is associated with increased level of SAP."

The mean and standard deviation of the biochemical parameters (blood fluoride, serum calcium and SAP) were taken from case study 2: Shivdaspura and Vanasthali Jaipur. More data points were generated assuming a normal distribution within the population for each of the parameters with sufficiently large sample size (100 subjects). The data points were then studied using the scatter plots shown in Figures 11.2 and 11.3.

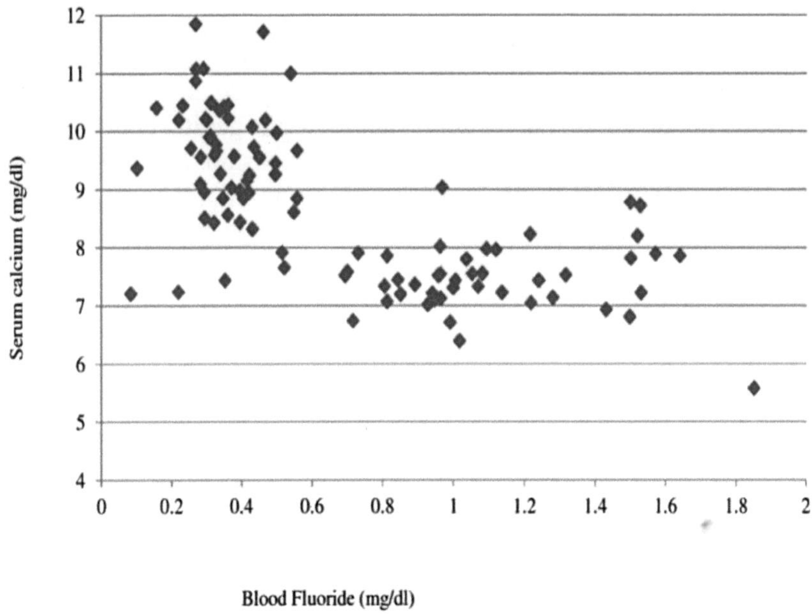

FIGURE 11.2 Scatter plot between data of serum calcium and blood fluoride generated by assuming a normal distribution of population and taking a sample size of 100.

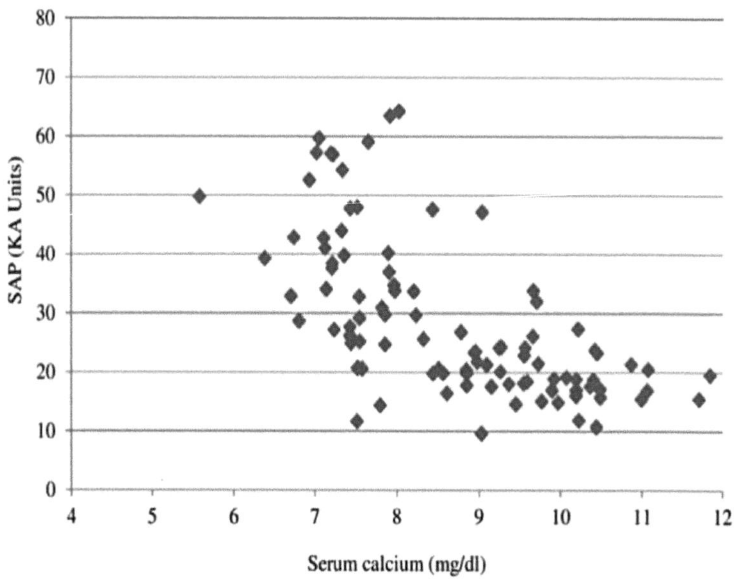

FIGURE 11.3 Scatter plot between data of serum alkaline phosphate (SAP) and serum calcium generated by assuming a normal distribution of population and taking a sample size of 100.

Figure 11.2 which shows the scatter plot of serum calcium and blood fluoride, reveals that serum calcium and blood fluoride are found to have a negative association (inverse relation) and the points getting clustered along a straight line depict the presence of a linear relationship. Figure 11.3 shows the scatter plot between serum calcium and blood fluoride and displays a negative association between the two variables. Although the points cluster along a line, they are not clustered quite as closely as they were for the scatter plot in Figure 11.2. The strength of these linear relationships was further explored using Pearson correlation (ρ). Following the null hypothesis was taken:

$$H_0 : \rho = 0 \ [\textit{There is no actual correlation}] \tag{11.7}$$

Using SPSS, the correlation matrix between serum calcium and blood fluoride and SAP was developed. Table 11.2 shows Pearson correlation coefficients between the three biochemical parameters maintaining the necessary pairs. Since the significance value was less than 0.05 in both the cases null hypothesis H_0 was rejected. Hence, $\rho = -0.673$ in the first case indicate blood fluoride has a strong negative correlation with serum calcium. While for other cases ρ = –0.520 denotes moderately strong negative correlation of SAP with serum calcium. Hence, the bivariate hypothesis postulated from the pathophysiology proves to be correct.

TABLE 11.2 Correlation Matrix of Pearson Correlation Coefficients between Serum Calcium and Blood Fluoride; and SAP and Serum Calcium

Serum Calcium (serum_Ca)	Blood Fluoride(blood_f)	SAP
Pearson Correlation (ρ)	–0.673	–0.520
Sig. (2-tailed)	0	0
N (sample size)	100	100

After a critical examination of the pre-treatment data of biochemical parameters of children from the villages of Jaipur, illustrated in case study 1, it was observed that the data could be used to interpret the fluctuations in biochemical parameters on the basis of the level of fluoride intake through drinking water. Biochemical parameters were classified into four categories by relating them to an ordinal variable 'water_fluoride_grade' with values 1, 2, 3, and 4 representing the fluoride levels in drinking water for villages of Ramsagarki Dhani, Rampura, Shivdaspura, and Rapuria respectively. Water_fluoride_grade =1 was used for drinking fluoride levels of 2.4 ppm, 2 for 4.6 ppm, 3 for 5.6, and 4 for 13.6 ppm. These data were then used as an input to MLP neural network analysis in SPSS v20. Tables 11.2–11.5 show the output of the MLP analysis.

Table 11.3 shows that 60.5% of the valid sample data was randomly picked and used for training the network while the rest 39.5% was utilized for testing. Table 11.4 lists all the biochemical parameters that were used as covariates in the input layer and describe the hidden and output layer (Table 11.6).

TABLE 11.3 Case Processing Summary of Neural Network Training and Testing Data

		N	Percent
Sample	**Training**	49	60.5%
	Testing	32	39.5%
Valid		81	100.0%

TABLE 11.4 Network Information of Input, Hidden, and Output Layers of Neural Network

Input Layer	**Covariates**	1	serum_ca
		2	SAP
		3	SIP
		4	SSA
		5	GAG
		6	SAA
		7	LAA
		8	serum_fluoride
		9	blood_fluoride
		10	urinary_fluoride
	Number of units[a]		10
	Rescaling method for covariates		Standardized
Hidden Layer(s)	Number of hidden layers		1
	Number of units in hidden layer 1[a]		4
	Activation function		Hyperbolic tangent
Output Layer	Dependent variables	1	water_fluoride_grade
	Number of units		4
	Activation function		Softmax
	Error function		Cross-entropy

[a]Excluding the bias unit.

TABLE 11.5 Model Summary (Cross Entropy Error and Percent Incorrect Predictions) of Training and Testing Data from the MLP Analysis

Training	Cross entropy error	2.511
	Percent incorrect predictions	0.0%
	Stopping rule used	1 consecutive step(s) with no decrease in error
	Training time	0:00:00.01
Testing	Cross entropy error	16.210
	Percent incorrect predictions	18.8%
Dependent variable: Water fluoride grade		

Error computations are based on the testing sample.

TABLE 11.6 Importance and Normalized Importance (Sensitivity Analysis) of Independent Variables from the MLP Analysis

Biochemical Parameter	Importance	Normalized Importance
GAG	0.168	100.00%
blood_fluoride	0.151	90.20%
serum_Ca	0.118	70.40%
serum_fluoride	0.1	59.60%
SSA	0.099	59.10%
SIP	0.087	51.90%
SAA	0.083	49.50%
urinary_fluoride	0.081	48.20%
SAP	0.079	47.40%
LAA	0.034	20.10%

Figure 11.4 displays the structure of a neural network with one hidden layer and four hidden units defined in the MLP neural network. Table 11.5 shows the model summary. The cross-entropy error, which the network tried to minimize during training, was 2.511. Moreover, the percentage of incorrect prediction was 18.8% in the testing samples. So, the percentage of correct prediction was 81.2%, which can be considered as good for physiological systems.

Table 11.6 and Figure 11.5 display normalized importance of various biochemical parameters generated from Independent variable importance analysis. This is a sensitivity analysis, based on the combined training and testing of samples. It computes and creates a table and a chart displaying the importance and normalized importance of each predictor in defining the neural network. It was observed that 'GAG' contributed the most in the neural network model construction, followed by 'blood fluoride,' 'serum calcium,' 'serum fluoride,' 'SSA,' etc.

Jha et al. [14] suggested that elevated content of GAG in bone and its reflection in serum can be an index to assess fluoride toxicity and detect fluorosis at very early stages, while the above MLP model has reinforced the fact that serum GAG is an important predictor of the level of drinking water fluoride, and stating vice versa the serum GAG values will be affected more significantly in comparison to other biochemical parameters with varying fluoride levels in drinking water.

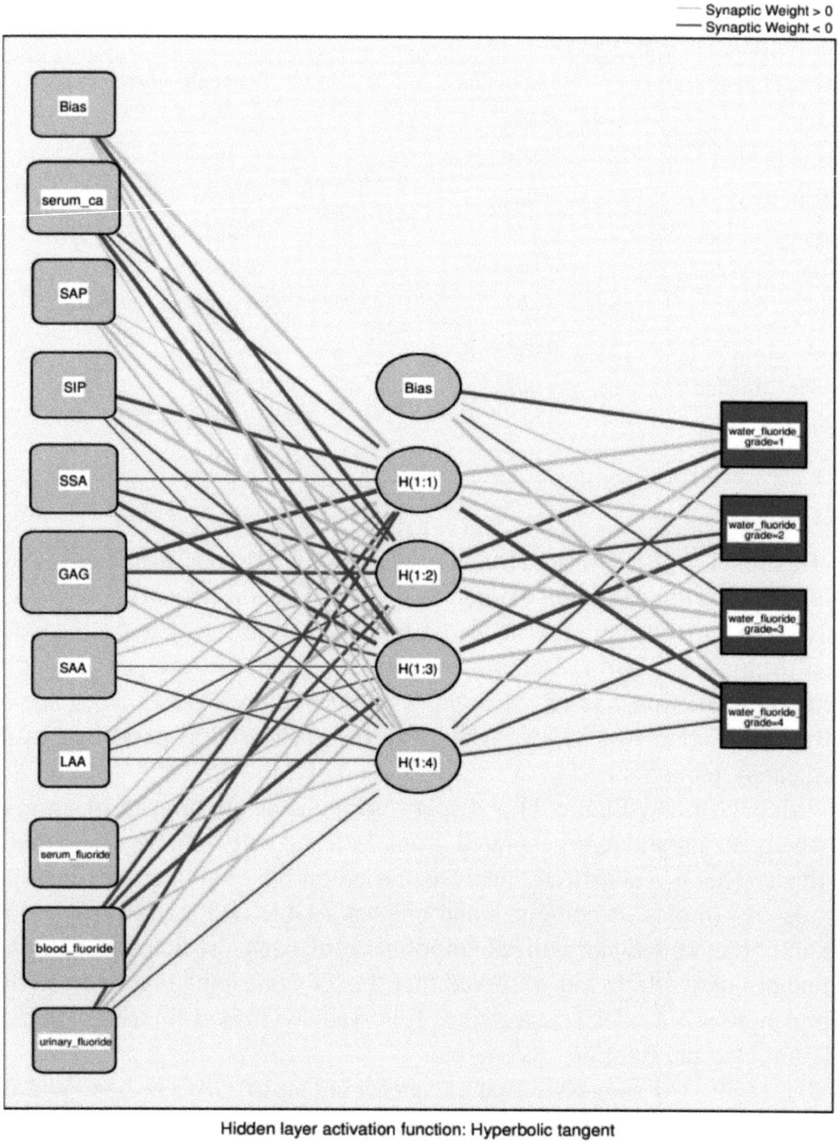

FIGURE 11.4 Depiction of structure of artificial neural network having various biochemical parameters in the input layer correlating with the drinking water fluoride level of the four target areas as depicted in case study 1 [2]; There is one hidden layer with four units and its activation function is hyperbolic tangent. The structure was produced by MLP analysis in SPSS.

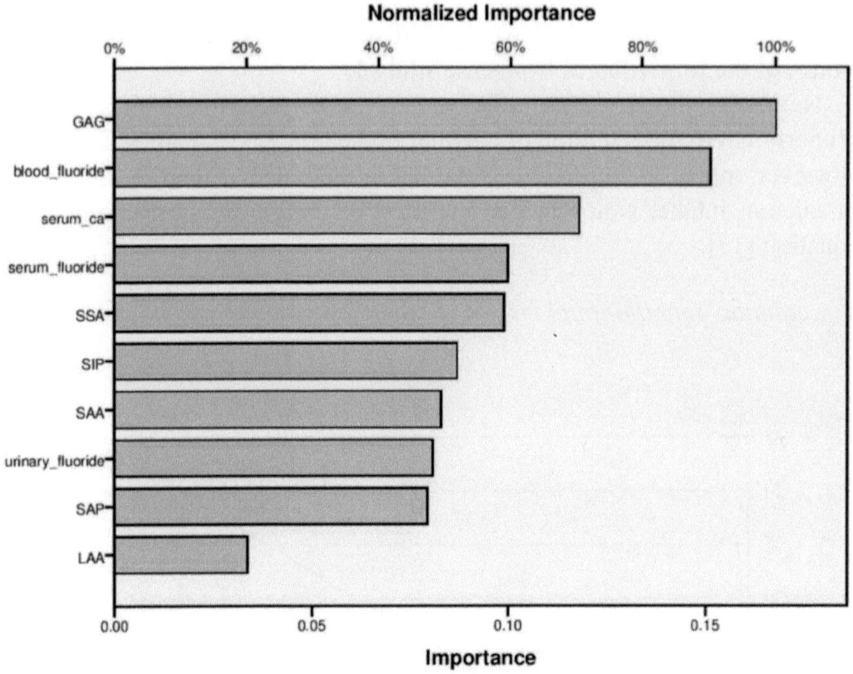

FIGURE 11.5 Bar chart of various biochemical parameters showing their normalized importance in correlating with drinking fluoride level and further in prediction of skeletal fluorosis; serum GAG was found to be significant. The chart was produced by MLP analysis in SPSS.

The observed importance of GAG can be further backed by the reasoning that bone and teeth, which are adversely affected by ingestion of high fluoride for prolonged duration, are made up of calcium and bone matrix [14, 15]. About 80–85% of bone matrix is made of collagen and the remaining 15–20% by glycosaminoglycans (GAG) and glycoproteins [16]. All three components of bone and teeth, i.e., collagen, GAG, and calcium are adversely affected by the ingestion of high fluoride through various food and water sources for a prolonged duration [14, 15]. Hence, serum GAG must be assessed as a key indicator of fluorosis.

The significant effect of serum GAG has been therefore used to develop RRM for calcium dose prediction at different water fluoride levels. Serum GAG was found to have the following strong linear relationship (R^2 = 0.9698) with fluoride concentration in drinking water.

$$serum\ GAG(mg/dl) = 2.2986 * Fluoride(mg/l) + 8.2875 \tag{11.8}$$

The above model was resultant of linear curve fitting of data in Figure 11.6 which plots the data of averages of serum GAG and drinking fluoride values of the four villages from case study 1.

Next, according to the proposed pathophysiology, serum GAG is inversely proportional to the retention of calcium in the bones, i.e., bone calcification. However, more calcium will be retained in the bones if there is an increase in calcium intake, which in the fluoride-free condition is governed by the equation [17]:

$$calcium\_retention(mg) = -78.5 + 0.44*calcium\_intake\ (mg) \tag{11.9}$$

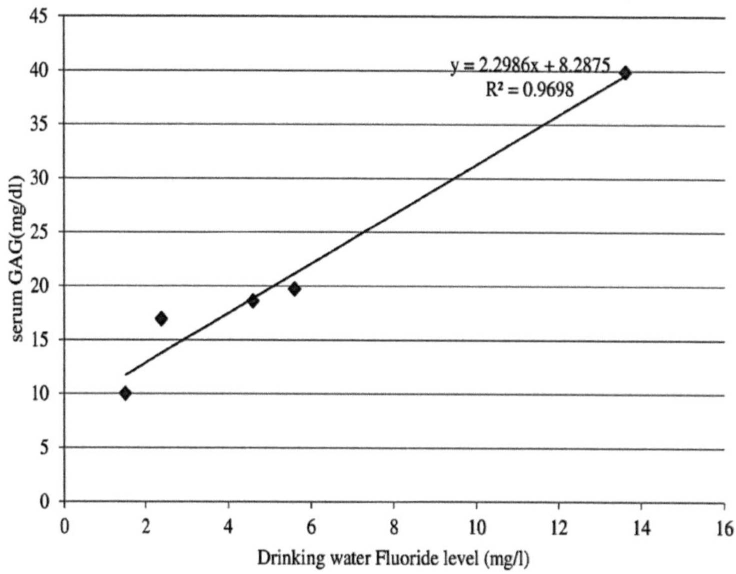

FIGURE 11.6 Plot showing the relationship between serum GAG and drinking water fluoride level fit to linear co-relation; the data used was derived from case study 1 [2].

Further, it was found that the serum GAG level in the human body in the normal conditions is in the range 9–11 mg/dl, and for calculations, it was assumed to be 10 mg/dl. The changes in the level of serum GAG because of the increase in the drinking fluoride level (Eqn. (11.8)), along with the varying calcium intake (Eqn. (11.9)) were used to calculate the reduced calcium retention in bones and the model (Eqn. (11.10)) was derived as follows:

$$Calcium\_retention(mg) = \frac{calcium\_intake(mg)}{2.2986 * Fluoride\left(\frac{mg}{l}\right) + 8.2875} * 4.4 - 78.5 \tag{11.10}$$

Calcium intake by the sample population of the four villages with varying drinking fluoride level (2.4 to 13.6 mg/l) ranged from 250 to 1010 mg/day [2]. When the calcium dietary intake is not enough to maintain the calcium homeostasis the value of retention was negative, which represents the process of leaching of bone, i.e., instead of bone calcification, calcium was leaching out from the bone. However, for the same level of water fluoride, if there is an increase in calcium intake, then calcium depletion would stop and retention would start. This critical calcium dose was plotted against the fluoride level (Figure 11.7) and the model (Eqn. (11.11)) was obtained by linear fitting:

$$Ca\_intake = 40.695(water\_F) + 158.45 \tag{11.11}$$

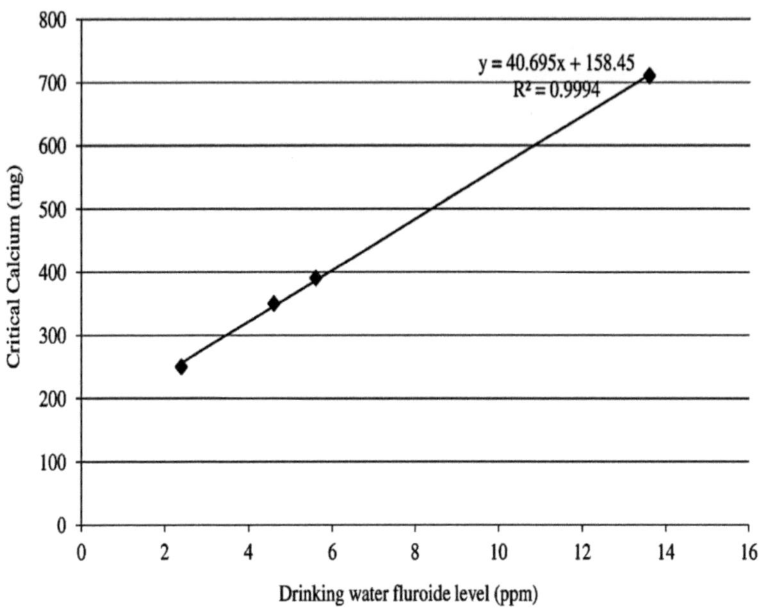

FIGURE 11.7 Plot of critical calcium and drinking water fluoride level.

Model Eqn. (11.11) predicts the supplemental dose of calcium to be taken at different water fluoride levels. The value of Ca dose at which retention starts is not the same as the required adequate Ca dose for a human being. The dosage predicted by RRM shows the amount of calcium required for maintaining calcium homeostasis within the body without the expense of calcium being leached out from the bones. However, for regular body growth and development Ca is a basic element and should be adequate in the diet to avoid other problems.

The above-developed model was extended for predicting the requirement of calcium to combat fluoride ingestion using data from the case study of Delhi by Susheela and Bhatnagar [13]. In this case, study two interventions were adopted to reverse fluorosis, first the fluorosis affected patients were advised to consume water with less fluorine, and second they were given calcium supplements simultaneously. The reported clinical pre-treatment data of blood and urinary fluoride were studied for various drinking fluoride levels and following correlations were fit from the reported data (as shown in Figures 11.8a and 11.8b):

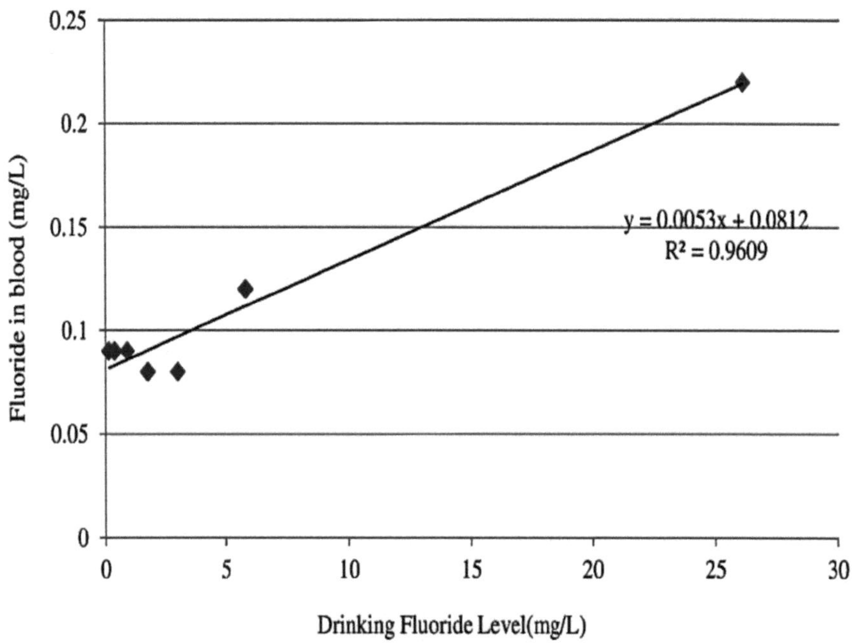

FIGURE 11.8(A) Plot of blood fluoride and drinking water fluoride fit to linear co-relation; the data was derived from case study 3 [13].

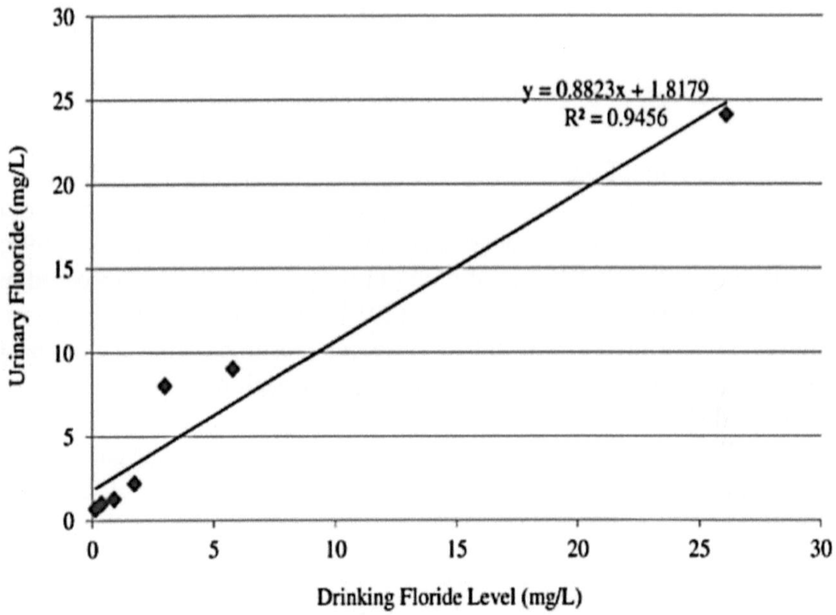

FIGURE 11.8(B) Plot between urinary fluoride and fluoride through drinking water fit to linear co-relation; the data was derived from case study 3 [13].

$$Blood\_F = 0.0053(DrinkingWater\_F) + 0.0812\ (R^2 = 0.9609) \quad (11.12)$$

$$Urinary\_F = 0.8823(DrinkingWater\_F) + 1.8179\ (R^2 = 0.9456) \quad (11.13)$$

Thereafter, the values of blood and urinary fluoride were estimated using co-relation 1 and 2 after the first intervention of each patient, i.e., after reducing the level of drinking fluoride. These values were found to be greater than the reported values by Susheela and Bhatnagar [13], because the reported values were taken after the second intervention, i.e., calcium supplementation. This difference in the values was hence used to calculate the effect of calcium in the treatment procedure by assuming that, apart from improving serum calcium levels the presence of extra calcium in daily intake directly inhibits the absorption of fluoride ions into the body, by forming insoluble complex of calcium fluoride [2]. Hence, the following two equations were used from the possible reactions occurring in serum based on its solubility product to find the extra calcium dose required due to fluoride toxicity.

$$Ca^{+2} + 2F^{-} \rightleftharpoons CaF_2 \tag{11.14}$$

$$Ca^{+2} + 2OH^{-} \rightleftharpoons Ca(OH)_2 \tag{11.15}$$

Along with the Eqn. (11.14) and (11.15) following three boundary conditions have been used: (1) According to the 7th edition of World Book Rush-Presbyterian-St. Luke's Medical Center Medical Encyclopedia (1995), World Book: 120–121. Volume of blood in body is 4.7 liters, (2) According to MedlinePlus Medical Encyclopedia (2013). A person urinates 2 liters in a day (3) 33% of the calcium intake dose gets absorbed from the stomach to serum [18]. The resulting "Delta Ca dose" for each patient with different water fluoride level was estimated and plotted (Figure 11.9) and Eqn. (11.16) was fitted linearly to the data to predict for the extra dose of calcium for different fluoride levels:

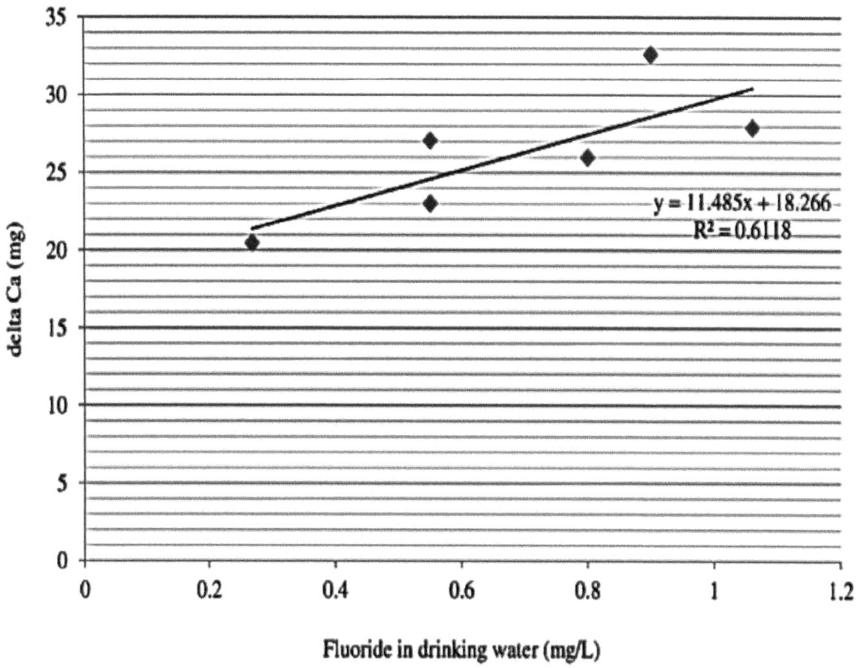

FIGURE 11.9 Plot between delta calcium and drinking water fluoride level fir to linear co-relation.

$$Delta\_Ca\_dose = 11.485 * (drinkingWater\_F) + 18.266 \ (R^2 = 0.6118) \tag{11.16}$$

However, apart from the above two reactions, calcium is responsible for many other reactions within the body so actual overall calcium dose requirement would be different from delta Ca dose. Gupta et al. [2] had proposed a thumb rule for treatment by Ca dosage stating "25 mg calcium dose is required per mg of fluoride intake." So, Figure 11.10 was plotted in order to draw a comparison among the calcium doses predicted by different models that have been developed so far. These include calcium dosage predicted by Eqn. (11.11) and Eqn. (11.16) for drinking fluoride level ranging from 1 to 15 ppm, and that recommended by Gupta et al. [2].

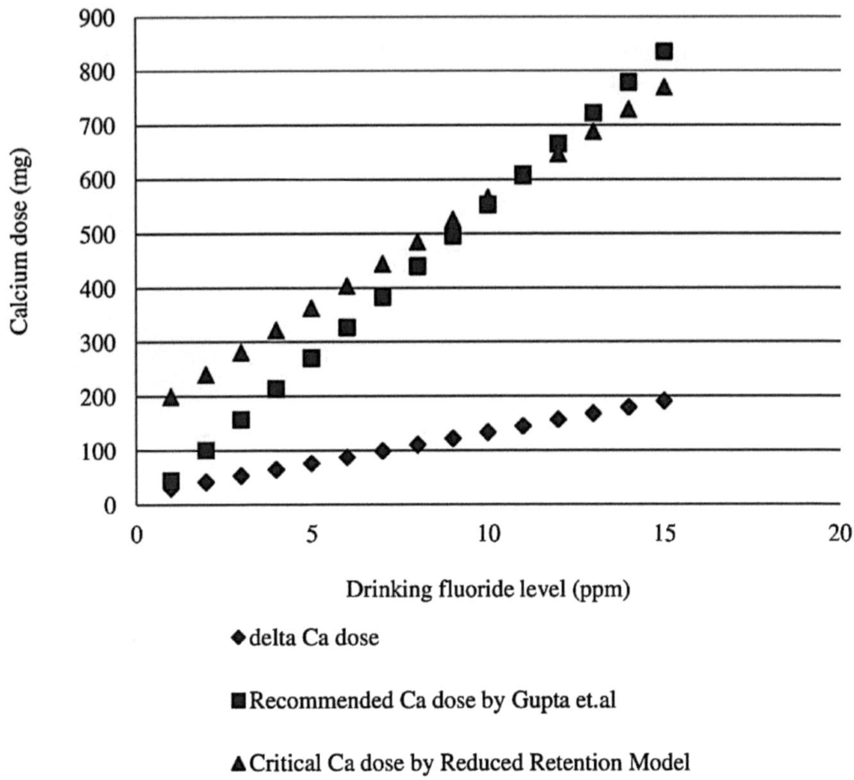

FIGURE 11.10 Plot between calcium doses predicted by delta calcium model, reduced retention model and recommended calcium dose by Gupta et al. [2] on y-axis and drinking water fluoride level on x-axis.

It was observed that RRM predicts values of calcium are almost comparable to the recommended values and the two curves almost overlap for drinking water fluoride level greater than 8 ppm. The deviations of delta Ca model can be accounted to the fact that Ca dose predicted by this model is in an addition to the Calcium requirement for normal bone growth. This is the value of Ca dose at which retention starts and hence should not be misinterpreted as the required adequate Ca dose for a human growth.

11.4.2 LR MODEL TO PREDICT SKELETAL FLUOROSIS

Bar chart shown in Figure 11.11 describes that even though the villages differ in the amount of fluoride consumed by its population through drinking water, the distribution of skeletal fluorosis looks similar, with less than 10% population in all the villages suffering due to skeletal fluorosis grade 2, around 30% from grade 1 and rest from grade 0. Grade 2, 1 and 0 presents the severity of fluorosis standing for severe, moderate, and mild respectively.

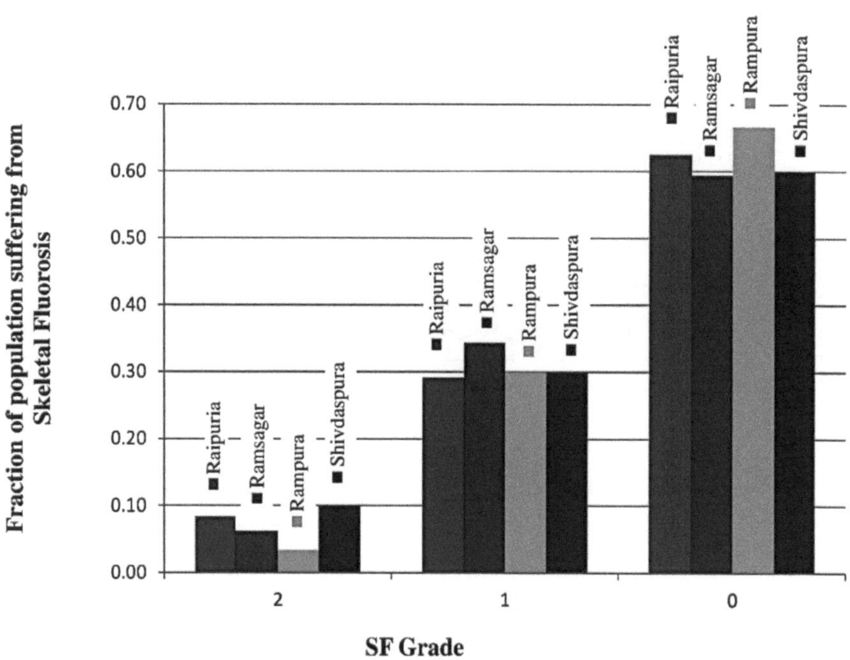

FIGURE 11.11 Bar chart depicting the severity of skeletal fluorosis (0, 1, and 2 indicate the grade of skeletal fluorosis showing mild, moderate, and severe level) among the population of four target villages of Jaipur depicted in case study 1 [2].

This result was further explored by developing a classification model to predict grade of skeletal fluorosis from a child's age, weight, and fluoride intake through drinking water resources. Backward logistic regression has been applied on the physical parameters' data obtained from case study 1. The SPSS v20 output is shown in Table 11.7.

From Table 11.7, it is turned out that none of the independent variables including 'age,' weight,' and 'fluoride_water' were significant enough ($p<0.5$) to develop a statistically significant model. Hence, all the variables got eliminated in step 4.

The model strengthens the result shown by bar graph in Figure 11.11 that grade of skeletal fluorosis is independent of fluoride intake through drinking water and also rules out any influence of age and weight of the subject. The reason behind this nature of result is due to limited data available. The values of daily dose of calcium, vitamin C, D, wider range of age and weight, levels of biochemical parameters within the body, etc., might be decisive in determining the severity of skeletal fluorosis.

TABLE 11.7 Variables Included in the Equation at Various Steps of Backward Logistic Regression*

		B	S.E.	Wald	df	Sig.	Exp(B)
Step 1[a]	age	0.227	0.187	1.476	1	0.224	1.255
	weight	–0.099	0.077	1.640	1	0.200	0.906
	fluoride_ water	–0.028	0.033	0.724	1	0.395	0.972
	Constant	0.274	0.999	0.075	1	0.784	1.315
Step 2[a]	age	0.231	0.180	1.647	1	0.199	1.260
	weight	–0.105	0.072	2.123	1	0.145	0.900
	Constant	0.041	0.942	0.002	1	0.965	1.042
Step 3[a]	weight	–0.030	0.033	0.820	1	0.365	0.971
	Constant	0.451	0.838	0.290	1	0.590	1.570
Step 4[a]	Constant	–0.279	0.250	1.238	1	0.266	0.757

[a] Variable(s) entered on step 1: age, weight, fluoride_water.

* The inapt variable gets removed in every subsequent step on the basis of Sig. value in backward logistic regression analysis.

11.5 CONCLUSIONS

RRM mathematically represented the biological changes as observed in the various biochemical parameters within the human body due to the prolonged intake of high fluoride which caused fluorosis. Delta Calcium model predicted the daily calcium dose requirement in order to prevent the harmful effects due to high fluoride intake as well as to reverse the fluoride deposition in individuals suffering from fluorosis.

The Pearson correlation coefficient between blood fluoride and serum calcium was -0.673 and between SAP and serum calcium was -0.520, both of which being negative demonstrate inverse correlation between the respective pairs. This lends support to the initial steps of proposed pathophysiology of fluoride intake in human body reported in this paper. A normalized importance chart was developed by MLP analysis of the biochemical parameters, which revealed significant importance of serum GAG in detection of onset of fluorosis and prediction of calcium dose. This novel finding was applied to develop RRM, which successfully predicted daily dose of calcium required to compete with fluoride intake through drinking water. These dosages were found to be in accordance with the medically recommended thumb rule of 25 mg calcium requirement per mg of fluoride intake. Further, the logistic regression analysis was applied for case study 1, which deduced that reported data of level of fluoride intake, age, and weight of the children were not sufficient enough to classify the perceived grade of skeletal fluorosis among the population of Ramsagar, Rampura, Shivdaspura, and Raipuria. This is due to the role of other parameters like serum calcium, serum fluoride, GAG, vitamin C, D, etc., against the development of skeletal fluorosis. In sum, this study gives an insight to contemporary mathematical modeling approach used to analyze the reported data of fluoride affected regions. Since, serum GAG was found to be a strong indicator of fluorosis; it may be considered as an early predictor for the onset of fluorosis and thus help in the medical management of this menace.

KEYWORDS

- **fluorosis**
- **multilayer perceptron**
- **parathyroid hormone**
- **Pearson product-moment coefficient**
- **reduced retention model**
- **serum GAG**

REFERENCES

1. George, S., Pandit, P., & Gupta, A. B., (2010). Residual aluminum in water defluoridated using activated alumina adsorption-modeling and simulation studies. *Water Res., 44*, 3055–3064.
2. Gupta, S. K., Khan, T. I., & Gupta, A. B., (2000). *Environmental Health Perspective of Fluorosis in Children*. LAP LAMBERT Academic Publishing GmbH & Co. KG, 201.
3. World Health Organization (WHO), (2011). *Guidelines For Drinking Water Equality* (4th edn., Vol. 4, pp. 178). Geneva.
4. Bureau of Indian Standards, (2009). *Draft Indian Standard Drinking Water-Specifications* (p. 7). Second revision of IS: 10500.
5. Shivaprasad, P., Singh, P. K., Saharan, V. K., & George, S., (2018). Synthesis of nano alumina for defluoridation of drinking water. *Nano-Struct. and Nano-Obj., 13*, 109–120.
6. Schwartz, P., Madsen, J. C., Rasmussen, A. Q., Transbol, I. B., & Brown, E. M., (1998). Evidence for a role of intracellular stored parathyroid hormone in producing hysteresis of the PTH–Calcium relationship in normal humans. *Clin. Endocrinol., 48*, 725–732.
7. Heath, D. A., (1998). Clinical manifestations of abnormalities of the calcium sensing receptor. *Clinical Endocrinology, 48*(3), 257.
8. Akesson, K., Lau, K. H. W., Johnston, P., Imperio, E., & Baylink, D. J., (1998). Effect of short-term depletion and repletion on biochemical markers of bone turnover in young adult women. *J. Clin. Endocrinol. Metab., 83*(6), 1921–1927.
9. Stauffer, M., Baylink, D., Wergedal, J., & Rich, C., (1973). Decreased bone formation, mineralization, and enhanced resorption in calcium-deficient rats. *Am. J. Physiol., 225*(2),–276.
10. Wright, K. R., & McMillan, P. J., (1994). Osteoclast recruitment and modulation by calcium deficiency, fasting, and calcium supplementation in the rat. *Calcified Tissue Int., 54*, 62–68.
11. Shortt, H. E., McRobert, G. R., Bernard, T. W., & Mannadi, N. A. S., (1937). Endemic fluorosis in the Madras presidency. *Ind. J. Med Res., 25*, 553–568.
12. Gupta, S. K., Gupta, R. C., & Seth, A. K., (1994). Reversal of clinical and dental fluorosis. *Indian Paediatrics., 31*(4), 439–443.
13. Susheela, A. K., & Bhatnagar, M., (2002). Reversal of fluoride induced cell injury through elimination of fluoride and consumption of diet rich in essential nutrients and antioxidants. *Mol. Cell. Biochem., 234, 235*, 335–340.
14. Jha, M., Susheela, A. K., Krishna, N., Rajyalaxmi, K., & Venkiah, K., (1982). Excessive ingestion of fluoride and its significance of sialic acid: Glycosaminoglycans in the serum of rabbit and human subjects. *J. Toxico. Clin. Toxicol., 19*(10), 1023–1030.
15. Susheela, A. K., (1979). Recent advances in research on fluoride toxicity and fluorosis. *ICMR Bulletin., 3*, 1–4.
16. Weatherell, J. A., Deutsch, D., & Robinson, C., (1975). Fluoride and relation to bone end tooth-In: Friedrich, K. C., & Hans, P. K., (eds.), *Calcium Metabolism, Bone, and Metabolic Bone Diseases, Proceedings of the X European Symposium on Calcified Tissues* (pp. 101–110). Hamburg (Germany), Springer-Verlag Berlin. Heidelberg. Newyork.
17. Indian Council of Medical Research (ICMR), (2010). *Nutrient Requirements and Recommended Dietary Allowances for Indians* (p. 165).
18. Challem, J., (2003). *User's Guide to Nutritional Supplements* (p. 19). Basic Health Publications.

CHAPTER 12

Diagnosis of *Mycobacterium tuberculosis* from Sputum DNA Using Multiplex PCR

R. GOPINATH,[1] SUTHANDIRA MUNISAMY,[2] M. JEYADEVASENA,[1] J. M. JEFFREY,[3] and ELANCHEZHIYAN MANICKAN[1]

[1]*Department of Microbiology, Dr. ALM PG IBMS, University of Madras, Taramani, Chennai – 600 113, Tamil Nadu, India, E-mail: emanickan@yahoo.com (E. Manickan)*

[2]*IRT Perundurai Medical College, Perundurai, Erode District, Tamil Nadu – 638053, India*

[3]*Department of Genetics, Dr. ALM PG IBMS, University of Madras, Taramani, Chennai – 600 113, Tamil Nadu, India*

ABSTRACT

The conventional techniques like the Acid-Fast Bacilli (AFB) staining and smear microscopy employed for the diagnosis of tuberculosis (TB) have a significant disadvantage of poor sensitivity and the gold standard diagnostic method employing culture is time-consuming. This study was conducted to isolate and identify *Mycobacterium tuberculosis* (MTB) from the sputum sample and to evaluate the efficacy of PCR as a modern diagnostic tool, for diagnosis of tuberculosis, especially in the smear-negative cases. The multiplex polymerase chain reaction (MPCR) offers a great promise as a diagnostic tool for TB owing to the rapid analysis, significant sensitivity, and specificity. Amplified nucleic acid hybridization assays such as MPCR have shown promising results. The aim of this study was to perform MPCR analysis using MTB genes namely PPE41, MPT53, LPQH, ESAT-6, and CFP-10, and analyze its efficiency in the diagnosis of TB. 252 sputum samples were collected from TB Hospital and they were subjected to Ziehl-Neelsen, culture analysis employing Lowenstein-Jensen (L-J) medium, and Multiplex PCR (MPCR) for the specific detection of mycobacterium DNA. The MPCR was performed for targeting genes of MTB. Of 252 cases, 102 samples showed positive for AFB, 124 samples were positive for culture method and 196 samples showed positive for MPCR. The overall results showed a sensitivity as well as specificity of the MPCR assay as 97.7% and 100%, respectively. Whereas the sensitivity and specificity in the case of culture was 89.2% and 98.1% and with AFB it was 63.3% and 92%, respectively. Analysis of the positive and negative predictive values of MPCR was found to be 91.1% and 99.3% when compared to the culture (78.2% and 87.1%) and AFB (71.2% and 82%), respectively. MPCR analysis performed by employing specified MTB primers was able to detect more TB positive cases when compared to the conventional AFB and L-J culture method for the detection of MTB. MPCR was found to be a more rapid, reliable, more sensitive, and highly specific diagnostic tool for the purpose of detection as well as the differentiation of MTB patients and this can serve as the potential diagnostic tool for the analysis of sputum samples.

12.1 INTRODUCTION

Tuberculosis (TB) is deemed as one of the leading cause for morbidity and mortality worldwide with an infection rate affecting approximately one-third of the population [1]. However, this is not the case when people predisposed to human immunodeficiency virus (HIV) get infected with TB. People with HIV-infection are 26 to 31 times more likely to procure a severe form of TB over a particular period of time when compared to HIV-negative persons [2, 3]. In 2015, the world had an estimated 10.4 million new TB cases. Over half of these were among men (5.9 million), and women constituted over a third (3.5 million). Ten percent of the cases were among children [1]. The situation is further aggravated due to the increasing occurrence of multi-drug resistant (MDR) and extremely drug-resistant (XDR) TB and with the significant emergence of HIV [4]. The laboratory diagnosis is extensively based on direct microscopy and culture for mycobacterium. Direct microscopic examination for detection of AFB has very low sensitivity and often is less specific [5, 6]. L-J culture test though being regarded as a gold standard technique is a labor intensive and time consuming procedure. Thus, the detection of MTB in clinical samples harboring small numbers of the organism still pose itself as a major challenge

[7]. As an alternative to these classical methods, new nucleic acid-based technologies have shown promises as a more rapid, sensitive, and specific means of detection and identification of Mycobacteria [8]. The need for a rapid diagnostic tool has led to the development of molecular methods for the strategic detection of *Mycobacterium tuberculosis.* The polymerase chain reaction (PCR) in the field of Molecular Biology has proven to be a useful technique in the diagnosis of TB infection. The problems associated with PCR method is the possibility of obtaining false-positivity in results due to the cross contamination of clinical specimens with *M. tuberculosis* DNA product with the PCR laboratory, and the inability of the PCR method to notice the minute difference between viable and nonviable organisms [9, 10]. The single gene target can result in false negativity of the results while more reliable results can be obtained by utilization of more than one target gene for amplification [11]. The multiplex PCR (MPCR) is the PCR in which several target genes are amplified simultaneously and a more sensitive and specific tool of diagnosis for *Mycobacterium tuberculosis* infection [12]. The aim of the present study was to evaluate the sensitivity and specificity of MPCR by utilizing MTB genes namely PPE41, MPT53, LPQH, ESAT-6, and CFP-10 for the diagnosis of TB patients.

12.2 MATERIALS AND METHODS

12.2.1 STUDY POPULATION

A total of 252 sputum samples were collected from IRT, Perundurai Medical College, Perundurai, Erode District, Tamil Nadu. Written consents were obtained from each patient after adequate furnishing of information and the approval for the study protocol was obtained from the Institutional Human Ethical Committee (No: UM/IHEC/16-2013-I).

12.2.2 SMEAR MICROSCOPY (SM)

Two drops of the sample pellet decontaminated by NaOH-NALC method was employed in smear microscopy (ZN staining), in accordance to the standard protocol [13]. The grading of the AFB results was done according to the WHO/International Union Against TB and Lung Disease guidelines and scored as "0" for the absence of AFB, scanty 1–9, 1+, 2+, or 3+ [13].

12.2.3 SAMPLE PREPARATION FOR DNA EXTRACTION

Commercially available Norgen Biotek Sputum DNA extraction kit (Canada) was utilized for the extraction of DNA from sputum samples.

12.2.4 DNA AMPLIFICATION OF MYCOBACTERIUM TUBERCULOSIS (TB)

The extracted DNA samples were subjected to MPCR amplification using appropriate primers of five MTB genes (Table 12.1). Amplifications were carried out using Eppendorf™ Thermal Cycler (Germany) consisting of initial denaturation at 95°C for 5 min followed by 32 cycles of denaturation at 95°C for 30 sec, annealing at 61°C for 30 sec and extension at 72°C for 45 sec. The amplifications were followed by final extension at 72°C for 5 min. The amplified products were then stored at a temperature of 4°C till the detection by electrophoresis.

TABLE 12.1 Sequences of Primers Used in Multiplex PCR

Gene	Sequence	PCR Product Size (bp)
PPE41	5'-AC GGATCCATGCATTTCGAAGCG-3'	604
	5'-AG GAATTCAGTGTCTGTACGCG-3'	
MPT53	5'-ATG GAT CCA TGA GTCT TCG CCT G-3'	539
	5'-ATG AAT TCG GAC GTC AGC GCA GC-3'	
LPQH	5'-ATG GAT CCG TGA AGC GTG GAC TG-3'	497
	5'-ATG AAT TCG GAA CAG GTC ACC TCG-3'	
ESAT-6	5'-TAAGGATCCATGACAGAGCAGGAGTG-3'	288
	5'-GCGAATTCTGCGAACATCCCAGTG-3'	
CFP-10	5'-TAA GGA TCC ATG GCA GAG ATG AAG AAC-3'	303
	5'-GCG AAT TCG AAG CCC ATT TGC GAG GA-3'	

12.2.5 DETECTION OF AMPLIFIED PRODUCTS

The MPCR products were then analyzed by electrophoresis at 100 V on 2% agarose gels stained with Ethidium bromide (0.5 μg/ml) and visualized under Gel documentation system (GELSTAN, India). The PCR products were compared with *H37Rv* positive control and PCR grade water and patient negative sample as negative control.

12.2.6 STATISTICAL ANALYSIS

The Sensitivity and specificity along with the positive predictive and negative predictive value of the MPCR were evaluated by utilizing online MedCalc and this data was compared with microbiological tests.

12.3 RESULTS AND DISCUSSION

In this study performed, a sample population of 252 sputum samples were collected and enrolled for the study. These samples were subjected to Ziehl-Neelsen staining, LJ culture and MPCR to detect MTB. Out of the sample population, 102 were found to be positive for AFB (Figure 12.1), 124 were positive for LJ culture (Figure 12.2) whereas MPCR amplification showed positive in 196 samples (Figure 12.3). MPCR showed higher number of positive result than compared to other gold standard techniques.

The statistical analysis of AFB smear examination showed a sensitivity of 63.3% and a specificity of 92%. The overall sensitivity and specificity of the culture is 89.2% and 98.1% whereas for MPCR it is 97.7% and 100% respectively.

The positive as well as the negative predictive values were found to be 71.2% and 82% for AFB; 78.2% and 87.1% for culture and 91.1% and 99.3% for MPCR, respectively (Table 12.2).

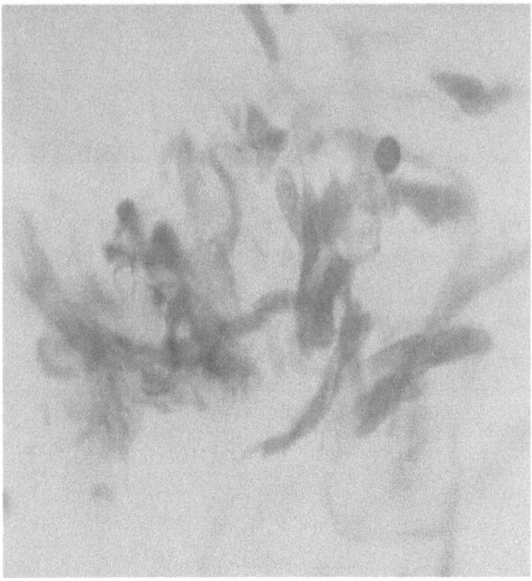

FIGURE 12.1 Ziehl-Neelsen staining.

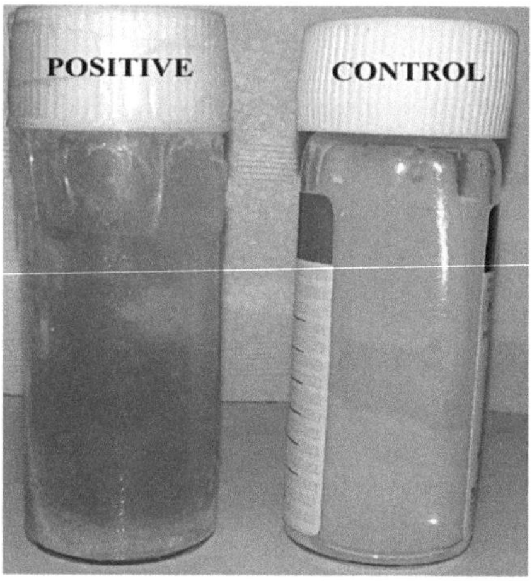

FIGURE 12.2 LJ culture slant.

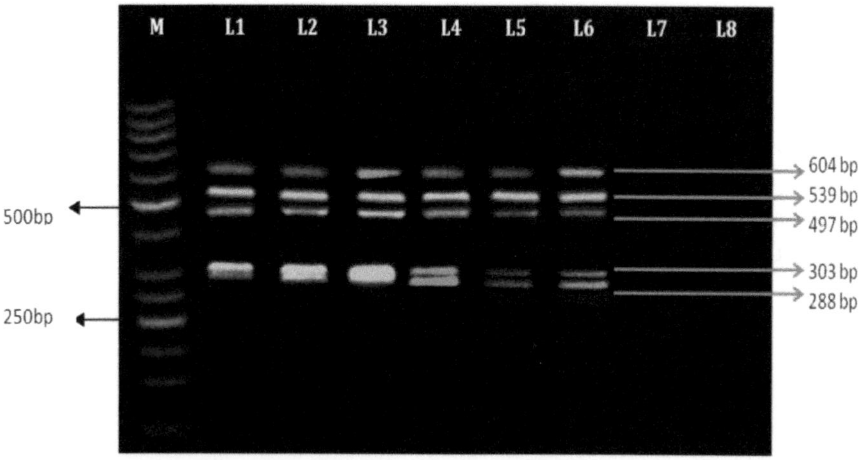

FIGURE 12.3 Multiplex PCR products of five MTB genes. Lane L1-H37Rv positive control; L2 to L6-patient sample showing the amplification of MTB genes PPE41 at 604bp, MPT53 at 539bp, LPQH at 497bp, CFP-10 at 303bp and ESAT-6 at 288bp; L7-negative control; L8-negative sample of patient.

TABLE 12.2 Comparison of MPCR Results with AFB and Conventional L-J Medium

Method	Sputum Samples from TB Patients N = 252			
	Sensitivity	Specificity	PPV	NPV
AFB smear Microscopy	63.3%	92%	71.2%	82%
LJ Culture	89.2%	98.1%	78.2%	87.1%
Multiplex PCR	97.7%	100%	91.1%	99.3%

TB is a leading threat to the health in both the developing as well as developed countries. In India, the diagnosis of TB is primarily based on the clinical features, histopathology, and demonstration of acid-fast bacilli and isolation of MTB in culture. These techniques have several drawbacks, such as lack of rapidity and inadequate sensitivity and specificity [14]. DNA amplification by PCR provides a rapid and sensitive method for the detection of MTB complex from clinical samples and cultures in the Mycobacteria laboratory routine. The conventional method of culture on the LJ media is regarded as the gold standard technique for the detection of Mycobacterium growth [15]. PCR due to higher rapidity and sensitivity can facilitate the earlier and accurate identification of the causative organism of TB and thus can very much be feasible for retreatment, prevention, and control of this chronic infection. PCR as a diagnostic tool had been widely used for many years for the purpose of detecting Mycobacterial DNA. However, PCR with one type of primer can cause ambiguity in the detection of MTB-DNA and hence should be verified with other primers specific for MTB [16]. The species-specific PCR assays developed for *M. tuberculosis* have been invalidated due to false-negative results owing to the absence of specific target sequences such as mtp40 in some MTB strains [17]. The common drawback associated with the PCR assays is the risk of obtaining false positivity in results due to contamination or presumably due to the presence of killed or dormant bacilli in the samples (Beige et al., 1999). The multiplex-PCR proposed in this study was based on the 2–3 amplification of the Mycobacterial genomic fragments [18]. This result was in accordance to study by Negi et al. [12] which reported a sensitivity of 73.6%with MPCR in smear-negative culture-positive which is significantly higher than the majority of studies reporting it at around 60% or even less in the study by Sarmiento et al. [19].

MTB detection by PCR assays is significantly more sensitive and rapid when compared to that of the culture and microscopy with their sensitivities ranging between 36% and 100% and their specificities ranging between 85% and 100% using various PCR targets such as the IS6110, MPB64, hup-B, TRC4, GCRS, Pab gene, etc., [20–22]. Deshpande et al. [23] stated in their study that the IS6110 PCR assay was feasible in terms of the sensitivity (82.4%) and specificity (75.9%) as compared to the culture results which revealed 68.6% sensitivity and 79.3% specificity. Therefore, it has been analyzed that the PCR based on IS6110 is a supportive method for the precise diagnosis of clinically diagnosed TB. Most reports of studies involving the IS6110-based detection have claimed sensitivities of about 75% and specificities of approximately 100%. Boondireke et al. (2007) has stated in their study that they obtained a sensitivity of 92.1% and specificity of 98.2%in MPCR.

MTB have proved its application as a reliable technique for diagnostic purposes. In our study, we found an overall sensitivity and specificity of 97.7% and 100% respectively in the MPCR analysis. MPCR using MTB specific primers detected more TB cases compared to that of conventional culture method. To the best of our knowledge, although MPCR assay has been utilized for TB diagnosis. It is obvious that further evaluation is required for further improvisation of the MPCR protocol for routine diagnostic purposes.

12.4 CONCLUSION

MPCR using MTB primers was able to pick up more TB patients compared to conventional L-J culture methods for the detection of MTB. MPCR using MTB genes is a highly sensitive and specific tool in the diagnosis of TB patients. MPCR using five gene primers is a rapid, reliable, and highly sensitive and specific diagnostic tool for the purpose of detecting and differentiating of the MTB patients and will be useful in diagnosing sputum samples. In future, MPCR with more than three target genes may have an important role in strengthening the diagnosis of TB.

ACKNOWLEDGMENTS

We thank UGC-UPE Phase II Biomedical Research, the University of Madras for providing the financial support for the accomplishment of this project. GR was a project fellow in this project.

KEYWORDS

- **extremely drug resistant**
- **human immunodeficiency virus**
- **multi-drug resistant**
- **multiplex PCR**
- ***Mycobacterium tuberculosis***
- **polymerase chain reaction**
- **tuberculosis**

REFERENCES

1. WHO, (2017). *Bending the Curve-Ending TB: Annual Report.*
2. Sudre, P., Ten, D. G., & Kochi, A., (1992). Tuberculosis: A global overview of the situation today. *Bull. World Health Organ., 70,* 149–159.
3. World Health Organization, (2014). *Global Tuberculosis Report 2014*. Geneva Switzerland.
4. World Health Organization, (2011). *Global Tuberculosis Control: Report 16*. Geneva.
5. Chan, C. M., Yuen, K. Y., Chan, K. S., Yam, W. C., Yim, K. H., & Ng, W. F., (1996). Single-tube nested PCR in the diagnosis of tuberculosis. *J. Clin. Pathol., 49*, 290–294.
6. Narita, M., Matsuzono, Y., Shibata, M., & Togashi, T., (1992). Nested amplification protocol for the detection of *Mycobacterium tuberculosis*. *Acta Paediatr., 81*, 997–1001.
7. Kharibam, S., Farooq, U., & Nudrat, S., (2016). Molecular detection of mycobacterium tuberculosis complex from clinical sputum samples in patients attending tertiary care centres in Uttar Pradesh province of India. *Acta Medica. International, 3*(1), 102–106.
8. Eisenach, K. D., Cave, M. D., Bates, J. H., & Crawford, J. T., (1990). Polymerase chain reaction amplification of a repetitive DNA sequence specific for *Mycobacterium tuberculosis*. *J. Infect. Dis., 161*, 977–981.
9. Yoneda, M., Fukui, Y., & Yamanouchi, T., (1965). Extracellular proteins of tubercle bacilli. V. Distribution of alpha and beta antigens in various mycobacteria. *Biken. J., 8*, 201–203.
10. Wiker, H. G., Harboe, M., & Lea, T. E., (1986). Purification and characterization of two protein antigens from the heterogeneous BCG85complex in *Mycobacterium bovis* BCG. *Int. Arch Allergy Appl. Immunol*., *81*, 298–306.
11. Kusum, S., Aman, S., Pallab, R., Kumar, S. S., Manish, M., & Sudesh, P., (2011). Multiplex PCR for rapid diagnosis of tuberculous meningitis. *J. Neurol., 258*, 1781–1787.
12. Negi, S. S., Anand, R., Basir, S. F., Pasha, S. T., Gupta, S., & Khare, S., (2006). Protein antigen b (Pab) based PCR test in diagnosis of pulmonary and extra-pulmonary tuberculosis. *Indian J. Med. Res., 124*, 81–88.
13. World Health Organization, (1998). *Laboratory Services in Tuberculosis Control: Part 2.*
14. Shukla, I., Varshney, S., Sarfraz, Malik, A., & Ahmad, Z., (2011). Evaluation of nested PCR targeting IS6110 of *Mycobacterium tuberculosis* for the diagnosis of pulmonary and extra-pulmonary tuberculosis. *Biol. Med., 3*, 171–175.
15. Clarridge, J. E., Shawaz, R. M., Shinnick, T. M., & Plikaytis, B. B., (1993). Large scale use of PCR for detection of mycobacterium tuberculosis in routine mycobacteriology laboratory. *J. Clin. Microbiol., 31*, 2049–2056.
16. Asthana, K. A., & Madan, M., (2015). Study of target gene IS 6110 and MPB 64 in diagnosis of pulmonary tuberculosis. *Int. J. Curr. Microbiol. App. Sci., 4*(8), 856–863.
17. Weil, A., Plikaytis, B. B., Butler, W. R., Woodley, C. L., & Shinnick, T. M., (1996). The mtp40 gene is not present in all strains of *Mycobacterium tuberculosis*. *J. Clin. Microbiol., 34*, 2309–2311.
18. Rodriguez, J. G., Mejia, G. A., Del Portillo, P., Patarrayo, M. E., & Murillo, L. A., (1995). Species-specific identification of *Mycobacterium bovis* by PCR. *Microbiology, 141*, 2131–2138.
19. Sarmiento, O. L., Weigle, K. A., Alexander, J., Weber, D. J., & Miller, W. C., (2003). Assessment by meta-analysis of PCR for diagnosis of smear-negative pulmonary tuberculosis. *J. Clin. Microbiol., 41*, 3233–3240.

20. Bhanu, N. V., Singh, U. B., Chakravorty, M., Suresh, N., Arora, J., Rana, T., Takkar, D., & Seth, P., (2005). Improved diagnostic value of PCR in the diagnosis of female genital tuberculosis leading to infertility. *J. Med. Microbiol., 54*, 927–931.
21. Sankar, S., Kuppanan, S., Balakrishnan, B., & Nandagopal, B., (2011). Analysis of sequence diversity among IS6110 sequence of *Mycobacterium tuberculosis*: Possible implications for PCR based detection. *Bioinformation, 6*, 283–285.
22. Jain, A., (2011). Extra pulmonary tuberculosis: A diagnostic dilemma. *Ind. J. Clin. Biochem., 26*, 269–273.
23. Deshpande, P. S., Kashyap, R. S., Ramteke, S. S., Nagdev, K. J., Purohit, H. J., & Hatim, F. D., (2007). Evaluation of the IS6110 PCR assay for the rapid diagnosis of tuberculous meningitis. *Cerebrospinal Fluid Res., 4*, 10.
24. Beige, J., Lokies, J., Schaberg, T., Finckh, M., Fischer, M., & Mauch, H., (1995). Clinical evaluation of a mycobacterium tuberculosis PCR assay. *J. Clin. Microbiol., 33*, 90–95.
25. Boondireke, S., Mungthin, M., Tan-ariya, P., Boonyongsunchai, P., Naaglor, T., Anan, W., Sompong, T., & Saovanee, L., (2010). Evaluation of sensitivity of multiplex PCR for detection of *Mycobacterium tuberculosis* and *Pneumocystis jirovecii* in clinical samples. *J. Clin. Microbiol., 48*(9), 3165–3168.
26. Microscopy Publication Number. World Health Organization, Geneva, Switzerland.
27. World Health Organization Annual Report, (2017).

CHAPTER 13

Anti-HSV and Cytotoxicity Properties of Three Different Nanoparticles Derived from Indian Medicinal Plants

K. VASANTHI, G. REENA, G. SATHYANARAYANAN, and ELANCHEZHIYAN MANICKAN

Department of Microbiology, Dr. ALM PG IBMS, University of Madras, Taramani, Chennai, Tamil Nadu, India

ABSTRACT

Currently, the medical field kept their step in nanotechnology which employs the nanoparticles for treating the diseases. The study highlighted the use of nanotechnology in the virus aqueous extracts from the medicinal plants and their effect against herpes virus. Aqueous extract of *Punica granatum* (*P. granatum*) (Peels, Juice), *Camellia sinensis (C. sinensis)*, *Nilavembu Kudineer Chooranam* (NKC), and *Acalypha indica (A. indica)*. Three different nanoparticles (NP) such as silver, gold, and bimetallic were synthesized from the above aqueous extracts and characterized. Anti-HSV (both 1 and 2) activities of these nanoparticles were done by CPE assay. Toxicity of the extracts was determined by MTT assay. Among the tested nanoparticles *P. granatum* peels silver NP (PgPSNP) exhibiting a potent anti-HSV activity followed by *P. granatum* juice silver NP > *C. sinensis* Silver NP > NKC Silver NP > *A. indica* silver NP. Neither bimetallic nor gold NP exhibited significant anti-HSV activity. Except *P. granatum* other extracts showed more toxicity. This study indicated that *P. granatum* peels silver NP (PgPSNP) showed the maximum anti-HSV activity besides its minimal toxicity observed. Thus, PgPSNP is a novel anti-HSV drug which is worth pursuing.

13.1 INTRODUCTION

Herpes simplex viruses (both types, HSV-1 and -2) are pathogenic to humans with a worldwide morbidity. HSV has a capacity to hide into our neurons and can reactivate, causing frequent recurrent infections in some patients, while most people experience few recurrences. Infections with HSV-1 and HSV-2 are highly prevalent. HSVs infected more than 3.7 billion people under the age of 50–60. The WHO estimated that over 500 million people are infected with HSV-2 worldwide with approximately 20 million cases annually [3]. Also, the global rates of either HSV-1 or HSV-2 are between 60% and 5% in adults. In India, the seropositivity has been reported to be 33.3% for HSV-1 and 16.6% for HSV-2 [3].

At present, there is no complete cure for this virus. Treatment focuses on getting rid of sores and limiting outbreaks. The medication includes acyclovir (ACV), famciclovir, valacyclovir, and these can help infected individuals reduce the risk of spreading the virus to others. Chronic use of anti-herpes virus drugs results in severe side effects and drug-resistant viruses. Use of ACV is unsuitable for pregnant women and neonates because of incorporation of drug into the host DNA and yields adverse effects. Also, effective vaccines against HSV infection are not yet identified. Furthermore, the available therapeutic vaccines does not protect the patients from recurrences because failure to stimulate the antibody-specific responses against HSV virus. Therefore, there is an urgent requirement of alternative agents to cure and prevent the HSV infection also these agents should be cheap, readily available, and less toxic.

Natural therapy is an alternative to allopathic medication which exploits the least side effects. Currently, there was an unlimited plants and herbs resources, therefore their phyto-chemicals were useful in concerned. Wide range of global population also prefers the use of natural products in treating and preventing the medical problems. But there is a few number of studies have used known purified plant chemicals, very few screening programs have been initiated on crude plant materials. Also, the drug delivery into the human body is very poor while using the plants as choice of drug [5].

Nanotechnology is an emerging area in pharmaceutical and medical field. Physicochemical and biological properties of materials would be varying fundamentally from their bulk part at their nanometric scale; this is due to the size-dependent quantum effect. The nanoparticles such as gold and carbon are surface-functionalized, have unique dimensions, and controlled drug release, thus can be used in the drug delivery [20–22]. On this basis the present study was highlighted the use of nanotechnology in the synthesis of drug from the medicinal plants and their effect against herpes virus.

13.2 MATERIALS AND METHODS

13.2.1 PLANT MATERIALS

The plants used for this study namely: *Punica granatum (P. granatum), Camellia sinensis (C. sinensis), Nilavembu Kudineer Chooranaum (NKC),* and *Acalypha indica (A. indica)* were collected from Chennai and some were purchased commercially from the stores (Figure 13.1). All these plants were taxonomically identified by the Department of Botany, University of Madras. Peels of *P. granatum* and leaves of the *A. indica* were shade dried, ground into a uniform powder using a blender, and stored at 4°C. Fruits of the *P. granatum* were grinded and the juice was lyophilized. Leaves of Camellia sinensis were purchased from commercial source available at Chennai. Dried leaves were ground into a uniform powder using a blender and stored at 4°C in refrigerator. NKC was purchased from the TAMPCOL and the dried leaves were ground into a uniform powder using a blender and stored at 4°C in refrigerator.

13.2.2 PREPARATION OF AQUEOUS EXTRACT

About 10 grams of the dried *C. sinensis*, NKC, and Peels of *P. granatum*, and *A. indica* powder was soaked in 100 ml of distilled water overnight at room temperature. It was then cotton filtered to remove the coarse particles and then filtered through Whatman No. 1 filter paper. Then the extracts was passed through 0.45 and 0.2 micron membrane filter and the water content of extract is removed by lyophilization and stored at 4°C until use. Juice of the *P. granatum* were filter by Whatman filter followed by 0.45 μm and 0.2 μm, lyophilized, and stored at 4°C until use.

13.2.3 SYNTHESIS OF NANOPARTICLES FROM THE EXTRACTS

Synthesis of gold nanoparticles was done as described previously by Tiwari (2011). Briefly, lyophilized powder of Plant extracts were reconstituted with 1 ml of sterile distilled water and mixed with 0.002 M of chlororauric acid (SRL Cat. No) (HAuCl4) in dark conditions with a preincubation at 90°C. After incubation the color of the solution were turned its color to ruby pink (Figure 13.2) indicates the gold nanoparticle formation. According to Klaus (2001) synthesis of silver Nanoparticles was done. Briefly, lyophilized powder of plant extracts were reconstituted with 1 ml of sterile distilled water and mixed with 20 ml of 10^{-3} M $AgNO_3$ (SRL Cat. No: Cat. No for (HAuCl4)- 12023, Cat. No for (AgNO3) – 94118) (99.99%) aqueous solution and kept at room temperature. After 1 hour the color of the solution were changed from colorless to honey brown (Figure 13.2) indicating the formation of silver nanoparticles and this is confirmed by UV-visible spectroscopy and other methods. Synthesis of bimetallic nanoparticles (Silver-Gold) were done according to the Pal et al. Briefly, lyophilized powder of plant extracts were reconstituted with 1 ml of sterile distilled water and mixed with equal amount of 10^{-3} M $AgNO_3$ and 0.002 M of chlororauric acid and incubated at room temperature. After incubation the color of the solution were turned its color in the combination of ruby pink and honey comb color (Figure 13.2).

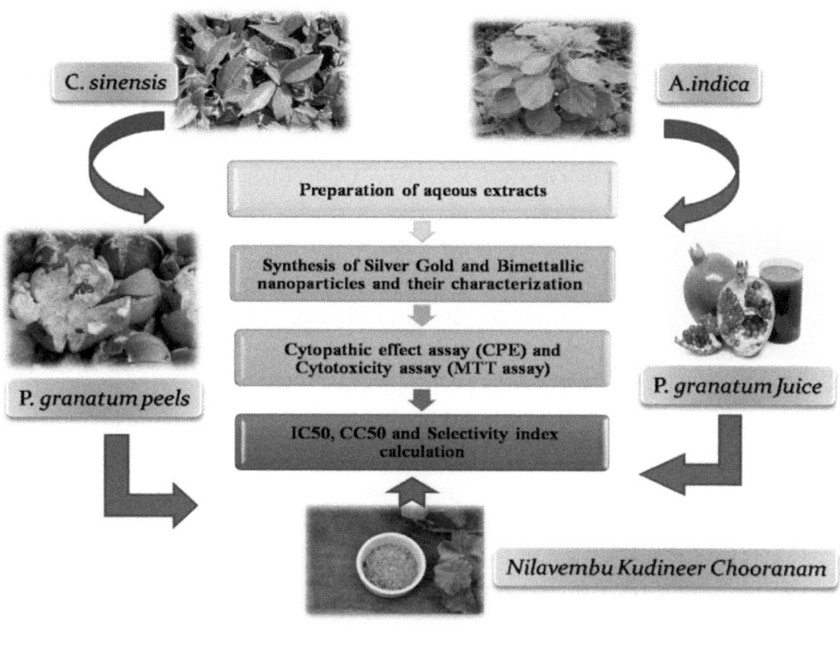

FIGURE 13.1 Plants used in this study and methodology.

FIGURE 13.2 Synthesis of three different nanoparticles from medicinal plants, nanoparticles of (a) NKC, (b) *P. granatum* peels, (c) *C. sinensis*, (d) *P. granatum* juice, (e) *A. indica.*

13.3 CHARACTERIZATION OF NANOPARTICLES

The newly synthesized nanomaterials was characterized using ultra violet visible (UV-Vis) spectrophotometer, high resolution scanning electron microscopy (HR-SEM), high resolution transmission electron microscopy (HR-TEM), FTIR spectra, X-ray diffraction (XRD) patterns (Figure 13.2).

13.3.1 IN VITRO STABILITY TEST

In vitro stability studies were performed with aqueous solutions of nanoparticles (0.5 ml) and 0.5 ml of 10% NaCl, 0.2 M cysteine, 0.2 M histidine, 0.5% HSA and 0.5% BSA solutions. The stability of nanoparticles will also investigated in phosphate buffer at pH 4, 5, 6, 8, 9, and 10 respectively. The stability of the nanoparticles will be measured by using a UV-Vis spectrum after 24 hrs and 360 hrs.

13.3.2 IN VITRO ANTI-HSV TESTING

1. **Cell Culture and Virus:** African green monkey kidney (Vero) cells were grown in DMEM (Sigma) supplemented with 10% heat-inactivated Fetal bovine serum (Gibco BRL Co., Germany) and 1% antibiotics (Penicillin [100 IU/ml] and Streptomycin, [100 μl/

ml] Himedia) at 37°C in a humidified atmosphere of 5% CO_2. Wild-type HSV-1 (753166) and HSV-2 (753167) strains obtained from Dr. S. Rajarajan (Department of Microbiology and Biotechnology, Presidency College, Chennai). Virus stocks were propagated in Vero cells and titer was calculated from plaque numbers (15×10^4/ml) for HSV-1 and PFU (20×10^4/ml) for HSV-2 according to the method of Reed and Muench (1938).

2. **Standard Antivirals:** ACV (Sigma) was used as standard antiviral drug (1 mg/ml) dissolved in PBS and was serially diluted with respect to test compounds at a concentration of 500 μg/ml (1:2) to 3.9 μg/ml (1:256).

13.3.3 CYTOPATIC EFFECT ASSAY (CPE ASSAY)

Vero cells were plated on 96-well plate, at a density of 10,000 cells/well for 70–80% confluency. Cells was infected with HSV-1 and HSV-2 and incubated for 90 min at 37°C. Varying concentrations of the different nanoparticles were added to the virus infected 96 well plates and incubate for 7 days at 37°C in 5% CO_2 environment. Cytopathic effects (CPE) and their reduction were observed for 7 days (Figure 13.3). Cell control and Virus control, Drug control were put along with the assay.

13.3.4 MTT ASSAYS TO MEASURE THE CYTOTOXICITY OF EXTRACT

MTT [3-(4,5-dimethylthiazol-2-yl)-2,5-diphenyltetrazolium bromide] assay was used for the screening of extracts toxicity using standard protocol. Briefly, verocells were incubated at 37°C on 96-well plates at a density of 10,000 cells/well, with 5% CO_2 in a humidified atmosphere with 10% Dulbecco's Modified Essentials Medium. After 24 hours, nanoparticles were added on monolayer (70–80% confluency) to final concentrations of 50 μg to 10 mg. The plate was incubated for further 5 day under the same conditions mentioned above. 200 μl MTT (Sigma-Aldrich, Catalogue No. M2003) solutions (5 mg/ml in phosphate buffer) was added to each well and incubated at 37°C for 4 hours. The MTT solution was decanted off, and Formosan was extracted from the cells with 250 μl of DMSO in each well. Color was measured with a 12-well ELISA plate reader at 550 nm (Figure 13.4). Toxicity Control used as 1% Triton X-100 (Qualigens, catalog No. 10655). All MTT assays were repeated three times. Cell viability was calculated using the following formula:

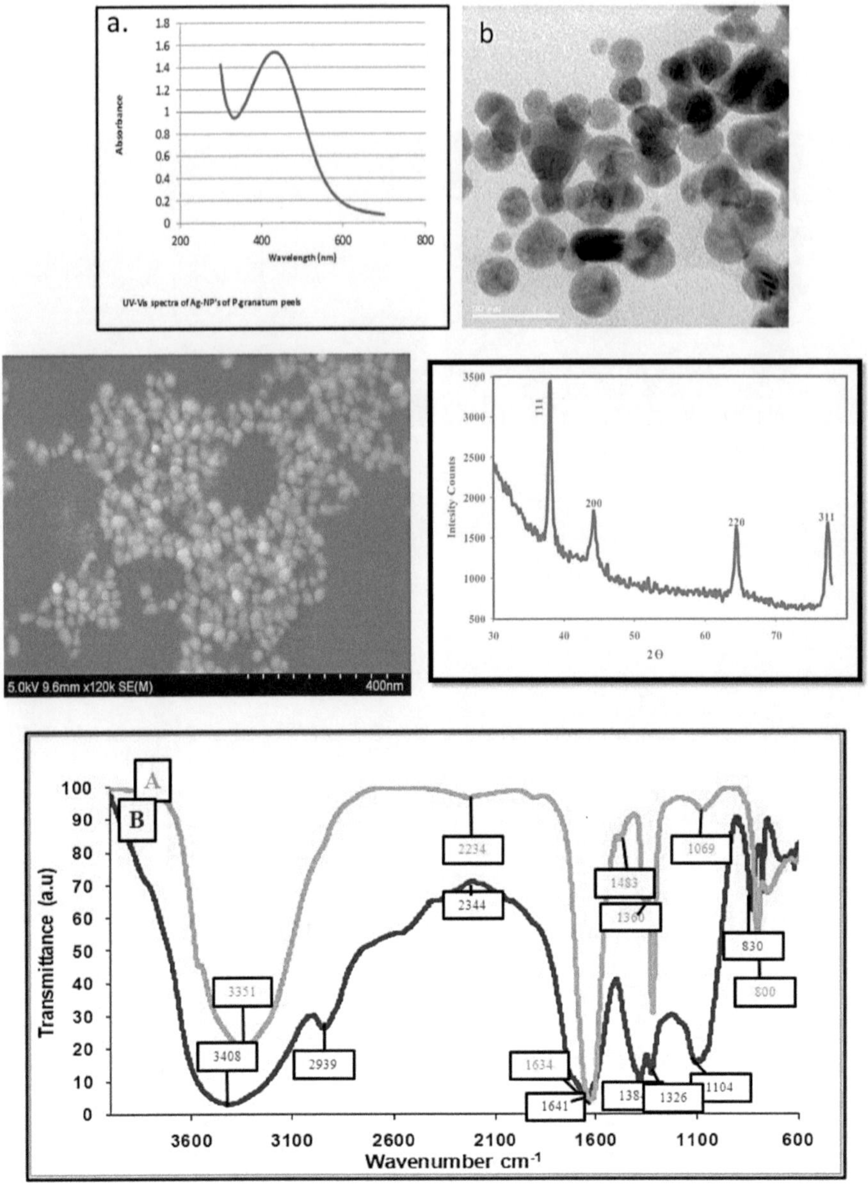

FIGURE 13.3 Characterization of nanoparticles. (a) UV-Vis reading of *P. granatum*,(b) TEM image of *P. granatum* peels of silver Np's,(c) SEM image of *P. granatum* peels silver Np's,(d) XRD of *P. granatum* peels silver Np's, (e) FTIR spectra of *P. granatum* peel synthesized silver nanoparticles and aqueous extract of *P. granatum* peel.

$$Cell\ viability(\%) = \frac{(OD\ value\ of\ treated\ well - OD\ value\ of\ the\ blank)}{(OD\ value\ of\ untreated\ well - OD\ value\ of\ the\ blank)}$$

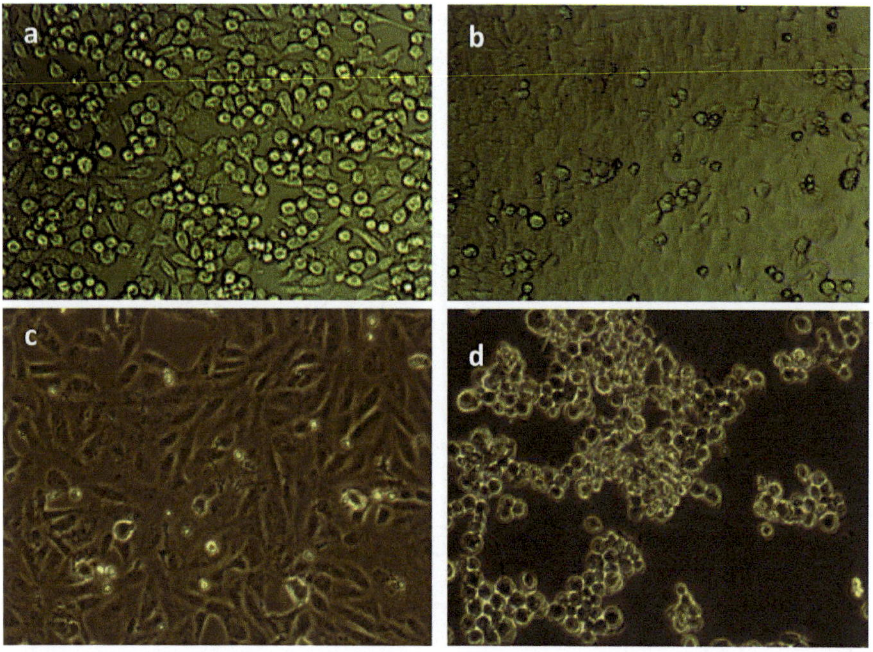

FIGURE 13.4 Antiviral assay, (a) CPE of Vero cell line, (b) reduction in CPE by *P. granatum* peels silver Np's, (c) cell control, and (d) virus control.

13.4 RESULTS

HSVs are unique viruses that target sensory neurons and after the first primary infection of HSV, which heals without a trace. However, the virus finds its way to dorsal root ganglion where they remain latent until the next recurrence episode. HSV type 1 (HSV-1) stays latent in the trigeminal ganglia and HSV type 2 (HSV-2) remain latent in the lumbo-sacral ganglia. During latency, the virus remains dormant and the only transcription found is latency associated transcripts (LAT). At the time of recurrence, the virus multiplies and exit through motor neurons and cause painful recurrent lesions at the neurodermatome where the nerves supply. The mechanism of latency and recurrence is poorly understood though the stress, radiation, etc., are attributed for those mechanisms. This kind of latent infections and periodic recurrences are the hallmarks of HSV infections. Nanoparticles have an extraordinary property to deliver a drug without losing its ability.

In this study, we evaluated the anti-HSV activity of three different nanoparticles (gold, silver, and bimetallic) which was derived from the four different plants. Figure 13.3a shows the UV-spectrophotometric reading of *P. granatum* silver nanoparticle showing 420 nm. Figure 13.3b shows the TEM image of *P. granatum* silver Np showing 20 nm of each particle. Figure 13.2c SEM showing surface morphology of *P. granatum* peels silver nanoparticles, showing well-separated spherical nanoparticle. Figure 13.2d shows the XRD analysis, structural information about crystalline metallic nanoparticle. XRD analysis showed intense peaks corresponding to (111), (200), (220), and (311) Bragg's reflection based on the face-centered cubic structure of silver nanoparticles. The broadening of Bragg's peaks indicates the formation of nanoparticle and its crystalline nature. There is no other peak in the XRD pattern, which confirmed the stability and high purity of silver nanoparticles. XRD pattern of newly synthesized silver nanoparticles showed intense peaks of (111), (200), (220), and (311) which confirmed the monophasic nature of pure Ag with face-centered cubic symmetry. Figure 13.3e shows FTIR Spectra of *P. granatum* peel synthesized silver nanoparticles and aqueous extract of *P. granatum peel*. The spectrum of *P. granatum extract* displayed important peaks at 3408, 2939 (O-H stretch hydroxyl of phenols or alcohols), 1641 (C=O stretch of amides), 1384, 1326 (N=O bend of Nitro groups), 1104 (C-O stretch of Carboxylic acid) and 830 cm^{-1} (C-H stretch of Aromatics). On the other hand, FTIR spectra of aqueous extract of *P. granatum* synthesized silver nanoparticles exhibited peaks at 3351, 1634, 1360, 1320, 1069, and 800 cm^{-1}. This confirms the involvement of these functional groups during formation of silver nanoparticles. The hydroxyl and carboxylic acid from the polyphenols and amino acid residues of proteins have ability to produce silver nanoparticles.

Table 13.1 shows a dose-response analysis of gold, silver, and bimetallic nanoparticles of aqueous extracts of *P. granatum* (Peels, Juice), *C. sinensis*, Nilavembu Kudineer Chooranam, and *A. indica* for both HSV-1 and HSV-2. Experiments showed that strong anti-HSV activity (both 1 and 2) were observed for silver nanoparticles of *P. granatum* peels extracts (1 μg/ml for HSV-1 and 3.9 μg/ml for HSV-2) followed by Silver nanoparticles of Juice (7.8 μg/ml for HSV-1 and 15.9 μg/ml for HSV-2) revealed by the complete reduction of CPE on vero cells (Table 13.1). However, the anti-HSV activity of nanoparticles derived from other aqueous plant extract was much milder (Table 13.1). Positive control (ACV) does not show CPE. MTT assay was done with vero cell line for all the nanoparticles (Figure 13.6). Except *P. granatum* nanoparticles, other nanoparticles had cytotoxicity at the various concentrations tested on vero cell line and their percentages were calculated using the formula. Selectivity of the nanoparticles was founded against HSV-1 and 2. Based on the results, it was found that the PgPSNP showed highest selectivity index (SI) (155 for HSV-1 and 150.1 for HSV-2), therefore the promising antiviral activity against HSV-1 and HSV-2, is followed by PgJSNP (33.21 for HSV-1 and 29.1 for HSV-2) whereas nanoparticles showed moderate Sis (Figures 13.4–13.10 and Table 13.1).

FIGURE 13.5 MTT assay after 48 hours incubation.

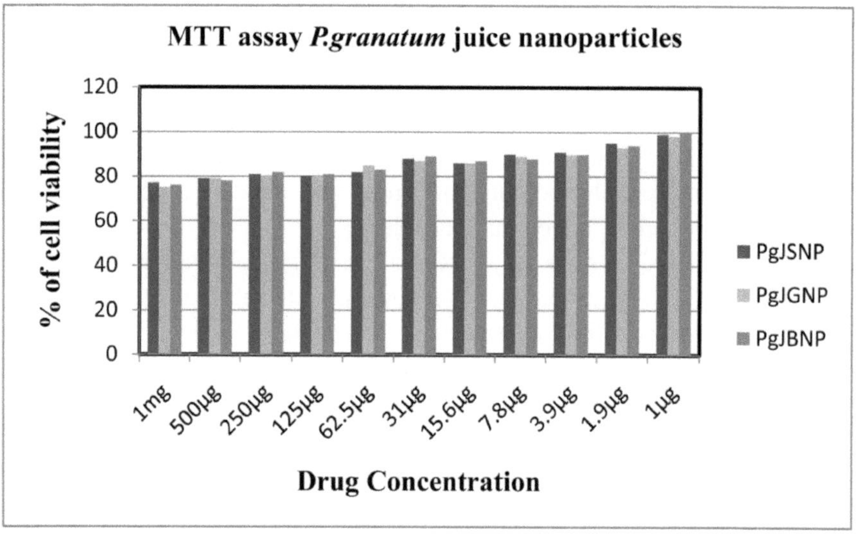

FIGURE 13.6 Percentage of cell viability for *P. granatum* juice nanoparticles using MTT assay at 48 hrs.

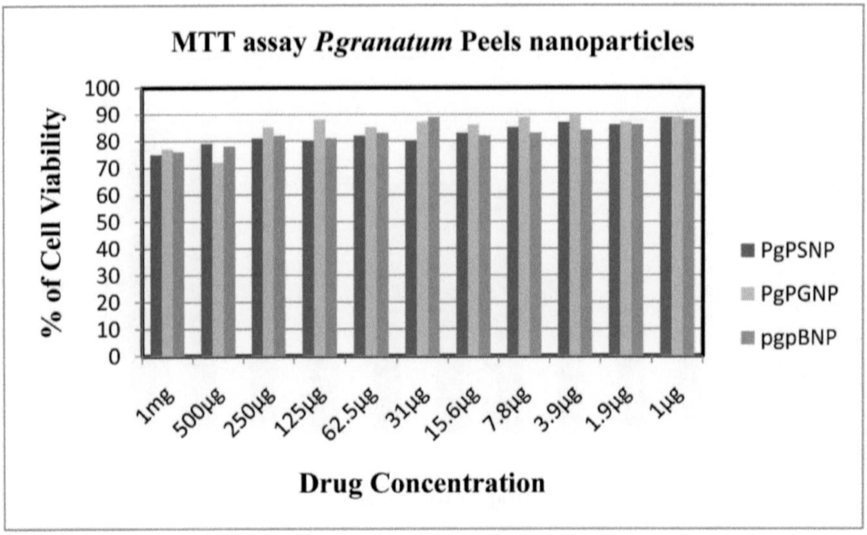

FIGURE 13.7 Percentage of cell viability for *P. granatum* peels nanoparticles using MTT assay at 48 hrs.

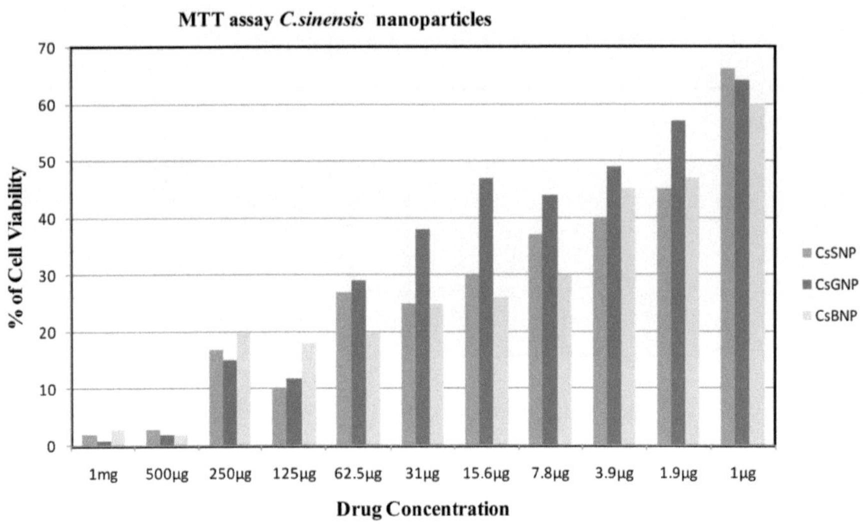

FIGURE 13.8 Percentage of cell viability for *C. sinensis* nanoparticles using MTT assay at 48 hrs.

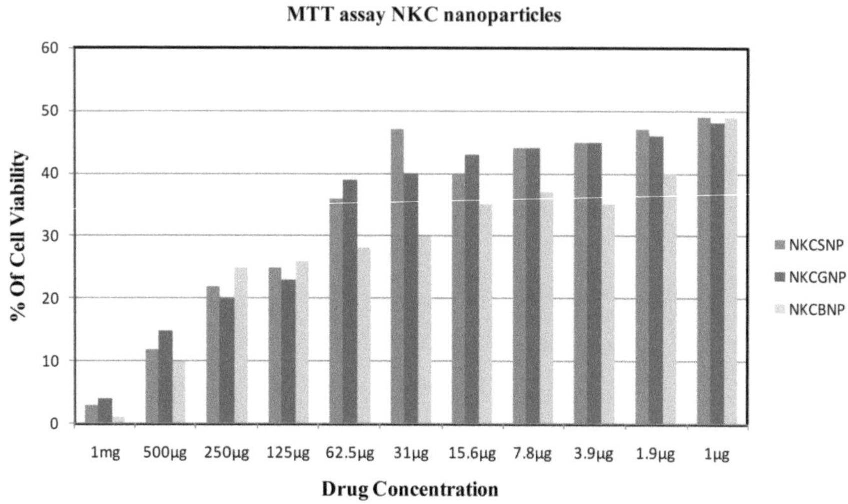

FIGURE 13.9 Percentage of cell viability for NKC nanoparticles using MTT assay at 48 hrs.

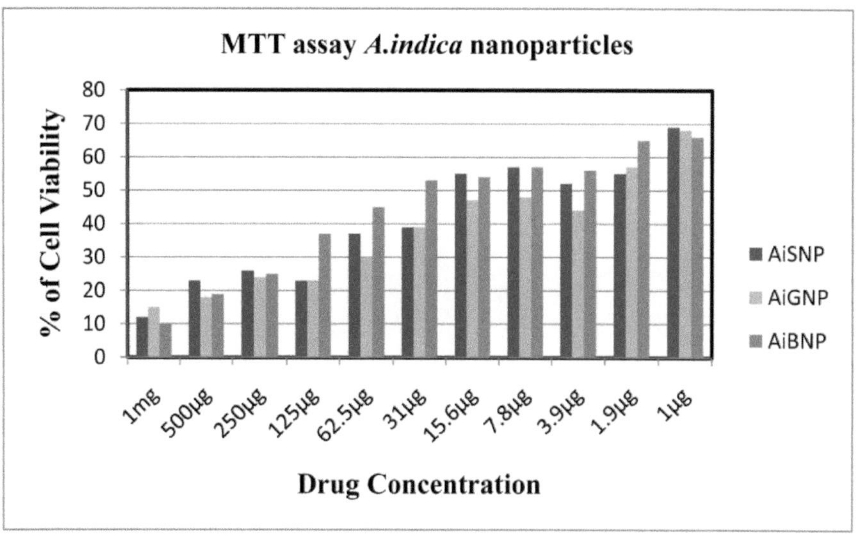

FIGURE 13.10 Percentage of cell viability for *A. indica* nanoparticles using MTT assay at 48 hrs.

TABLE 13.1 Antiviral Activity and Cytotoxicity Effect of Nanoparticles

Sl. No.	Nanoparticles	Cytotoxicity CC_{50} (μg/ml)	Antiviral Activity IC_{50} (μg/ml)		Selectivity Index	
			HSV-1	HSV-2	HSV-1	HSV-2
1.	*P. granatum* peel silver nanoparticle (PgPSNP)	453	2.93	3.01	155	150.1
2.	*P. granatum* peel gold nanoparticle (PgPGNP)	416.08	42.09	45.70	9.89	9.10
3.	*P. granatum* peel bimetallic nanoparticle (PgPBNP)	418.04	34.39	39.78	12.16	10.51
4.	*P. granatum* juice silver nanoparticle (PgJSNP)	353	10.63	12.09	33.21	29.1
5.	*P. granatum* juice gold nanoparticle(PgJGNP)	456	47.80	47.98	9.54	9.50
6.	*P. granatum* juice bimetalllic nanoparticle(PgJBNP)	398	50.67	56.80	7.85	7
7.	*C. sinensis* silver nanoparticle (CsSNP)	415	89.12	96.34	4.66	4.31
8.	*C. sinensis* gold nanoparticle (CsGNP)	402	100.65	120.87	4	3.3
9.	*C. sinensis* bimetallic nanoparticle (CsBNP)	437	95.69	107.04	4.6	4.1
10.	Nilavembu Kudineer Chooranaum Silver nanoparticle (NKCSNP)	429	226.10	224	1.9	2
11.	Nilavembu Kudineer Chooranaum Gold nanoparticle (NKCGNP)	312	250	223.41	1.2	1.4
12.	Nilavembu Kudineer Chooranaum Bimetallic nanoparticle (NKCBNP)	395	212	206.93	1.9	2
13.	*A. indica* silver nanoparticle (AiSNP)	453	160.04	178	2.8	2.5
14.	*A. indica* gold nanoparticle (AiGNP)	378	198.23	178.02	1.9	2.1
15.	*A. indica* bimetallic nanoparticle (AiBNP)	409	145	178.96	2.8	2.3
16.	Acyclovir (ACV)	250	1.5	2	167	125

13.5 DISCUSSION

HSVs infection is common in worldwide. After the primary infection the virus, remain latent in the host body till the end. There is no exact treatment for this infection. Currently using drugs not completely eliminating the virus from the human body, it is giving timely cure from the infection. Extensive and long term clinical use of anti-herpes virus agents like ACV, and its derivatives results in severe side effects and drug-resistant viruses. Further, ACV is reported to incorporate into the cellular DNA, yielding adverse drug reactions and thus, unsuitable for pregnant women and neonates. Moreover, the major determinants of effective immunity against HSV infection are not yet identified. Furthermore, the therapeutic vaccines failed to induce antibody-specific responses to protect recipients from recurrences. Therefore, there is an urgent need for cheap, readily available, less toxic alternate agents to control and prevent HSV infection and its transmission [11].

Nanotechnology is an emerging area in the drug development. It has a huge capacity and thus used drug delivery system. Since the host cells were actively absorbs the nanoparticles than the other larger micro molecules, the drug delivery system is effectively achieved by the nanoparticles [12]. Limited number of studies revealed the approach of metal-based nanoparticles such as tin, silver, gold, and zinc oxide nanoparticles in herpes infection treatments. Reduction of the progeny viruses with weak cytotoxicity was due to the interaction between silver nanoparticles and HSV-2 by *in vitro* [16]. Formation of the bond by the nanoparticles between the glycoprotein membrane of HSV-2 and the receptor of the host inhibits the entry of virus in the cell [16]. Modified tannic acid with silver nanoparticles has been reported in reduction of HSV-2 infection in both *in vitro* and *in vivo* [17]. The antiviral activities of the nanoparticles were greatly influenced by the particle size and the dose of the formulation. Because the smaller-sized nanoparticles were characterized by the production of cytokines and chemokines which was greatly useful for anti-viral response [17].

Discovery of water-soluble gold nanoparticles were useful in preventing the herpes virus infection by interacting the viral attachment and penetration [19]. Baram-Pinto et al. developed a non-toxic formulation such as gold-based mercapto ethane sulfonated nanoparticles. These nanoparticles can be widely used for topical application, due to its non-toxic formulation it can blindly use as therapeutic and prophylactic application. The mechanism behind this discovery was blocking of viral attachment to the cell and thus preventing the cell-to-cell spread of virus [20]. Another nanoparticle such as zinc oxide nanoparticles also developed to inhibit the entry of HSV-2 virus and preventing the cell to cell spreading in the vaginal lining by its negatively charged surface [18].

Our study underlines the capacity of the nanoparticles which was derived from the plants extract against the HSV. Thus, our nanoparticles such as PgPSNP and PgJSNP revealed anti-HSV activity due to the presence of more active compounds punicalagin. Our also study supports the previous reporting, punicalagin (active compound) of the extract and juice of *P. granatum* were responsible for the antiviral activity. Thus, the fractionation of these bioactive compounds against the viral activity should be explored further [14, 15]. The active compounds in the remaining extracts might be less in quantities to inhibit the virus particles. Elucidation of those active compounds from those plants might be providing the new and effective antiviral agents. This study clearly indicated that *P. granatum* peels silver NP (PgPSNP) showed the maximum anti-HSV activity besides its minimal toxicity observed. Thus, PgPSNP is a novel anti-HSV drug which is worth pursuing.

KEYWORDS

- **cytopathic effects**
- **cytotoxicity**
- **herpes simplex-1**
- **herpes simplex-2**
- **nanoparticles**
- **plant extracts**

REFERENCES

1. Steiner, I., & Benninger, F., (2013). Update on herpes virus infections of the nervous system. *Curr. Neurol. Neurosci. Rep., 13*, 414.
2. Aggarwal, A., & Kaur, R., (2004). Seroprevalence of herpes simplex virus1 and 2 Antibodies in STD clinic patients. *Indian J. Med. Microbiol., 22*, 2446.
3. Bogaerts, J., Ahmed, J., Akhter, N., Begum, N., Rahman, M., Nahar, S., et al., (2001). Sexually transmitted infections among married women in Dhaka, Bangladesh: Unexpected high prevalence of herpes simplex type 2 infection. *Sex Transm. Infect., 77*, 1149.
4. Mathiesen, T., Linde, A., Olding-Stenkvist, E., & Wahren, B., (1988). *Specific IgG Subclass Reactivity in Herpes Simplex Encephalitis.*

5. Jassim, S. A. A., & Naji, M. A., (2003). Novel antiviral agents: A medicinal plant perspective. *Journal of Applied Microbiology, 95*, 412–427.
6. Bradley, H., Markowitz, L. E., Gibson, T., et al., (2014). Seroprevalence of herpes simplex virus types 1 and 2-United States, 1999–2010. *J. Infect. Dis., 209*, 325–333.
7. Luker, G., Bardill, J., Prior, J., Pica, C., Piwnica-Worms, D., & Leib, D., (2002). Primary infection with HSV-1 and reactivation from latency. *J. Virol., 76*, 12149–12161.
8. Corey, L., Adams, H. G., Brown, Z. A., & Holmes, K. K., (1983). Genital herpes simplex virus infections: Clinical manifestations, course, and complications. *Ann. Intern. Med., 98*, 958–972.
9. Forsgren, M., Skoog, E., Jeansson, S., Olofsson, S., & Giesecke, J., (1994). Prevalence of antibodies to herpes simplex virus in pregnant women in Stockholm in 1969, 1983, and 1989: Implications for STD epidemiology. *Int. J. Sex. Transm. Dis. AIDS., 5*, 113–116.
10. Cowan, F. M., Johnson, A. M., Ashley, R., Corey, L., & Mindel, A., (1996). Relationship between antibodies to herpes simplex virus (HSV) and symptoms of HSV infection. *J. Infect. Dis., 174*, 470–545.
11. Bag, P., Chattopadhyay, D., Mukherjee, H., Ojha, D., Mandal, N., Sarkar, M. C., et al., (2012). Therapeutic vaccines failed to induce antibody-specific responses. *Journal of Virology, 9*, 98.
12. Yokoyama, M., (2005). Drug targeting with nano-sized carrier systems. *J. Artif. Organs., 8*, 77–84. doi: 10.1007/s10047-005-0285-0.
13. Khan, M. T., Ather, A., Thompson, K. D., & Gambari, R., (2005). Extracts and molecules from medicinal plants against herpes simplex viruses. *Antiviral Res., 67*, 107–119.
14. Lin, L. T., Chen, T. Y., Chung, C. Y., Noyce, R. S., Grindley, T. B., McCormick, C., et al., (2011). Hydrolyzable tannins(chebulagic acid and punicalagin) target viral glycoprotein-glycosaminoglycan interactions to inhibit herpes simplex virus 1 entry and cell-to-cell spread. *J. Virol., 85*, 4386–4398.
15. Lu, J., Wei, Y., & Yuan, Q., (2007). Preparative separation of punicalagin from pomegranate husk by high-speed countercurrent chromatography. *J. Chromatogr B Analyt. Technol. Biomed Life Sci., 857*, 175–179.
16. Hu, R. L., Li, S. R., Kong, F. J., Hou, R. J., Guan, X. L., & Guo, F., (2014). Inhibition effect of silver nanoparticles on herpes simplex virus 2. *Genet. Mol. Res., 13*, 7022–7028.
17. Orlowski, P., Tomaszewska, E., Gniadek, M., Baska, P., Nowakowska, J., Sokolowska, J., Nowak, Z., Donten, M., Celichowski, G., Grobelny, J., et al., (2014). Tannic acid modified silver nanoparticles show antiviral activity in herpes simplex virus type 2 infection. *PLoS One, 9*, e104113.
18. Antoine, T. E., Hadigal, S. R., Yakoub, A. M., Mishra, Y. K., Bhattacharya, P., Haddad, C., Valyi-Nagy, T., et al., (2016). Intravaginal zinc oxide tetrapod nanoparticles as novel immuno protective agents against genital herpes. *J. Immunol., 196*, 4566–4575.
19. Sarid, R., Gedanken, A., & Baram-Pinto, D., (2012). *Pharmaceutical Compositions Comprising Water-Soluble Sulfonate-Protected Nanoparticles and Uses Thereof.* U.S. Patent 20120027809 A1.
20. Baram-Pinto, D., Shukla, S., Gedanken, A., & Sarid, R., (2010). Inhibition of HSV-1 attachment, entry, and cell-to-cell spread by functionalized multivalent gold nanoparticles. *Small, 6*, 1044–1050.

21. Angshuman, P., Sunil, S., & Surekha, D., (2007). Preparation of silver, gold, and silver-gold bimetallic nanoparticles in w/o micro emulsion containing TritonX-100. *Colloids and Surfaces A: Physicochemical and Engineering Aspects, 302*(1–3), 483–487.
22. Piotr, O., Andrzej, K., Emilia, T., Ranoszek-Soliwoda, K., Agnieszka, W., Jakub, G., Grzegorz, C., et al., (2018). Antiviral activity of tannic acid modified silver nanoparticles: Potential to activate immune response in herpes genitalis. *Viruses, 10*(10), 524.
23. Tiwari, P. M., Vig, K., Dennis, V. A., & Singh, S. R., (2011). Functionalized Gold Nanoparticles and Their Biomedical Applications. *Nanomaterials, 1*(1), 31–63.
24. Klaus, T., Joerger, R., Olsson, E., & Granqvist, C. Gr., (1999). Silver-based crystalline nanoparticles, microbially fabricated. *Proc Natl Acad Sci USA, 96*, 13611–13614
25. Reed, L. J., & Muench, H., (1938). A Simple method of estimating fifty per cent endpoints. *American Journal of Epidemiology, 27*(3),493–497.

CHAPTER 14

A Positive Correlation Between IgG2 Antibody Preponderance and Clearance of HSV During Active HSV Infections

K. VASANTHI,[1] G. SATHYA NARAYANAN,[1] PUGALENDHI,[2] and ELANCHEZHIYAN MANICKAN[1]

[1]*Department of Microbiology, Dr. ALM PG IBMS, University of Madras, Taramani, Chennai, Tamil Nadu, India*

[2]*Voluntary Health Science, Taramani, Chennai, Tamil Nadu, India*

ABSTRACT

Herpes simplex virus (HSV) is a human pathogenic virus that causes infection worldwide. We have studied the levels of total immunoglobulins and HSV specific immunoglobulins then the results were correlated with the latency. In this study, we enrolled 323 subjects, and blood was collected from 183 HSV active infections (94 were HSV-1 positive and 89 were HSV-2 positive), recently healed (58 HSV-1 and 48 HSV-2) and 140 age and sex controlled negative individuals. HSV positivity of the collected samples was confirmed by PCR. Total IgG and HSV specific IgG were tested by ELISA. The mean total serum IgG of HSV-1 cases was 554 pg/ml and HSV-2 cases was 696 pg/ml. Interestingly during active infection, there was a preponderance of IgG1 (mean 185 pg/ml for HSV-1 and 195 pg/ml for HSV-2) compared to IgG2 (mean 35 pg/ml of HSV-1 and 28 pg/ml of HSV-2). However, among the recently healed individuals, the total IgG remains similar but there was an up-regulation of IgG2 that was noticed in the ranges of 185 pg/ml HSV-1 and 205 pg/ml HSV-2. Interestingly the IgG1 was the vice versa (30 pg/ml for HSV-1 and 32 pg/ml for HSV-2) importantly, healthy controls had negligible anti-HSV antibodies. From this study, we could confirm that presence of IgG1 antibodies is non-protective against HSV infections as seen among the non-healers. However, the presence of IgG2 is protective as noticed among the recent healers, it is long known that Th-1 cytokine is associated with IgG2 antibody production and Th-2 cytokines promote IgG1. Taken together it becomes much clearer that during HSV infections Th-1 cells are protective and Th-2 cells are detrimental to the host.

14.1 INTRODUCTION

HSV infections are ubiquitous and cause significant morbidity worldwide with the capacity to hide in our neurons for life (latent infection). Infections with HSV-1 and HSV-2 are highly prevalent and it is estimated that more than 3.7 billion people under the age of 50 to 67% of the population are infected with herpes simplex virus (HSV) both type 1 and 2 (HSV-1, HSV-2). The WHO annually estimating over 500 million people are infected with HSV-2 worldwide with approximately 20 million cases. Also, the global rates of either HSV-1 or HSV-2 are between 60% and (5% in adults. In India, the seropositivity has been reported to be 33.3% for HSV-1 and 16.6% for HSV-2 [3]. HSV establishing the lifelong persistence of its infection in the human host by its successful hiding into the neurons of peripheral ganglia and while reactivating it was liberated out from those neurons and cause recurrent infection [7].

Primary infection occurs at the early days of childhood. Once the virus infects the epithelial cells it finds its way to innervating neurons and migrates centripetally towards the CNS and reaches dorsal root ganglion. In the dorsal root ganglion, it remains almost inert (latent infection) and subsequently reactivated due to some undefined stimulus and reaches the same neurodermatome and causes chronic recurrent infections. Those undefined, nonspecific stimulus are stress, catamenia (menstrual cycle), fever, exposure to sunlight, and trauma [9].

During HSV infection the level of the immunoglobulins (IgGs) were increased up to four times than the normal level. These antibodies were serving as markers for the detection of HSV infection. Antibodies for the HSV infection usually appear in 4–7 days after the infection and extend its peak in 2–4 weeks. IgM antibodies were subjected to appear transiently and are followed by IgG and IgA antibodies during the primary infection. These antibodies were memorized lifelong in the host with the minor fluctuations. High levels of antibodies were produced when the primary infection are more sever or more recurrence infection. Although the levels of the IgGs were increased during the HSV infection but the role and status of HSV infection during the recurrent condition was remain unclear. Each IgGs increase or decrease were directly correlating cell-mediated immune response; therefore, it is significant to examine the IgGs levels during infectious status and in normal status. For example, decreasing of IgG1 was associated with recurrent infections. The depression in IgG levels might contribute to the reactivation of HSV in subjects with recurrent infection. The study of IgG and their subclasses' role during the viral infection was most important to understand the pathogenesis of the virus. Thus the present study is carried to find out the levels of total and HSV specific IgGs (IgM, IgG, and their subclasses) in serum during the active and recent herpes infection comparing with healthy subjects.

14.2 MATERIALS AND METHODS

14.2.1 STUDY POPULATION

In this study, we enrolled 429 individuals of which 183 were HSV active infection (94 were HSV-1 positive and 89 were HSV-2 positive), 106 were recently healed subjects (58 HSV-1 and 48 HSV-2) and 140 were HSV negative (Figure 14.1). Serum was obtained from serum was obtained from the above said subjects. Clinical diagnosis of HSV Positive patients was made by the department of skin and venereal diseases, Voluntary Health Services, Chennai. Age-wise classifications and gender-wise classification were done in each of the three groups. Human ethical clearance was obtained from the Institutional Human Ethical Committee (UM/IHEC/02-2017-I). A proforma was made and informed consent from each participant was obtained. The ages of the study population ranged from 1–80 years (mean, 41 years).

14.2.2 BLOOD COLLECTION

Whole blood was collected from the outpatients (Voluntary Health Services, Chennai) who had lesions on the genital area and around the mouth and from healthy individuals. Then the samples were transported to the institution using an appropriate sample container box.

14.2.3 DNA EXTRACTION

1. **Isolation of DNA from Blood Samples:** Viral DNA was extracted from blood using the NucleoSpin Blood Mini kit (LOT No. 1607/001)

according to manufacture instructions and stored at –20°C for further processing.

2. **Quantification of DNA:** Quantity and quality of the DNA was detected using NanoDrop UV spectrophotometer (Thermo Scientific, USA) at 260 and 280 A.

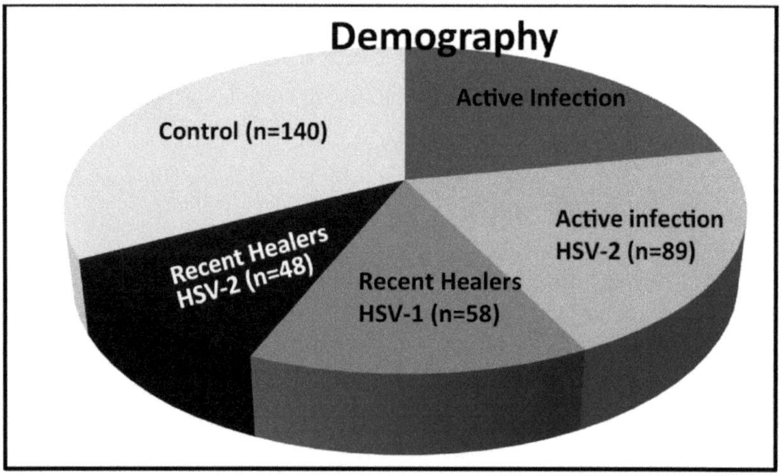

FIGURE 14.1 Study population.

14.2.4 PCR AMPLIFICATION FOR CONFIRMATION OF HSV

PCR was performed in targeting Glycoprotein D gene of HSV-1 and 2 for blood samples using primers described by Kessler et al. (2000). Briefly, PCR was done using a 50 µl master mix containing 5 µl of template DNA, 1 mM of each primer (Table 14.1), 1 µl of dNTPs, 2 units of Taq polymerase enzyme and 5 µl of 10X reaction buffer. PCR conditions included an initial denaturation at 95°C for 1 min followed by 35 cycles at 95°C for 30 sec (3 min during cycle 1), 55°C for 30 sec and 72°C for 30 sec and a final extension of 72°C for 5 mins and cooled down to 4°C. Nested amplification was done with a 5 µl aliquot from the first run; 50 µl of Taq PCR master mix solution; 36 µl of double-distilled, DNase-free water; and a 1 mM concentration of each nested primer by the same cycle protocol as described above. Each amplification run contained one negative and one positive control. The negative control consisted of blank reagent and water. For the positive control, total cellular nucleic acid was extracted from virus stocks which were grown in the vero cell line.

TABLE 14.1 Details of Primers Used in the Study

Primer (Initial PCR)	Sequence (5` to 3`)	Expected Amplicon Size
HSV R1F	5'TGCTCCTACAACAAGGTC3'	200 bp
HSVR1R	5'CGGTGCTCCAGGATATAA3'	
Primer (Nested PCR)	Sequence (5` to 3`)	Expected Amplicon Size
HSVR2F	5'ATCCGAACGCAGCCCCGGCTG3'	142 bp
HSVR2R	5'TCTCCGTCCAGTCGTTTATCTTTC3'	

14.2.5 PCR AMPLICONS-GEL VISUALIZATION

The PCR amplicons were resolved in 2% agarose with ethidium bromide (10 mg/ml) by electrophoresis for ~20 min at 100 V and gel was analyzed using the Carestream Gel documentation system (Figure 14.2).

14.2.6 SERUM ISOLATION

A clot from the whole blood has been removed by centrifuging at 1500 g for 10 min in a refrigerated centrifuge. The resulting supernatant was designated serum. Following centrifugation, the liquid component (serum) has been immediately transferred into a clean polypropylene (PP) tube using a Pasteur pipette. The samples were maintained at 2–8°C while handling and immediately analyzed, avoiding freeze-thaw cycles because this is detrimental to many serum components.

14.2.7 DETERMINATION OF ANTIBODIES LEVELS

The HSV confirmed (PCR positive) samples were classified as age-wise and gender-wise. Similar steps were done for healthy individuals to confirm HSV negativity. For the detection of total IgGs, levels (IgM, IgG) from the above serum samples EIA kits were obtained from ABCAM (Cat. No. ab137982-IgM, ab195215-IgG). For the detection of HSV, specific IgGs levels (IgM, IgG) from the above serum samples EIA kits were procured from CALBIOTECH, USA, and MYBIOSOURCE (USA) (Cat. No. H1M4908, H2M4989, H1G5010, H2G4881). For the detection of IgG subclasses from the above demography, the EIA kits were obtained from

Invitrogen-ThermoScientific (Cat. No. 991000). The analysis of immunoglobulins levels on serum samples was performed according to the manufactures instructions. Then the plates were read at 450 nm in an ELISA plate reader. A standard curve was plotted by linear regression analysis using the values of the standard and the values of the unknown samples were calculated from the graph.

14.2.8 DATA ANALYSIS

Obtained data were analyzed for statistical significance by Student's-t test.

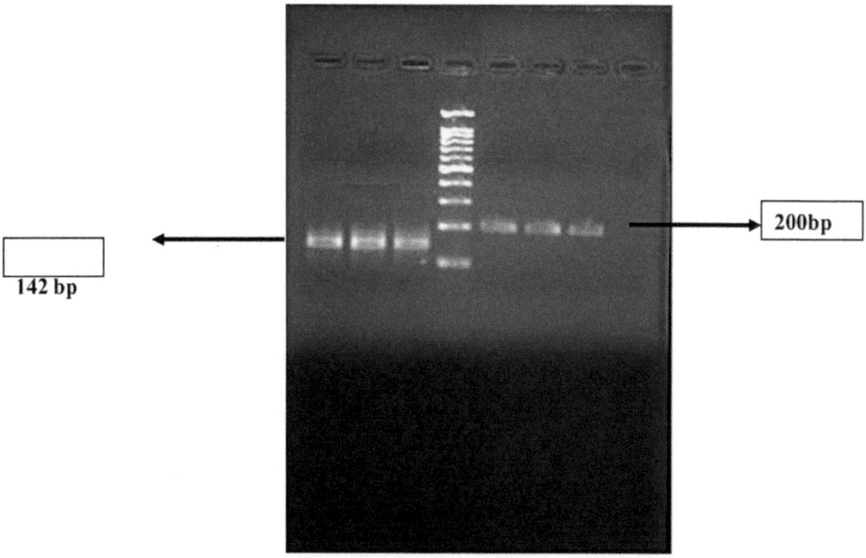

FIGURE 14.2 The HSV positivity of the clinical samples by PCR (in which 200 bp are the first-round PCR product and 142 bp is the second round PCR product).

14.3 RESULTS AND DISCUSSION

14.3.1 RESULTS

1. **Serum Total Immunoglobulin (IgG) and HSV Specific Immunoglobulin Levels among HSV Positive and Control Individuals:** Human total IgGs estimation of healthy control subjects in serum

showed (IgG-579, IgM-402 pg/ml). When the same estimation for done for HSV-1 and HSV-2 positive subjects showed similar levels (Human total IgGs for active infection subjects: IgG-554, IgM-362, pg/ml for HSV-1 and IgG-696, IgM-493 pg/ml for HSV-2, respectively) and (Human total IgGs for recent healed subjects: IgG-536, IgM-348, pg/ml for H SV-1 and IgG-528, IgM-475 pg/ml for HSV-2, respectively) (Figure 14.3). Human HSV specific IgGs estimation of healthy control subjects in the serum showed (IgG-10, IgM-7 pg/ml). When the similar estimation were observed for the active infection and recent healed subjects there was an increased level of IgG and IgM (for active infection subjects: IgG-328, IgM-312 pg/ml for HSV-1 and IgG-376, IgM-346 pg/ml for HSV-2, respectively) and (for recent healed subjects: IgG-306, IgM-296 for HSV-1 and IgG-354, IgM-309 for HSV-2, respectively) (Figure 14.4). This increment was statistically significant ($p<0.001$; Students *t*-test) when compared with healthy controls. This is the important key point were noted among HSV cases, and such increment was not observed in total serum IgGs. These increments might indicate the active and past infection but there is an important observation linked with this increase that is, there was a positive correlation with respective IgG antibody positivity and declination of Th-1 cytokine productivity (data not shown). Such association was not observed with IgM antibody positivity. From this data, we propose that HSV specific IgG and IgM were significantly increased among the HSV infected individuals and such triggers may provoke reactivation, recrudescence, and recurrent herpetic infections. Also, there is a decline of Th-1 cytokines due to an imbalance in HSV specific IgGs which switch on the Th-1 cytokine production. Taken together it could be assumed that Th-1 cytokines prevent HSV reactivation. Reduction or failure of Th-1 cytokine secretion due to an imbalance of HSV specific IgG may be a trigger for HSV reactivation and recurrent infection.

2. **Serum IgG1vs IgG2 Levels among HSV Positive and Control Individuals:** Based on the increase on the HSV specific IgG we further extended our study and estimated the levels of HSV specific IgG subclasses (IgG1 which promotes the production of Th-2 cytokine and IgG2 which promotes the Th-1 cytokine). We further observed the levels of HSV Specific IgG subclasses such as IgG1 vs. IgG2 during active and recent healed cases. Interestingly we noticed during active infection there was a preponderance of IgG1 (mean 185 pg/ml for

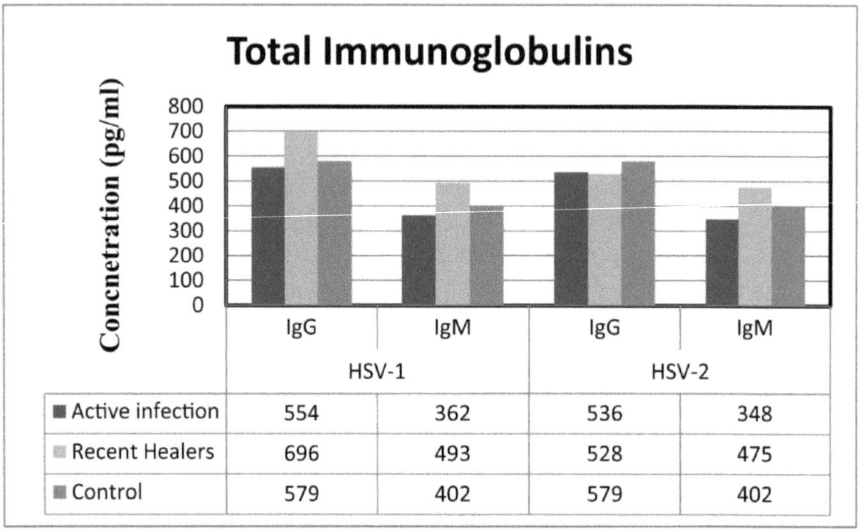

FIGURE 14.3 Shows the overall serum total immunoglobulin levels of patients with active HSV infections, recent healers, and healthy controls. Total IgG, IgM immunoglobulin levels of the subjects were estimated by ELISA.

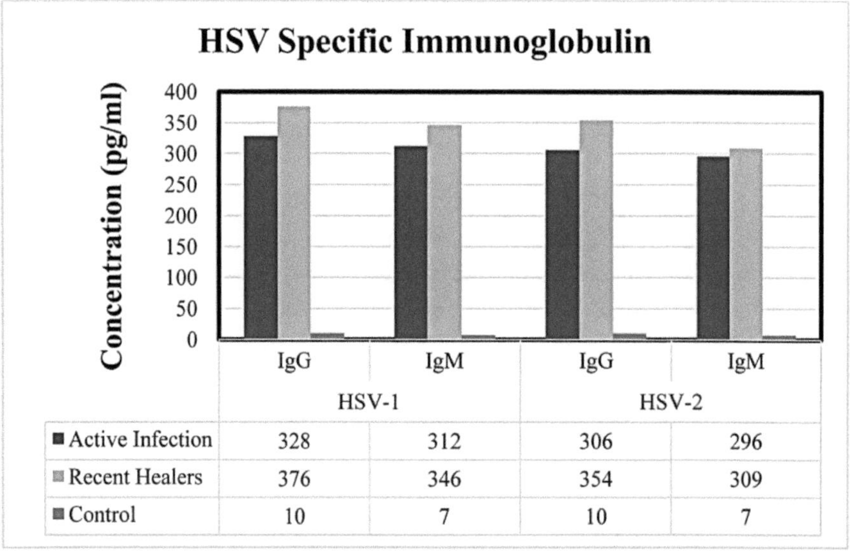

FIGURE 14.4 Shows the overall serum HSV specific immunoglobulin levels of patients with active HSV infections, recent healers, and healthy controls. IgG, IgM HSV Specific immunoglobulins levels of the subjects were estimated by ELISA.

HSV-1 and 195 pg/ml for HSV-2) compared to IgG2 (mean 35 pg/ml of HSV-1 and 28 pg/ml of HSV-2). However, among the recently healed individuals, there was an up-regulation of IgG2 was noticed in the ranges of 185 pg/ml HSV-1 and 205 pg/ml HSV-2. Interestingly the IgG1 was the vice versa (30 pg/ml for HSV-1 and 32 pg/ml for HSV-2) importantly, healthy controls had negligible anti-HSV antibodies (Figure 14.4). From this data, we propose that the IgG1 is responsible for the Th-2 cytokines production and the IgG2 is responsible for the Th-1 cytokines which prevent the HSV recrudescence. Taken together it could be assumed that diminishing in the production of IgG2 leads to reduction or failure of Th-1 cytokine secretion which may be a trigger for HSV reactivation and recurrent infection (Figure 14.5).

3. **Correlation of Gender or Age and Impairment of IgG and IgA among HSV Patients:** As discussed above there was a correlation observed between anti-HSV IgG1 and IgG2 levels and impairment of Th-1 cytokines and recurrence of HSV infections. Next, we wanted to elucidate whether such correlation existed between gender and age of the patients with a change in IgGs production or HSV recurrence. As shown in the Figures 14.5–14.8, there were comparable levels of Total IgGs vs. HSV specific IgG and IgG1 vs. IgG2 were produced irrespective of the gender or age and there was no significant difference ($p>0.01$) within HSV groups. This suggests that there were no such correlations or bias existed between gender or age and IgGs produced or HSV recurrence.

14.3.2 DISCUSSION

During HSV, infection cell-mediated immunity performs the major clearance of the virus which is inside the infected cells. In contrast, humoral immunity with the help of antibodies clears the entire virus in the circulation. Among the antibodies, IgM and IgG cleared the viruses during the early phase and late phase, respectively. IgA antibody also helps in the clearance of the virus among the IgG antibodies two subclasses namely IgG1 and IgG2 were found to be associated with susceptibility or protection, respectively. In our study, we found the patients with active HSV infection there was a preponderance of IgG1. I contrast patients among there was a recovery we found a predominance of IgG2. We and others had shown that patients having active HSV infection had a Th1 polarization and who were recovered was found to be more towards Th2 cytokines. The current data corroborates the previous observation.

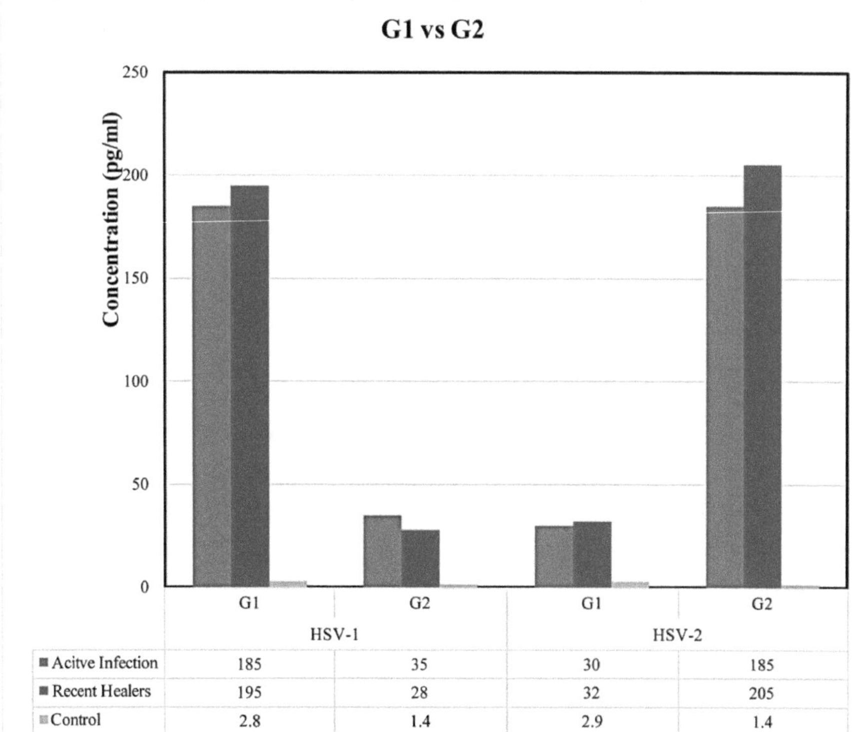

FIGURE 14.5 Shows the serum G1 and G2 levels of patients with active HSV infections, recent healers and healthy controls.

In our study, the IgG subclass evaluation showed that HSV patients with active infection had more IgG1 antibody during both HSV-1 and HSV-2 infection. However, among the just recovered patients the antibody subtypes as completely to Th1 phenotype, i.e., IgG2. In a report by Coleman et al. found that among the 157 patients they found a preponderance of IgG1 in a mouse model. In their study, other subtypes namely IgG2, IgG3, and IgG4 were not detected [5]. In a similar study, Mckendal and Woo found that 51% of the animals showed IgG1 whereas other subclassed of IgG were at the minimal level in a BABLc mouse model [6]. Ishizaka et al. found that IgG2A was more protective than IgG1 against HSV-2. On this, findings commonly observed that IgG2A was more protective against both HSV-1 and HSV-2 suggesting that Th1 was protective [4].

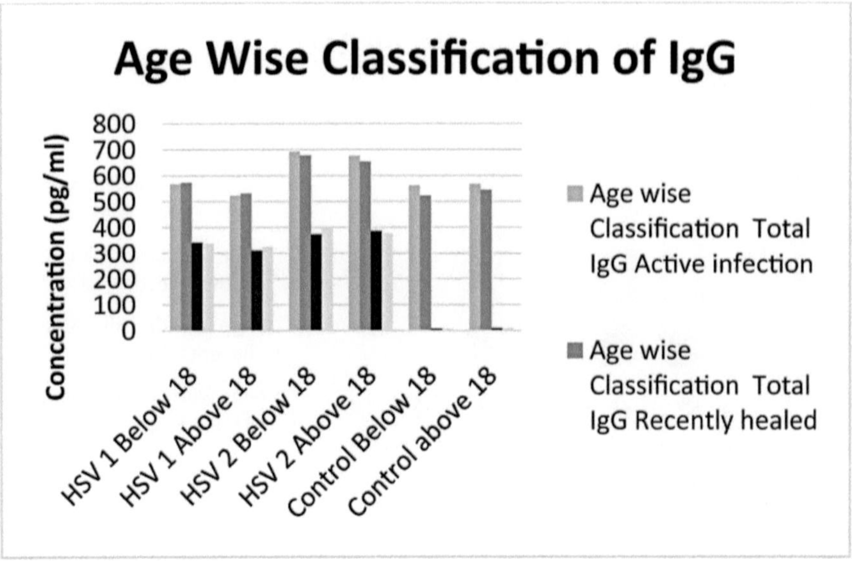

FIGURE 14.6 The Age-wise classification of HSV positive subjects. The study population was segregated based on their, i.e., <18 years versus >18 years, and the level of IgG are plotted.

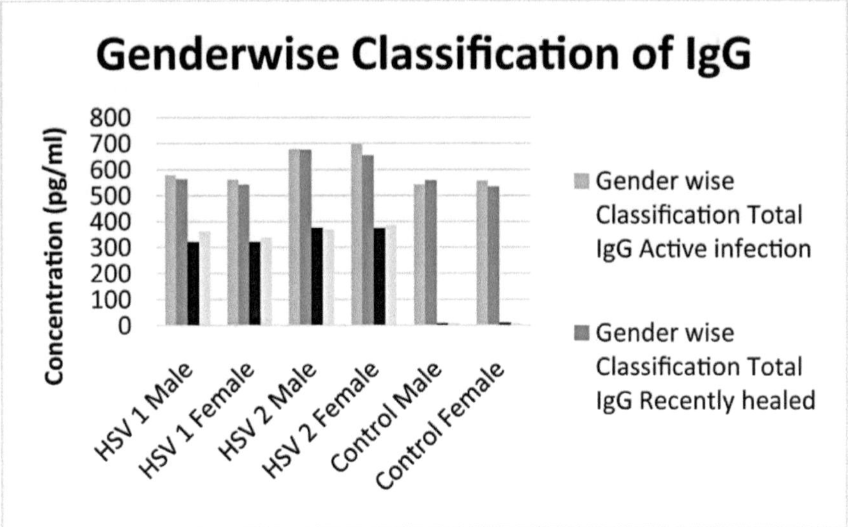

FIGURE 14.7 The gender-wise classification of HSV positive subjects. The study population was segregated based on their sex, i.e., male versus female and the level of IgG are plotted.

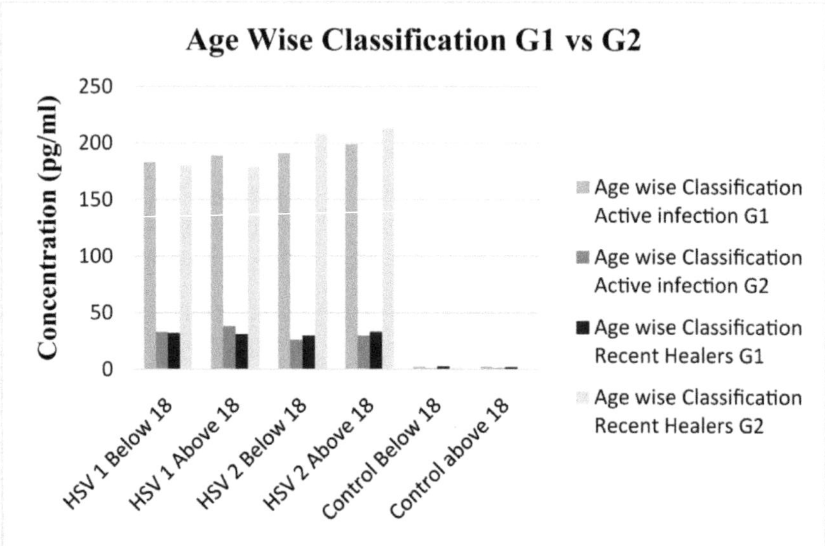

FIGURE 14.8 The age-wise classification of HSV positive subjects. The study population was segregated based on their age, i.e., <18 years versus >18 years and the level of each G1, G2 are plotted.

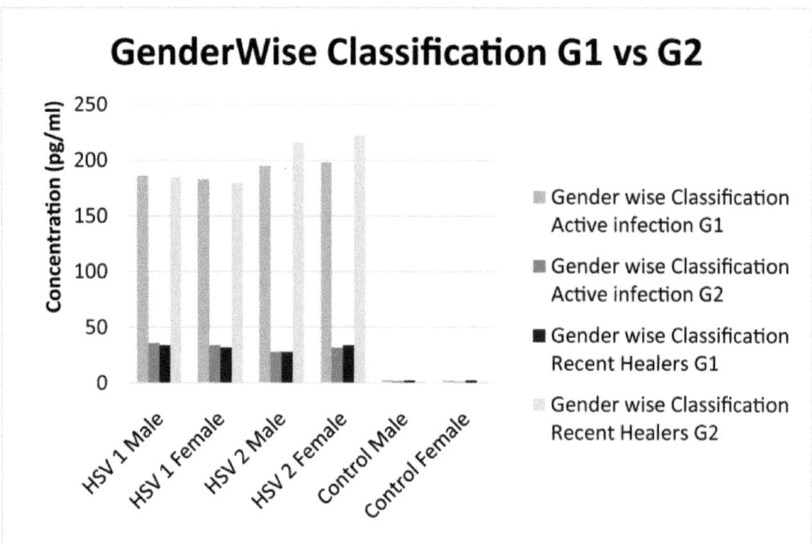

FIGURE 14.9 The gender-wise classification of HSV positive subjects. The study population was segregated based on their sex, i.e., male versus female and the level of each G1 and G2 are plotted.

Our study endorses these previous reports but n humans. Our study is a well-structured study because we not only evaluated the IgG subclasses among active HSV infection but also among recent healers to obtain the complete antibody profile. It is well known that IgG2 is associated with Th1 responses and IgG1 is commonly associated with susceptibility of HSV infection. We have already demonstrated that Th1 cytokines a protective against HSV infection and latency and Th2 cytokines were resulting in pathological consequences. We have also reported that the reduction in Th1 cytokines resulted in the HSV reactivation and recurrent HSV infection [8] and here we are showing an associated antibody isotype which was prevalent in the protected individuals. Our study has thrown more light on the spectrum of immune events at the humoral immune response level.

14.4 CONCLUSION

HSV causes both lytic and latent diseases to humans. Immediately after the primary infection, HSV establishes a chronic latent infection in various ganglia including trigeminal, facial, and vagus ganglia by hiding into them. This dormant virus then awaits a trigger to reactivate it. Once the triggering signals are sensed by the virus it undergoes an early phase followed by waves of reactivation which cause lesions and blisters that are so painful. While stress, trauma, menstrual period, and radiation has been described as predisposing factors for HSV recrudescence. In this paper, we propose a connection between IgGs and the impairment of Th-1 cytokines. This connection may be one of the predisposing factors for the recurrence of HSV infection. Here we found that a decrease of the IgG2 IgG diminishes the Th-1 cytokines production which could provoke the virus and cause recurrent herpetic lesions.

KEYWORDS

- **cytokines**
- **herpes simplex virus**
- **herpetic lesions**
- **immunoglobulin**
- **latency**
- **latent diseases**

REFERENCES

1. Steiner, I., & Benninger, F., (2013). Update on herpes virus infections of the nervous system. *Curr. Neurol. Neurosci. Rep., 13*, 414.
2. Aggarwal, A., & Kaur, R., (2004). Seroprevalence of herpes simplex virus1 and 2 Antibodies in STD clinic patients. *Indian J. Med. Microbiol., 22*, 2446.
3. Bogaerts, J., Ahmed, J., Akhter, N., Begum, N., Rahman, M., Nahar, S., et al., (2001). Sexually transmitted infections among married women in Dhaka, Bangladesh: Unexpected high prevalence of herpes simplex type 2 infection. *Sex Transm. Infect., 77*, 1149.
4. Sally, T. I., Priscilla, P., Jillian, S., & Eric, M. M., (1995). IgG subtype is correlated with efficiency of passive protection and effect or function of anti-herpes simplex virus glycoprotein d monoclonal antibodies. *The Journal of Infectious Diseases, 172*, 1108–1111.
5. Marie, C. R., Andre, J. N., Susan, C. W., Donald, J. P., Charlotte, M. B., & Charles, B. R., (1985). IgG Subclass antibodies to herpes simplex virus. *The Journal of Infectious Diseases, 151*(5).
6. Robert, R. M., & Wayne, W., (1988). Murine IgG subclass responses to herpes simplex virus type 1 and polypeptides. *J. Gen. Virol., 69*, 847–857.
7. Khaldin, A. A., & Baskakova, D. V., (2007). Epidemiological aspects of diseases caused by the herpes simplex virus (Review). *Consilium. Medicum., 2*, 27–30.
8. Vasanthi, K., Dhanalakshmi, M., Pugalendhi, V., & Elanchezhiyan, M., (2017). Reactivation from latency and recurrence of HSV infections due to impaired Th-1 cytokines. *Human Journals Research Article, 10*(1).
9. Halford, W. P., Gebhardt, B. M., & Carr, D. J., (1996). Persistent cytokine expression in trigeminal ganglion latently infected with herpes simplex virus type 1. *J. Immunol., 157*(8), 3542–3549.
10. Kessler, H. H., Muhlbauer, G., Rinner, B., Stelzl, et al., (2000). Detection of Herpes Simplex Virus DNA by Real-Time PCR. *J Clin Microbiol. 38*(7), 2638-2642.

CHAPTER 15

Effects of Thermal Cycling on Surface Hardness, Diametral Tensile Strength and Porosity of an Organically Modified Ceramic (ORMOCER)-Based Visible Light Cure Dental Restorative Resin

P. P. LIZYMOL

Scientist F and in Charge, Division of Dental Products, Biomedical Technology Wing, Sree Chitra Tirunal Institute for Medical Sciences and Technology, Poojappura, Thiruvananthapuram, 695012, Kerala, India, Tel.: +91-471-2520221, Fax: +91-471-2341814, E-mails: lizymol@rediffmail.com; lizymol@sctimst.ac.in (P. P. Lizymol)

ABSTRACT

Dental restorative material placed in the tooth cavity is exposed to cyclic changes of temperature. The present study aimed to find the effect of thermal cycling on surface hardness, diametral tensile strength and porosity of a visible light cure composite.

15.1 INTRODUCTION

Dental restorative polymer composite materials based on polymerizable bisphenol-A glycidyl methacrylate (Bis-GMA) monomers [1, 2] and quartz/radiopaque glass fillers has been the most popular materials used in dentistry, since Bowen [1, 2] introduced (Bis-GMA) in the 1960s. Though they have good aesthetic and physical properties [3], attempts including few structural variations in the organic matrix of dental composites are going on to improve the clinical performance of restorative materials [4–14]. Among these modifications, urethane dimethacrylates (UDMAs), [4] urethane tetra-methacrylates, [5] organically modified ceramics (ORMOCERS) [6,-13], and bioactive materials [14] are included.

ORMOCERS are very promising materials. Although ORMOCERS are very promising, few investigations [6] have confirmed the potential of ORMOCERS as biomaterials or low-contraction materials applied to teeth restoration. The two composites (Definite® (Degussa AG, Hanau, Germany) or Admira® (Voco GmbH, Cuxhaven, Germany) available in the market based on ORMOCER technology. Admira composite contained 78% inorganic particles (barium and aluminum silicate) with an average size of 0.7 μ and the organic fraction composed of 65.5%, conventional organic dimethacrylates such as BisGMA, and UDMA along with 34.5% of triethylene glycol dimethacrylate (TEGDMA). The concept of ORMOCER [7] is to combine properties of organic polymers with glass-like materials to generate new/synergistic properties. The processing steps are based on sol-gel type reaction. Our previous studies reported [8–12] the development of a noncytotoxic and biocompatible organically modified ceramic composite with lower polymerization shrinkage compared to a composite containing BisGMA.

VOCO, the Dentalists [13] claimed that they have developed a new radiopaque light-curing composite, Admira Fusion which based on a nanohybrid silicate and ORMOCER technology with 1.25% polymerization shrinkage (volumetric)) is suitable for posterior and anterior restoration. Though the popularity of tooth-colored polymeric restorations is increasingly in dentistry, about half of the restorations have to be removed or replaced due to the failed restorations. The poor clinical performance of the restorative composite is due to various reasons. Polymerization shrinkage in the oral cavity, low monomer conversion, presence of residual monomer, cytotoxic effects of leachants, intensity of light, exposure time, shelf life of material food habits, bacterial infection and the handling characteristics have significant effect on the durability of restoration in the oral cavity. Current research [15] in dental materials reached from the traditional bioinert materials for restorative purposes to replace the decayed tooth to bioactive materials has a therapeutic function. As per the reported studies of Maktabi et al. [16], when the dental composite is placed in the oral cavity incrementally, the performance of the light-curing procedure is low, which leads to the formation of biofilm and secondary caries. Studies by Shimokawa CA [17] showed that microhardness of bulk-fill resin-based composites (RBCs) has a positive correlation on the performance of the light-curing unit by using four different light-curing units (LCUs. Microhardness is directly related to the monomer conversion.

Our previous studies [18] showed that the selection of photo-initiator has a significant effect on monomer conversion and clinical performance. Compared to CQ/amine, photo-polymerization of BisGMA is found to be more efficient with TPO. Both exposure and storage times were important variables in CQ/amine, but not in TPO. Free radicals generated by CQ/ Amine showed more radiative and nonradiative energy loss compared to TPO photolysis. The better monomer conversion of TPO based system reduces the adverse toxicological effects due to chemical and mechanical degradation.

The clinical durability of composite restorative materials is significantly affected by cyclic temperature changes. The composite restorative material has to expose to cyclic temperature changes in the oral environment during normal eating and drinking probably from low temperature as 5°C to about 55°C. During this process, the materials may undergo degradation and decrease in properties.

As per the Reported studies [19], the observed changes in the marginal gap of crowns made from light-cured resin during thermal cycling are comparatively less than autopolymerized resin. They suggested that the improved characteristics of the light-polymerized material may improve the longevity of provisional crowns during clinical applications.

The effect of thermal cycling on degradation of the commercially available visible light cure dental restorative material (Admira®) was evaluated in terms of Vickers hardness number (VHN) and DTS. We have carried out an accelerated thermal cycling test on cured dental composites. Though we have carried out the aging studies of various dental restorative materials [12], changes on porosity, and physico-mechanical properties of visible light cure dental restorative composites based on ORMOCER technology has not reported.

MicroCT (μCT) evaluation was carried out to find out whether thermal cycling has any effect on internal structure of the cured composite. Micro-CT is a non-destructive 3D imaging technique [20] that can be used to inspect the internal structures of small objects with high spatial resolution and unprecedented speed. In the present study, thermally cycled samples were examined using μCT technique to evaluate the increase in porosity, which can also be the reason for loss of properties. The change in porosity was correlated with the decrease in VHN and DTS.

15.2 EXPERIMENTAL

15.2.1 MATERIALS

Commercially available dental resin Admira (VOCO, Cuxhaven, Germany) was used for the studies.

15.2.2 PREPARATION OF SAMPLES

Samples with 6 mm (dia) x and 3 mm (thick) prepared as per the reported procedure [8] was cured of 40 seconds and stored at 37°C for 24 hours and used for thermal cycling.

15.2.3 THERMAL CYCLING

The prepared samples were allowed to expose cyclic temperature changes in distilled water from 5°C to 55°C in a dental thermal cycler (Willytec, Germany). The dwell time used was 15 sec at 5°C and 55°C. A 15 sec time was given as the drain time at 22°C. Samples were subjected to 500, 1000, 1500, and 2000 cycles. The thermo cycled samples were used for microCT, microhardness, and DTS evaluation. For microCT evaluation, the same sample before and after thermal cycling was evaluated at each cycles up to 2000 cycles of thermal cycling.

15.2.4 MICROARCHITECTURAL ANALYSIS

A micro Computed Tomography (MicroCT) system (μCT 40, ScancoMedical, Bassersdorf, Switzerland) was used to non-destuctively image and quantifies the 3D microstructural morphology of each sample. The samples were scanned using with 45 [KeV] energy, 177 [μA] intensity, and 12 μm slice thickness. 2D images reconstructed using the Isotropic slice data generated by the system were used for the qualitative analysis and to get the 3D images. Porosity (%) and total pore volume were calculated using the equations:

$$[1-Bv/Tv] \times 100 \text{ and } Tv-Bv$$

where, Tv is the total volume and Bv is the bone volume (volume excluding the pores).

15.2.5 EVALUATION OF SURFACE HARDNESS (VHN)

Vickers microhardness tester (Model HMV2, Shimadzu, Japan was used to evaluate the surface hardness [Vickers hardness (VHN)] of each side of samples. A load of 100 gm was applied for 15 sec. and the measurement was done as per the reported procedure [8] Vickers hardness was calculated from the equation:

$$H_V = 0.1891\ F/d^2$$

where H_V = hardness number; F = Test load (N); d = mean diagonal length of the indentation (mm).

The mean and standard deviation for six measurements was recorded.

15.2.6 EVALUATION OF DTS

The diametral tensile strength (DTS) was determined as described before [8] using a Universal Testing Machine (Instron, Model 1011, UK). DTS was calculated using the following equation:

$$DTS = 2P/\pi DL$$

where, P is the load at break in Newtons, D is the diameter and L is the thickness of the specimen in mm. Statistical analysis was carried out using ANOVA (analysis of variance) single factor to determine significant changes ($P < 0.05$).

15.3 RESULTS AND DISCUSSION

Dental restorative material is expected to remain lifelong in the oral environment, which is a thermal cycling system with temperature variation from 5 to 55°C, which changes with the food habits. Variation in temperature can make irreversible changes in properties of the restorative material which adversely affects the clinical performance.

The variation of surface hardness (VHN) of Admira with thermal cycling given in Figure 15.1 shows that hardness decreases with thermal cycling up to 1000 cycles ($P = 2.35E\text{-}14$) and increases after 1000 cycles.

Compared to samples without thermal cycling (0 cycles) significant decrease in VHN is observed up to 2000 cycles. P-value obtained in statistical analysis is given in Table 15.1.

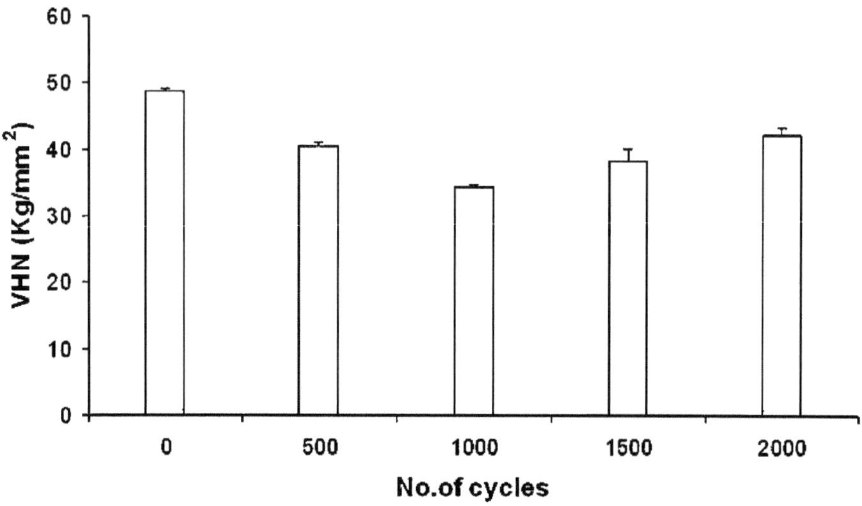

FIGURE 15.1 Effect of thermal cycling on surface hardness (VHN).

TABLE 15.1 Effect of Thermal Cycling on VHN; Analysis of Variance (P Value)

VHN 0 CYCLES	VHN 500 CYCLES	2.22E-11
VHN 0 CYCLES	VHN 1000 CYCLES	2.35E-14
VHN 0 CYCLES	VHN 1500 CYCLES	1.36E-07
VHN 0 CYCLES	VHN 2000 CYCLES	4.32E-08
VHN 500 CYCLES	VHN 1000 CYCLES	5.68E-11
VHN 1000 CYCLES	VHN 1500 CYCLES	0.000688
VHN 1500 CYCLES	VHN 2000 CYCLES	0.00084

The variation in surface hardness with thermal cycling indicated the initial surface softening in the aqueous medium after prolonged exposure (> 500 cycles). Surface hardness is related to the amount of water taken up, the greater the uptake the weaker the resultant swollen polymer. But later long term exposure after 1000 cycles indicated that the surface is supersaturated with the adsorbed moisture and further exposure has no deteriorating effect. The increase in VHN after 1000 cycles may be due to thermally induced cross-linking at the surface. Hardness and monomer conversion are directly related [6, 10]. The greater the monomer conversion, the more will be the hardness. Hardness values have been used [10] as an indirect measure of degree of conversion and degradation in aqueous medium. Figure 15.2 shows that DTS of Admira has no significant change up to 500 cycles. After 500 cycles, DTS decreases with increase in cycles.

Micro-CT nondestructively and reproducibly provided 3D representations of structural characteristics. Microarchitecture parameters including volume fraction, strut density, strut thickness, and degree of anisotropy were calculated using 3D stereology. Figure 15.3 shows the threshold inversion which represents the porosity after deducting the Admira from the scanned image using the attached software (Micro-CT evaluation programme version 6.0).

Figure 15.4 shows porosity distribution of Admira from 0–2000 cycles. Up to 500 cycles, no change in porosity (%) and pore volume (Figure 15.4) was observed. Pore size is found to be 24 μm up to 500 cycles. After 500 cycles both pore size and pore volume are found to increase. Figure 15.4 shows that pore size is 90 μm after 500 cycles. Total porosity (%) and pore volume of Admira (Table 15.2) are found to increase with increase in thermal cycling. Increase in total porosity (%) and pore volume may be due to the dissolution of filler particles and uncured resin. This increase in pore size may be the reason for decrease in DTS.

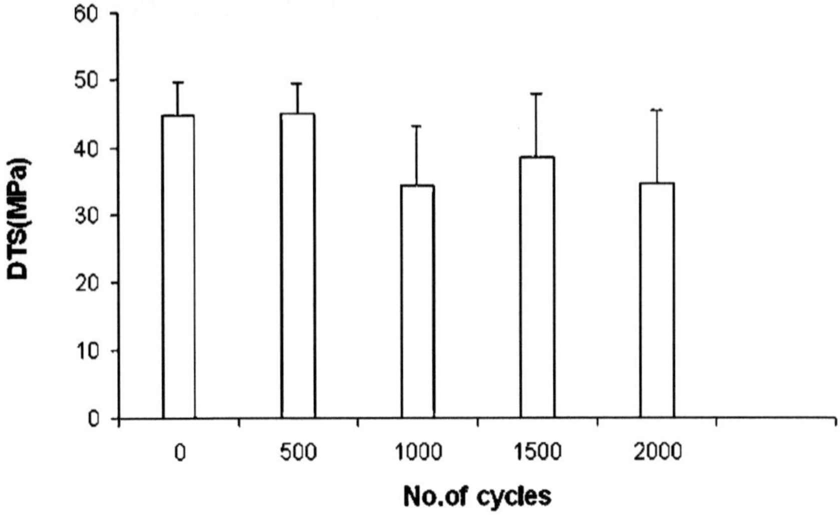

FIGURE 15.2 Effect of thermal cycling on DTS.

TABLE 15.2 Porosity and Total Pore Volume of Thermocycled Admira

Sample Code	Porosity (%)	Pore Volume (mm^3)
Admira-0	2.24	1.96
Admira-500	2.25	1.98
Admira-1000	3.40	3.00
Admira-1500	3.41	2.99
Admira-2000	3.57	3.14

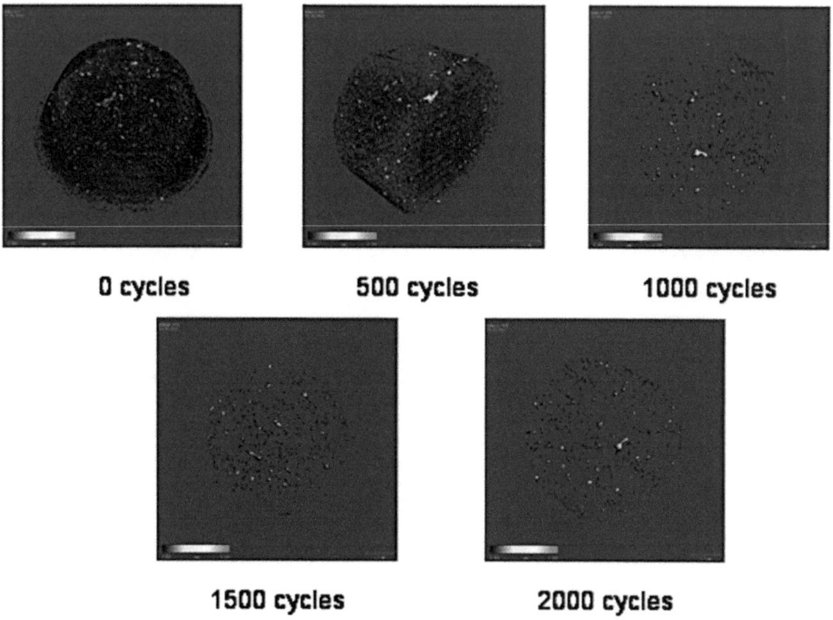

FIGURE 15.3 Effect of thermal cycling on porosity.

15.4 CONCLUSION

Effects of thermal cycling on DTS and VHN of a commercially available visible light cure dental restorative material were evaluated. Changes in DTS and VHN due to thermal cycling were examined and correlated to the presence of internal porosity with the help of micro-computed tomography images of the cured sample. The photo-cured composite samples were subjected to thermal cycling up to 2000 cycles from 5°C to 55°C in a dental thermal cycler and the properties were evaluated. ANOVA single factor showed that thermal cycling up to 500 cycles resulted in no significant change in DTS where a significant decrease in VHN. Pore size was not changed with thermal cycling up to 500 cycles as evidenced by microCT results. Analysis of images and data provided by μCT after thermal cycling shows that total porosity (%), pore-volume, and pore size of Admira are increased with thermal cycling (>500 cycles). This increase in porosity may be the reason for the decrease in DTS whereas surface softening may be the reason for the decrease in surface hardness of photo-cured dental composite during thermal cycling. The study showed that thermal cycling from 5 to 55°C caused irreversible changes in properties of the restorative material which adversely affects the clinical performance.

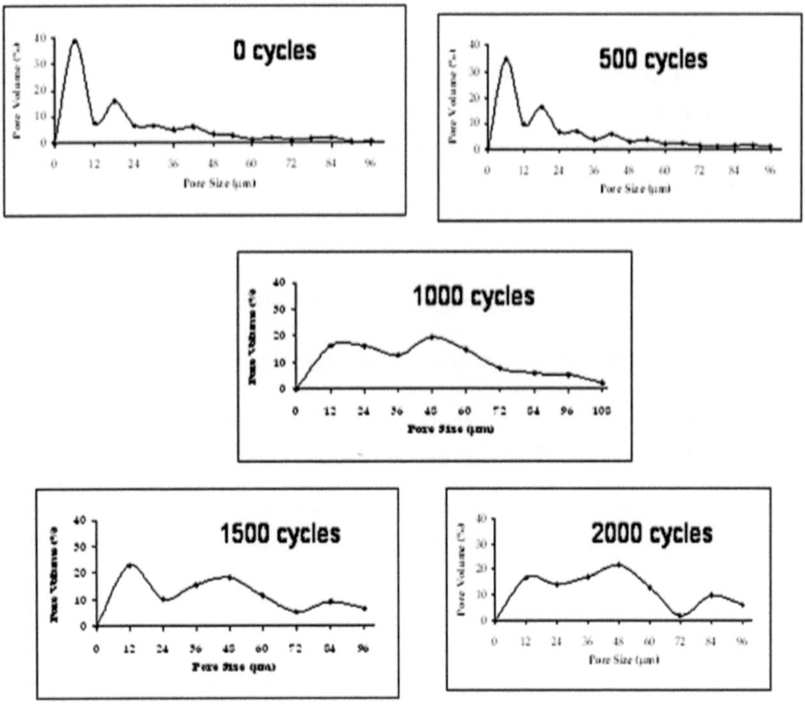

FIGURE 15.4 Effect of thermal cycling on pores distribution.

ACKNOWLEDGMENTS

The author wishes to thank Former and present Directors, Head, Biomedical Technology Wing, and Dr. Kalliyana Krishnan, Former Head, Department of Biomaterial Science and Technology, Sree Chitra Tirunal Institute for Medical Sciences and Technology for providing the facilities to carry out the project. Financial support from the Department of Science and Technology, India, under the scheme "Fast track proposal for young scientists" and technical assistance obtained from Mr. Arun Torris for micro-CT scanning are gratefully acknowledged.

KEYWORDS

- **analysis of variance**
- **diametral tensile strength (DTS)**
- **microcomputed tomography**

- **thermal cycling**
- **Vickers hardness number (VHN)**
- **visible light cure**

REFERENCES

1. Bowen, R. L., (1956). Use of epoxy resins in restorative materials. *J. Dent. Res., 35*, 361–369.
2. Bowen, R. L., (1963). Properties of a silica-reinforced polymer for dental restorations. *J. Am. Dent. Asso., 66*, 57–64.
3. Cecilia, P. T., Benedito, D. M. P., & Mônica, C. S., (2003). Wear of dental resin composites: Insights into underlying processes and assessment methods: A review. *J. Biomed Mat Res Part B: Applied Biomaterials, 65B*(**2**), **280–285.**
4. Kalliyanakrishnan, V., Lizymol, P. P., & Jacob, K. A., (2004). Synthesis and characterization of a urethane dimethacrylate resin with application in dentistry, *J. Polym. Mate., 21*(2), 137–144.
5. Kalliaynakrishnan, V., Lizymol, P. P., & Sindhu, P. N., (1999). Urethane tetramethacrylates: Novel substitutes as resin matrix in radio paque dental composites. *J. Appl. Polym. Sci., 74*, 735–746.
6. Fabio, F. S., Luis, C. M., Marysilvia, F., & Marcia, R. B., (2007). Degree of conversion versus the depth of polymerization of an organically modified ceramic dental restoration composite by Fourier transform Infrared spectroscopy. *J. Appl. Polym. Sci., 104*, 325–330.
7. Hass, K. H., (2000). Hybrid inorganic-organic polymers based on organically modified si-alkoxides. *Adv. Eng. Mater., 2*(9), 571–582.
8. Lizymol, P. P., (2010). Studies on new organically modified ceramics based dental restorative resins. *J. Appl. Polym. Sci., 116*, 509–517.
9. Lizymol, P. P., Mohanan, P. V., & Shabareeswaran, V. K. A., (2012). Biological evaluation of a new organically modified ceramic based dental restorative resin submitted journal of bioactive and compatible polymers. *J. Appl. Polym. Sci., 125*, 620–629.
10. Lizymol, P. P., (2004). Thermal studies: A comparison of the thermal properties of different oligomers by thermogravimetric techniques *J, Appl. Polym. Sci., 93*, 977–985.
11. Lizymol, P. P., (2010). Studies on shrinkage, depth of cure and cytotoxic behavior of novel organically modified ceramic based dental restorative resins. *J. Appl. Polym. Sci., 116*, 2645–2650.
12. Lizymol, P. P., & Kalliyana, K. V., (2009). Aging effects of dental restorative materials upon surface hardness. *J. Polym. Mater., 26*(2), 207–214.
13. VOCO, (2017). Game-changing restorative solutions. *British Dental Journal, 222*(10).
14. Vibha, C., & Lizymol, P. P., (2019). Synthesis and characterization of a novel radiopaque dimethacrylatezirconium containing pre-polymer for biomedical applications *Materials Letters, 237*, 294–297.

15. Ke, Z., Bashayer, B., Christopher, D. L., Michael, D. W., Mary, A. S. M., Yuncong, L., Mark, A. R., et al., (2018). Developing a new generation of therapeutic dental polymers to inhibit oral biofilms and protect teeth. *Materials, 11*, 1747–1764.
16. Maktabi, H., Ibrahim, M., Alkhubaizi, Q., Weir, M., Xu, H., Strassler, H., Fugolin, A. P. P., et al., (2019). Underperforming light curing procedures trigger detrimental irradiance-dependent biofilm response on incrementally placed dental composites. *J. Dentistry*. https://doi.org/10.1016/j.jdent.2019.04.003 (accessed on 24 June 2020).
17. Shimokawa, C. A. K., Turbino, M. L., Giannini, M., Braga, R. R., & Price, R. B., (2018). Effect of light curing units on the polymerization of bulk fill resin-based composites. *Dental Materials, 34*(8), 1211–1221.
18. Vaidyanathan, T. K., Vaidyanathan, J., Lizymol, P. P., Arya, S., & Kalliyanakrishnan, V., (2017). Study of visible light activated polymerization in BisGMA-TEGDMA monomer with type 1 and type 2 photoinitiators using Raman spectroscopy. *Dental Materials, 33*(1), 1–11.
19. Robert, J. D., Peter, K., Saul, W., & Vaidyanathan, T. K., (1999). Effects of occlusal loading and thermal cycling on the marginal gaps of light-polymerized and auto polymerized resin provisional crowns. *J. Prosthet. Dent., 82*, 161–166.
20. Angela, S. P. L., Thomas, H. B., Sarah, H. C., & Robert, E. G., **(2003).** Microarchitectural and mechanical characterization of oriented porous polymer scaffolds, *Biomaterials, 24*(3), 481–489.

Index

D

R

S

T